Environmental Division

AA009018

Held at the 2023 AIChE Annual Meeting

Orlando, Florida, USA
5-10 November 2023

ISBN: 978-1-7138-9288-5

Environmental Division

Held at the 2023 AIChE Annual Meeting

Orlando, Florida, USA
5-10 November 2023

The Global Home of Chemical Engineers

Printed from e-media with permission by:

Curran Associates, Inc.
57 Morehouse Lane
Red Hook, NY 12571

Some format issues inherent in the e-media version may also appear in this print version.

Printed with permission by Curran Associates, Inc. (2024)

For permission requests, please contact AIChE
at the address below.

AIChE
120 Wall Street, FL 23
New York, NY 10005-4020

Phone: (800) 242-4363
Fax: (203) 775-5177

www.aiche.org

Additional copies of this publication are available from:

Curran Associates, Inc.
57 Morehouse Lane
Red Hook, NY 12571 USA
Phone: 845-758-0400
Fax: 845-758-2633
Email: curran@proceedings.com
Web: www.proceedings.com

TABLE OF CONTENTS

ADVANCED TREATMENT TECHNOLOGIES FOR WATER I

ATMOSPHERIC CHEMISTRY AND PHYSICS: LABORATORY, MODELING, AND FIELD STUDIES

CO2 INDUSTRIAL, ENGINEERING AND RANDD APPROACHES

DESIGN AND ANALYSIS OF CARBON CAPTURE AND NEGATIVE EMISSIONS TECHNOLOGIES - EXPERIMENTAL

DESIGN AND ANALYSIS OF CARBON CAPTURE AND NEGATIVE EMISSIONS TECHNOLOGIES - MODELS

DESIGN AND ANALYSIS OF SUSTAINABLE CARBON CAPTURE AND EMISSIONS CONTROL TECHNOLOGIES I

DESIGN AND OPTIMIZATION OF INTEGRATED ENERGY SYSTEMS

FUNDAMENTALS AND APPLICATIONS FOR WASTE TREATMENT AND VALORIZATION I

FUNDAMENTALS OF ENVIRONMENTAL KINETICS AND FOOD, ENERGY, AND WATER SYSTEMS

POSTER SESSION: ENVIRONMENTAL DIVISION

REMEDIATION OF EMERGING CONTAMINANTS AND LEGACY COMPOUNDS

SUSTAINABILITY FUNDAMENTALS AND METRICS APPLICATIONS

SUSTAINABLE FUEL FROM RENEWABLE RESOURCES

WASTE PLASTIC - RECYCLE, REUSE AND REMEDIATION STRATEGIES I

WATER REUSE AND RECYCLING

EMERGING FRONTIERS IN ENVIRONMENTAL CHEMICAL ENGINEERING

ADVANCED TREATMENT TECHNOLOGIES FOR WATER II

DESIGN AND ANALYSIS OF SUSTAINABLE CARBON CAPTURE AND EMISSIONS CONTROL TECHNOLOGIES II

WASTE PLASTIC - RECYCLE, REUSE AND REMEDIATION STRATEGIES II

FUNDAMENTALS AND APPLICATIONS FOR WASTE TREATMENT AND VALORIZATION II

79a Superfine Activated Carbon-Functionalized Adsorptive Thin-Film Nanocomposite Membranes for Selective Pfas Removal from Water

Medha Kasula, Jasneet Pala, Milad Esfahani

Due to the widespread production and use of per- and poly-fluoroalkyl substances (PFAS), as well as their ability to move and persist in the environment; surveys conducted by the Centers for Disease Control and Prevention (CDC) show significant human exposure to PFAS. The membrane filtration and adsorption process are the two main effective separation and removal methods for PFAS removal from water, where each technique possesses a unique capability. Thin-film composite (TFC) nanofiltration (NF) membranes showed an acceptable (<90%) removal of PFAS. However, the presence of rejected PFAS in the concentrate stream needs another separation process. In this study, we report the fabrication of adsorptive thin-film nanocomposite (TFN) NF membranes functionalized with super fine (<100 nm) activated carbon (SFAC) for enhanced perfluorooctanesulfonic acid (PFOS) removal by adsorption during the water filtration. The NF membrane was fabricated on a poly (ether sulfone) (PES) support layer with a polyamide (PA) selective layer functionalized by immobilization of AC inside the PA via interfacial polymerization. The isotherm adsorption of SPAC showed 93% removal of PFOS in DI water. The bare NF membrane showed 12 ($L.m^{-2}.h^{-1}.bar^{-1}$) water permeability, around 60% NaCl rejection, more than 99% salt $MgSO_4$ rejection, and 90% PFAS removal. The functionalized SFAC-TFN membranes, showed similar PFOS removal, but the dominant mechanism appeared to be adsorption instead of rejection. SFAC-TFN-0.5 membranes (the TFN membrane functionalized with the optimal amount of 0.5 wt.% SFAC) showed a PFOS removal of 94 %, with more than 80% being adsorption and an increased permeability of 22 ($L.m^{-2}.h^{-1}.bar^{-1}$) compared to the bare TFC. The rejection of PFOS in presence of humic acid is also observed to study the selectivity of PFOS on the adsorptive membrane. In addition, the SFAC-TFN-0.5 showed acceptable re-generability without sacrificing the membrane performance.

79b Development of Cationic Hydrogel PAC Composites for Pfas Remediation in Aqueous Solutions

Maria Victoria Ximenes Klaus, Nicole Marguerite, J. Zach Hilt

Decades of production of per- and polyfluoroalkyl substances (PFAS), anthropogenic surfactants developed for a plethora of applications, have since plagued the environment as a result of their prevalence and persistent nature. Some PFAS have been connected to a myriad of harmful health impacts including dyslipidemia, birth defects in newborns, and increased risks to several types of cancers. Current treatments for PFAS remediation in the environment and water sources include granular activated carbon (GAC) adsorption, reverse osmosis and nanofiltration membrane separation, and anion exchange resins (AER) with each having advantages and disadvantages (e.g., limited selectivity, cost, etc.). Hence, the need for new cost-effective and efficient technologies for PFAS removal is of utmost importance. Novel materials based synthetic sorbents (i.e., hydrogels and hydrogel composites) have recently gained attention for remediation of these contaminants due to their high-water retention capacity and ability to be functionalized to target PFAS sorption. Herein, acrylamide-based hydrogel composites were synthesized using two cationic co-monomers, N-[3-(dimethylamino)propyl] acrylamide methyl chloride quaternary (DMAPAQ) and diallyldimethylammonium chloride (DADMAC), with powdered activated carbon (PAC) and characterized to determine their affinity for PFAS. Physicochemical characterization included Fourier-Transform infrared spectroscopy (FTIR) to identify chemical compositions, thermogravimetric analysis (TGA) to confirm PAC loading percentages, and aqueous swelling for expansibility behavior. Finally, aqueous sorption studies were performed to determine the affinity of two legacy PFAS compounds, perfluorooctanoic acid (PFOA) and perfluorooctanoic sulfonic acid (PFOS), and two short-chain analytes, perfluorobutanoic acid (PFBA) and perfluorobutanesulfonic acid (PFBS) to compare impacts of long-chain and short-chain compounds, as well as sulfonic versus carbonic functionality of PFAS. FTIR confirmed expected functionalization with observation of carbonyl and amine groups, and TGA analysis also confirmed the loading of various concentrations of PAC within the network. Congruently, the swelling ratio decreased with increasing crosslinking density and increased with the inclusion of cationic co-monomers, as expected. Finally, sorption of PFAS was analyzed via liquid chromatography tandem mass spectrometry (LM-MS/MS) where PFAS binding increased with increasing cationic functionality and PAC loading density.

79d The Role of Redox-Active Metallopolymers in the Selective Remediation of Water from Pfas Contaminants

Paola Baldaguez Medina, Valentina Ardila Contreras, Johannes Elbert, Markus Gallei, Xiao Su

Per- and polyfluoroalkyl substances (PFAS) are persistent and toxic compounds that are commonly found in water and pose a significant health risk to humans. The conventional methods for capturing PFAS from water suffer from a lack of selectivity, particularly at low concentrations. Here, we introduce electrosorption techniques as a modular, environmentally friendly, and easily scalable approach to enhance the selectivity of PFAS. We employed metallopolymers as electrode materials due to their attractive properties for energy storage and electronic devices. Specifically, we investigated five different metallopolymers with distinct functional groups and electron densities to determine their selectivity towards PFAS compounds. Our results demonstrate that metallopolymers with higher electron density exhibit greater uptake and regeneration efficiency. Notably, $PMAECoPF_6$ showed the highest affinity towards perfluorooctanoic acid (PFOA), with an uptake of over 90 mg-PFOA/g-adsorbent at 0.0 V vs. Ag/AgCl and a regeneration efficiency of over 85% at -0.4 V vs. Ag/AgCl in one hour of experimentation. Moreover, we observed that the adsorption capacity of metallopolymers was unaffected by the water matrices, including tap water, secondary water effluent, and various salt concentrations. The desorption kinetics further revealed that electrochemical input can increase the regeneration efficiency by >40% compared to the open circuit. Overall, we have developed a metallopolymers-based system that employs electrosorption techniques for selective capture and release of PFAS from water. Our findings provide a promising approach for future water remediation technology, thereby creating a water-energy nexus.

79e Two-Dimensional Molybdenum Disulfide Based Field Effect Transistor Sensor for Pfas Detection in Water

Md Mohidul Alam Sabuj, Mo Li, Maria Armanious, Meng-Qiang Zhao

Per- and poly-fluoroalkyl substances (PFAS), a large group of persistent anthropogenic organic chemicals widely applied in industrial chemical processes and consumer products, have become an emerging concern because of their adverse environmental effects and human health. Traditional technologies such as liquid chromatography-mass spectroscopy and gas chromatography-mass spectroscopy are often not appropriate for continuous monitoring of the environment due to the high expenses and requirement of well-equipped laboratories with highly trained personnel. Therefore, developing a simple, inexpensive, and scalable method for rapid and in-situ detection of PFAS molecules in the environment, especially in drinking water, is of primary importance.

Here, we report a novel field effect transistor (FET) based on two-dimensional (2D) molybdenum disulfide (MoS_2) for PFAS detection in water. High-density MoS_2 flakes were synthesized by chemical vapor deposition and transferred on the channel areas between source-drain electrodes. Then, a hexagonal boron nitride (h-BN) film was transferred to the MoS_2 layer as an encapsulation layer to improve stability by preventing chemical degradation upon exposure to air or water. A pyrene-based linker molecule was used to bond PFAS molecules to the FET channel area, assisted by the hBN films. The PFAS FET sensor can be fabricated by functionalizing the channel area with PFAS specified molecular imprinting polymer (MIP) probe molecules for selective detection. It was observed that adding a pyrene-based linker produces an n-type doping effect by increasing the electron concentration in the channel, which increases the electrical conductivity. Finally, the performance of the device was characterized by measuring the changes in electrical conductivity of the MoS_2 channel with different concentrations of PFAS molecules. Our results demonstrate that the FET sensor exhibits a linear response to PFAS in the ppb to ppt concentration range, with a high sensitivity of <0.01 ppb and good selectivity. This work provides an essential advancement in developing next-generation electronic sensors, which can be a cost-effective, rapid, and sensitive detection tool for in-situ PFAS monitoring in water.

79f Polyelectrolyte Nanofiltration Membranes for Base Recovery Processes

Joshua Livingston, G. Kane Jennings, Shihong Lin

Climate change and overconsumption threaten the availability of potable water, non-renewable minerals (phosphorous, nitrogen, etc.), and valuable metals (lithium, copper, rare earth minerals). Nanofiltration (NF) membranes address these challenges by enabling monovalent/divalent species separation with unrivaled precision. Base recovery processes have emerged as essential tools in resource recovery, targeting the removal of multivalent species from hydroxide-rich streams. Most notably, there is growing demand for membranes capable of separating CO_3^{2-} from hydroxide-rich CO_2 capture solvents and PO_4^{3-} from hydroxide-rich adsorbent regeneration solvents. The conventional NF membrane's polyamide active layer degrades following long-term exposure to alkaline conditions. In this study, we fabricate alkaline-resistant polyelectrolyte membranes by depositing alternating layers of polycation (poly(diallyl dimethylammonium chloride)) (PDADMAC) and polyanion poly(sodium 4-styrenesulfonate) (PSS). The PDADMAC/PSS bilayer is held together via electrostatic interactions that are maintained at extreme pH conditions. We demonstrate their promise for base recovery applications by testing CO_3^{2-}/OH^- and PO_4^{3-}/OH^- separation performance. The alkaline stability of polyelectrolyte and polyamide membranes is evaluated by exposing membranes to pH 13 KOH solutions, then testing performance periodically over four weeks. CO_3^{2-} and PO_4^{3-} rejection increase as the number of polyelectrolyte bilayers increases; five PDADMAC/PSS bilayers remove >99% of CO_3^{2-} while allowing complete passage of ^-OH at pH 13. Commercial polyamide NF membranes degrade rapidly following long-term exposure to alkaline conditions, but polyelectrolyte membranes maintain long-term performance. Ultimately, PDADMAC/PSS polyelectrolyte membranes purify ^-OH-rich streams, retain multivalent species for further treatments, and show superior alkaline resistance to polyamide membranes.

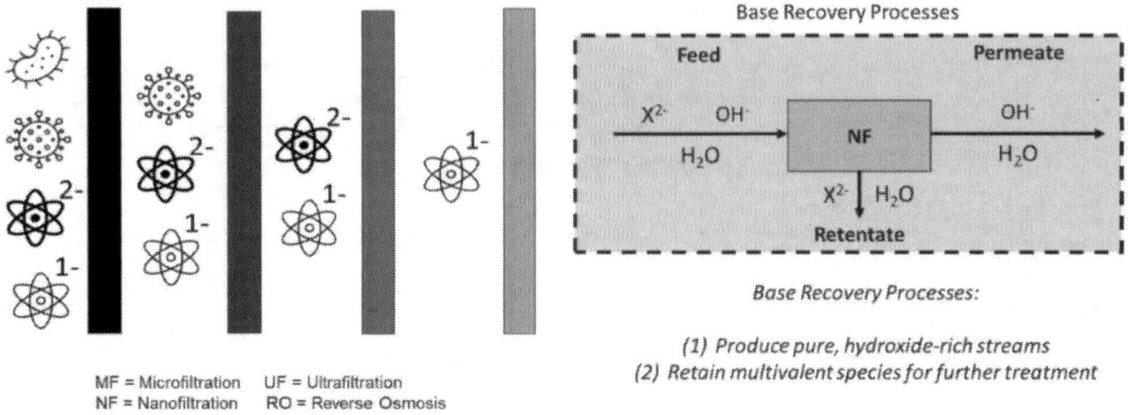

Scheme 1: Base recovery processes utilize nanofiltration membranes retain and concentrate multivalent species while permitting hydroxide to permeate through the membrane.

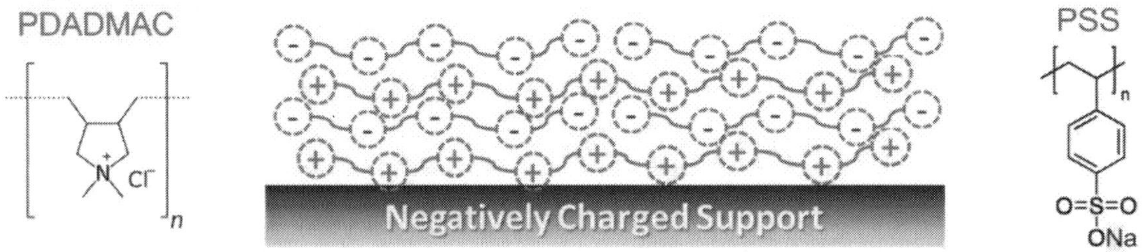

Scheme 2: Polyelectrolyte nanofiltration membranes maintain mechanical stability via the electrostatic interactions between alternating polycation and polyanion layers. Membranes terminated with the polyanion are negatively charged.

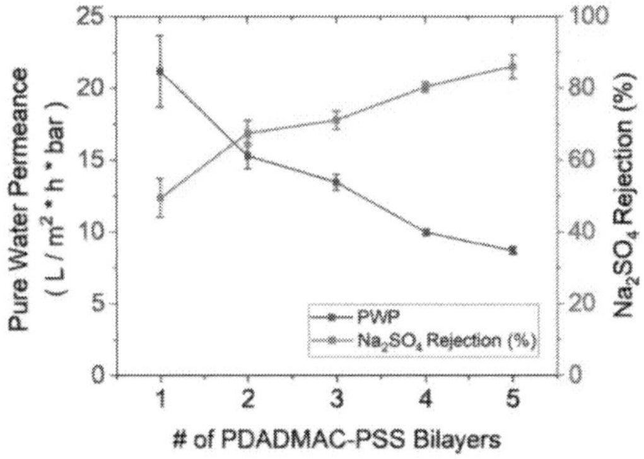

Figure 1: Na2SO4 rejection increases as the number of PDADMAC/PSS bilayers increases while the pure water permeance decreases.

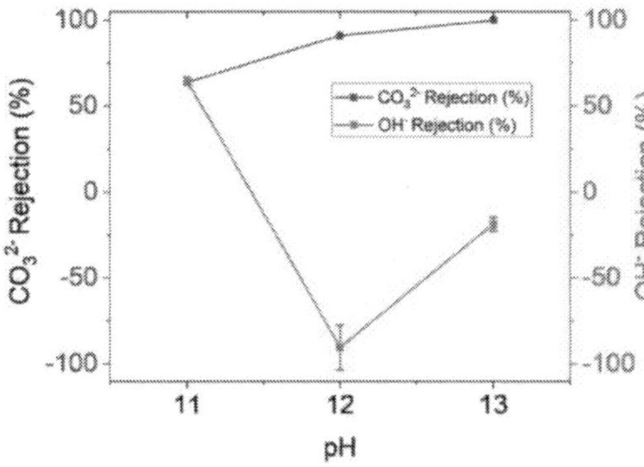

Figure 2: Carbonate rejection increases as the feed solution pH increases. Negative hydroxide rejection is observed when carbonate rejection increases.

79g Evaluation of a Janus Membrane for Enhanced Rejection of PFAS in Membrane Distillation

Hyunsik Kim, Mallikarjuna Nadagouda, Tae Lee

Treating per- and poly-fluoroalkyl substances (PFAS) contaminated water is an important research topic due to the potential health risks associated with PFAS exposure. Membrane distillation (MD), which utilizes hydrophobic microporous membranes that prevent the transport of non-volatile solutes, is a promising method for separating PFAS from challenging water matrices such as landfill leachate, wastewater, and reverse osmosis residuals. However, amphiphilic PFAS, such as perfluorooctanoic acid (PFOA), may establish hydrophobic interactions with the membrane surface leading to the adsorption of PFAS onto the membrane surface. PFAS adsorption presents a major challenge in the MD process, as it can result in both surface diffusion and pore wetting, which severely impact the separation efficiency. Here, we hypothesize that controlling hydrophobic-hydrophobic interactions is vital for enhancing the membrane's resistance to PFAS adsorption and surface diffusion. In order to verify the above, a Janus membrane was fabricated by solution-casting a blend of polyvinyl alcohol (PVA) and sulfosuccinic acid (SSA) onto the hydrophobic MD membrane, giving it both hydrophilic and hydrophobic properties. LC-MS/MS analysis was performed on both the feed and permeate samples to measure the extent of PFAS rejection. The electronic conductivity of the permeate as well as the impedance across the membrane were measured to monitor the progress of the pore wetting. A plate-and-frame membrane cell was employed to evaluate the benefits of a Janus membrane relative to a pristine polytetrafluoroethylene (PTFE) under high concentrations of PFOA and NaCl. The study demonstrated the membrane's remarkable capability to suppress pore wetting as evidenced by superior salt rejection (>99%) and a steady permeate flux under stressed operating conditions where the membrane retentate was continuously recycled. The current research implies a high potential for functionalized membranes for targeted treatment of PFAS from difficult-to-treat water matrices.

79h Graphene As a Rational Interface for Enhanced Removal of Cyanotoxins and Emerging Contaminants for Water Security

Jesse Roberts, Sarah Grace Zetterholm, Luke A. Gurtowski, Ashvin Fernando, Kevin Wyss, Angela Evans, Justin Puhnaty, Brianna Fernando, Audie Thompson, James M. Tour, Chris Griggs

Emerging contaminants represent a global environmental threat to ecosystems and drinking water supplies due to their potential adverse effects on human health. These contaminants, including cyanotoxins and micropollutants, are not effectively removed by traditional treatment methods, such filtration and UV disinfection. Graphene nanomaterials have recently emerged as a promising solution for efficient and sustainable removal of emerging contaminants from water due to their high surface area, cost effectiveness, and ability to adsorb many classes of contaminants.

Here, we show the direct use of graphene nanoplatelets (GnPs) as an adsorptive nanomaterial for the enhanced removal of cyanotoxins, including microcystin-LR, anatoxin a, and saxitoxin, as well as micropollutants such as DEET and DEP. Using batch adsorption experiments, we benchmarked GnPs against the commercially available granular activated carbon (GAC) to evaluate the efficiency of GnPs for adsorbing these contaminants from water.

Our experimental results demonstrated that the sp2 carbon network and pore characteristics (pore size and morphology) of graphene make GnPs a highly effective adsorbent for the removal of both cyanotoxins and micropollutants from water. Using the intraparticle-diffusion (IPD) model, we proved that the mesopore structure and surface interactions of graphene played a major role in increasing the rate of adsorption for all contaminants where adsorption reached equilibrium in under 30 minutes for GnPs. The surface properties also drastically increased the adsorption capacity for GnPs to over 100x greater than GAC's capacity when looking at microcystin-LR. Initial results with all other contaminants show that the adsorption capacity using GnPs can be up to two orders of magnitude higher than GAC's capacity.

Although noncovalent interactions typically rely on computational methods for mechanistic insight, novel NMR titrations were used to probe primary interactions between graphene and MC-LR adduct. This study proved that interactions between the aromatic ring on MC-LR and graphene sp2 orbitals are a dominant interaction. With rapid kinetics and adsorption capacities much higher than GAC, it is anticipated that graphene will offer a novel molecular approach for removal of toxins and emerging contaminants with aromatic systems.

✓ ¹H-NMR based visualization
✓ adsorption-desorption studies
✓ π-based interaction driven binding
✓ Utilization of graphene NPs for MCLR capture

71a Strategies to Understand Dust Transport and Screen for Potential Health Impacts

Kamaljeet Kaur, Reuben Attah, Kerry Kelly

Globally, arid and semi-arid lands cover approximately one third of the Earth's land surface, and our changing climate is expanding the extent of these lands. The Great Salt Lake (GSL) is a prime example of this phenomena. It is at historically low levels, with more than 750 mi^2 of exposed lakebed, some of which contain elevated levels of toxic metals, such as lead and arsenic. Under appropriate meteorological conditions, portions of this exposed lakebed produce substantial dust plumes, and the prevailing winds can transport this dust directly into the highly populated Wasatch Front. However, regulatory measurements capable of capturing dust (PM10, particulate matter smaller than 10 μm in diameter) are sparse, sparser than measurements of PM2.5 (PM smaller than 2.5 μm in diameter). Although low-cost sensors could supplement regulatory monitors, as numerous studies have shown for PM2.5 concentration, most of these sensors are not effective at measuring PM10 despite claims by sensor manufacturers. This study focuses on understanding dust transport using novel low-cost sensor measurement strategies and as well as screening results to understand the oxidative potential of dust from the GSL.

71b Observations of Atmospheric Conditions during the 2020 Solar Eclipse over South America

Konstantine Geranios, Alexander Chambers, Malachi Mooney-Rivkin, Jennifer Fowler, Matthew Bernards

Despite the challenges of conducting field work in an international setting during the global COVID-19 pandemic, a field project entitled "Atmospheric Gravity Wave Radiosonde Field Campaign for Eclipse 2020" was successfully conducted in Chile during the total solar eclipse (TSE) on December 14, 2020. The project collected radiosonde and surface measurements before, during, and after the TSE at Toltén and Villarrica, Chile, along the path of totality, to study the atmospheric impacts of the TSE with a focus on atmospheric gravity waves (momentum and energy transfer) generated by the TSE in the stratosphere.

High resolution radiosonde soundings were collected from both sites, with 48 hours of hourly high-altitude balloon launches centered around totality. Soundings profiled temperature, pressure, relative humidity, windspeed, and wind direction every 1 second (equivalent to ~5 m vertically) from the surface into the stratosphere. Additional surface measurements were collected using a Lufft weather station every 30 seconds over the duration of the campaign.

Unfortunately, the meteorological conditions surrounding the actual eclipse were not ideal. An atmospheric river (AR) was co-located with the TSE path before and during TSE on December 14, 2020, based on analysis of the integrated water vapor and integrated vapor transport. The AR brought downpours and thick cloud cover to both field sites. However, the data collected allowed for detailed analysis of the combinatorial impacts of the AR and TSE on solar irradiance, surface temperature, evolution of the planetary boundary layer, lapse rate, and generation of atmospheric gravity waves. These impacts will be discussed in detail.

Additionally, a nearly identical field campaign will be conducted in Lakeview, OR during the annular solar eclipse event on October 14, 2023. Radiosondes will be launched 24 hours prior to totality and 6 hours after totality in this campaign. Very preliminary results of the impacts of the eclipse on solar irradiance, surface temperature, evolution of the planetary boundary layer, lapse rate, and generation of atmospheric gravity waves will also be presented.

71c COSMO-RS Re-Parameterization for Accurately Predicting Deep Eutectic Solvent-Fluorinated Compound Systems.

Thomas Quaid, Toufiq Reza

Greenhouse Gases (GHG) are vaporized compounds that absorb infrared radiation (IR) and act as heat sinks in Earth's atmosphere. Among GHG, fluorinated compounds are considered the most potent with some species like R134a having global warming potential of 1,430, meaning it is 1,430 times more potent than CO_2. The aim of this research is to analyze the potential use of "A conductor like screening model" (COSMO) to evaluate a new class of green solvents called deep eutectic solvents (DES) to selectively capture R134a. DES are considered cheap, environmentally benign, and highly tunable as they are comprised of two or more components with nearly infinite combinations and compositions. They have been proven to be highly efficient at selectively absorbing CO_2, but little research has been done regarding other GHG's like R134a. Due to the lack of experimental research regarding DES and fluorinated compounds, and the complex nature of the solvent-solute interactions, a complete re-tuning of COSMO-RS is undertaken to make the software a viable option for screening DES on a large scale to make experimental work efficient. Through initial benchmarking for DES and R134a systems it was concluded COSMO-RS produced errors too significant for reliable computations. Thus, a deep study and reparameterization of the softwares variables including but not limited to hydrogen bond temperature dependence, vander Waals temperature dependence, gas phase transition scalar, and misfit factor are undertake to make the method suitable for such systems. The parameterization was done through literature values with similar systems and significant increase in accuracy has been achieved. The work is beneficial to all researchers who utilize COSMO as it shows the inner workings and method for training it to work with exotic systems.

71d Rapid Photolysis Decay of Gaseous Organic Nitrates Formed from Hydroxyl and Nitrate Radical Oxidations of α-Pinene and β-Pinene

Nga Lee Ng, Masayuki Takeuchi, Yuchen Wang

Photolysis of gaseous organic nitrates is crucial for understanding the formation and fate of air pollutants, such as nitrogen oxides (NO_x) and ozone (O_3). Monoterpenes are prevalent biogenic volatile organic compounds, contributing to the formation of organic nitrates; however, there is currently a lack of experimental constraint on the photolysis chemistry of monoterpene nitrates. Here, we conducted laboratory chamber experiments to investigate photolysis of organic nitrates formed from hydroxyl and nitrate radical oxidations of α-pinene and β-pinene. We directly measured their photolysis rates by chemical ionization mass spectrometry (CIMS). The chamber photolysis rate constants vary depending not only on the molecular formulae of organic nitrates but also on the precursor type and oxidation condition. While the photolysis rate constants of many detected C_{10} organic nitrates are on the order of 10^{-5} s^{-1} or larger, the other organic nitrates exhibit little to no decay behavior. The most photo-labile organic nitrate is $C_{10}H_{17}NO_5$ formed from nitrate radical oxidation of α-pinene and β-pinene with a chamber photolysis rate constant of 1.1 (± 0.1) and 1.3 (± 0.3) × 10^{-4} s^{-1}, respectively. We further determined that the ambient photolysis rate constant is as large as 6.4 (± 3.0) × 10^{-4} (corresponding photolysis lifetime of 0.43 ± 0.20 h). Compared to other loss processes (i.e., photooxidation and dry deposition) of gaseous organic nitrates, photolysis is a comparable or dominant loss process. These findings have important atmospheric implications regarding the role of monoterpene nitrates on the spatial distribution of NO_x and O_3 formation.

71e $PM_{2.5}$ bound Species Variation and Source Characterization in the Post-Lockdown Period of the Covid-19 Pandemic and Fireworks Period in National Capital of India

Mohd Faisal, Umer Ali, Vikram Singh, Mayank Kumar

The rapid growth of industrialization and increase in energy demand has led to the significant deterioration of air quality. Long-term Exposure to high $PM_{2.5}$ levels can result in acute and chronic health effects in adults and children. Therefore, knowledge and identification of aerosol sources can be very helpful for making quality control strategies, aerosol health impacts and climate change effects. Delhi is one of the world's most polluted cities, with numerous severe air pollution episodes and haze events in recent years, this forces us to investigate the sources of air pollution to make effective mitigation technology. This work analyzes the $PM_{2.5}$ Highly bound time-resolved non-refractory and elemental compositions to observe the lag effects of the COVID-19 lockdown in the capital city of India. Here, a comparative scenario between lockdown and unlock periods along with previous years' work is presented. The following eight source profiles are identified from the source apportionment of elements such as power plant (35.6%), dust (25.7%), biomass burning (16.2%), vehicular emission (10%), secondary chloride (4%), waste incineration (3.6%), coal combustion (2.8%) and industrial emission (2.2%). The apportion sources of organic aerosol (OA) yielded two primary factors, primary organic aerosol (POA) and biomass burning organic aerosol (BBOA), and two secondary factors, oxygenated organic aerosols 1 and 2 (OOA-1 and OOA-2). The two oxygenated factors, OOA-1 and OOA-2 dominated the total OA mass concentration with 67.5% and 22% contribution, followed by POA and BBOA with 6.5% and 4%, respectively. In this work, the POA was found to be lower as compared to previous studies, and it was about 72% lower, as reported by Shukla et al. (2021). The significant reduction might be attributed to restrictions imposed by the government in the lockdown period. Also in order to study the short-term air quality degradation with very high-intensity episodic event , especially the worsening of air quality in the form of ambient particulate matter (PM) pollution, which has been observed to link as a result of the firework displays (FW) due to nationwide celebrations in different countries like the Diwali Festival in India, Independence Day in the United States, the Lantern and Spring Festivals in China, Guy Fawkes Night in the United Kingdom, and the New Year's Eve celebrations around the world. Fireworks activity is behind the substantial increase in particulate concentration during Diwali in Delhi and the concomitantly enhanced atmospheric metal concentration. Therefore, a comprehensive understanding of the impact of fireworks on the atmosphere is crucial to developing a mitigation strategy. Accordingly, this study focuses on the comparative analysis of fireworks events in Diwali across 2019, 2020, and 2021. The current analysis studies the impact of Diwali fireworks for three consecutive years on Delhi's composition, concentration, and ambient $PM_{2.5}$ sources. The measurement of elements in $PM_{2.5}$ is performed using automated ambient metals monitor with sufficient resolution to capture the time-variation during the fireworks event. Dispersion normalized PMF (DN-PMF) was also performed using the time-dependent ventilation coefficient to factor out the impact of meteorology across the three years. The average ventilation coefficient was 3672, 2874, and 2000 m^2 /s in 2019, 29 2020, and 2021, respectively. Nine different sources were identified for all three years. The study apportioned over 85% of elemental mass to the firecrackers during Diwali.

71f Eliminating Soot Emissions from Jet Fuel Combustion

Georgios Kelesidis, Amogh Nagarkar, Una Trivanovic, Sotiris E. Pratsinis

Soot from jet fuel combustion in aircraft engines contributes to global warming through the formation of contrail cirrus clouds that make up to 60 % of the total radiative forcing from aviation. Here, the elimination of such emissions is explored through gas injection (containing 0 - 25 vol % O_2) at the exhaust of enclosed spray combustion of jet fuel that emulates nicely aircraft soot emissions (Trivanovic et al., 2022). It is shown that injecting gas containing 5 vol % of O_2 enhances the formation of polyaromatic hydrocarbons that adsorb on the surface of soot. This increases its number density and volume fraction by 25 and 80 %, respectively. Further increasing the O_2 concentration to 20 or 25 vol %, however, enhances oxidation (Kelesidis & Pratsinis, 2019) and nearly eliminates soot emissions from jet fuel spray combustion, reducing their total number density and volume fraction by 87.3 or 95.4 and 98.3 or 99.6 %, respectively (Fig. 1)! The reduction of soot emissions attained here is up to 60 % higher than that by blending jet fuel with alternative ones. So, injection of air just after the aircraft engine exhaust can reduce the radiative forcing from their emissions by 50 %.

References:

Kelesidis, G.A., Trivanovic, U., and Pratsinis, S.E. 2022. High-throughput generation of aircraft-like soot. Aerosol Sci. Technol. 56:732–743, doi.org/10.1080/02786826.2022.2070055.

Kelesidis, G. A., and Pratsinis, S. E. 2019. Estimating the internal and surface oxidation of soot agglomerates. Combust. Flame 209: 493-499, doi.org/10.1016/j.combustflame.2019.08.001.

Figure 1. Volume fraction of aviation surrogate soot as a function of the O_2 concentration in the gas injected at the exhaust of jet fuel combustion.

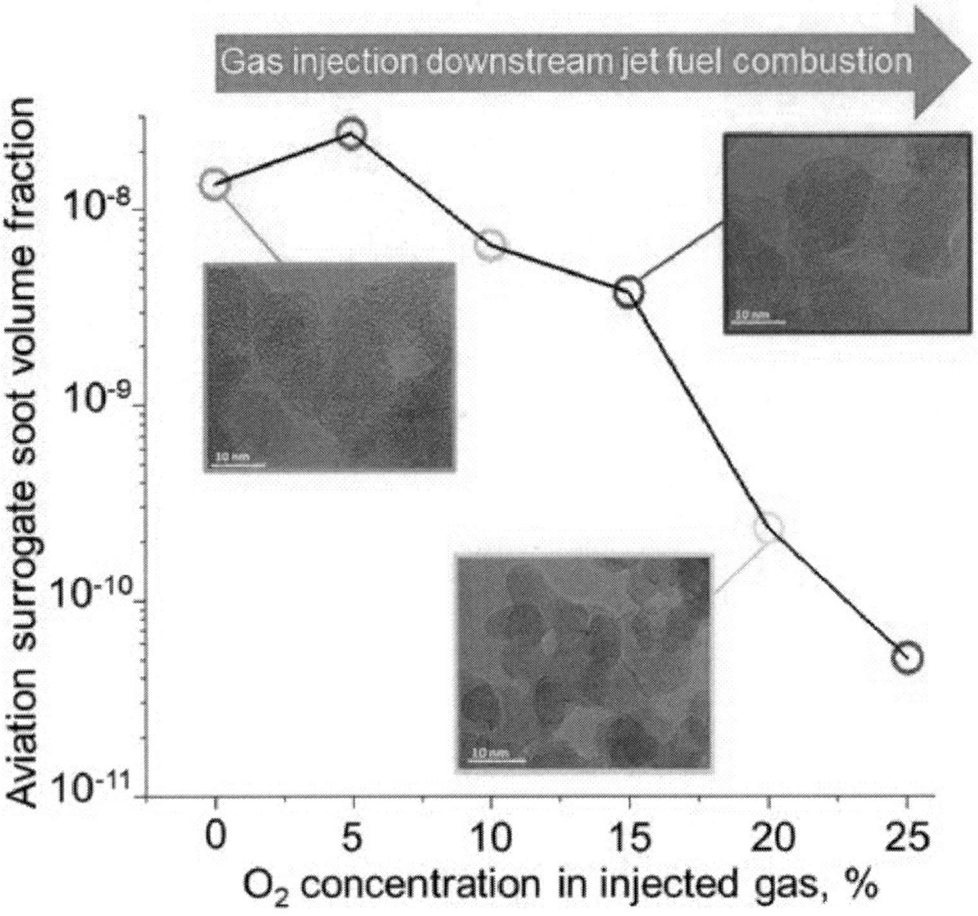

Gas injection downstream jet fuel combustion
Aviation surrogate soot volume fraction
10^{-8}
10^{-9}
10^{-10}
10^{-11}
0
5
10
15
20
25
O_2 concentration in injected gas, %

71g "I Can't Breathe:" the Invisible Slow Violence of Breathing Politics in Minneapolis

Marwa El-Sayed, Heather O'Leary, Scott Parr, Deondre Smiles

Following the death of George Floyd in Minneapolis, his utterance, "I can't breathe," reverberated internationally as the world population grappled with the twin specters of life-threatening COVID-19 respiratory morbidities and mounting years under increasingly polarized racist regimes. Despite crisis fatigue, national and international outpourings of solidarity trended on social and mainstream media. However, in this moment, the legacy of structural and slow violences against the living, breathing Minneapolis–St. Paul communities of color were obscured. This study addresses transdisciplinary breathing politics in this mid-sized American city to integrate atmospheric indicators (concentrations of criteria pollutants including particulate matter and gaseous pollutants), traffic indicators (Minnesota Department of Transportation permanent traffic monitoring station data), and social indicators (community responses in newspaper and Twitter archives), ultimately making visible how Floyd's utterance reflects much deeper patterns of stratified urban public health risks and socio-environmental airscape politics.

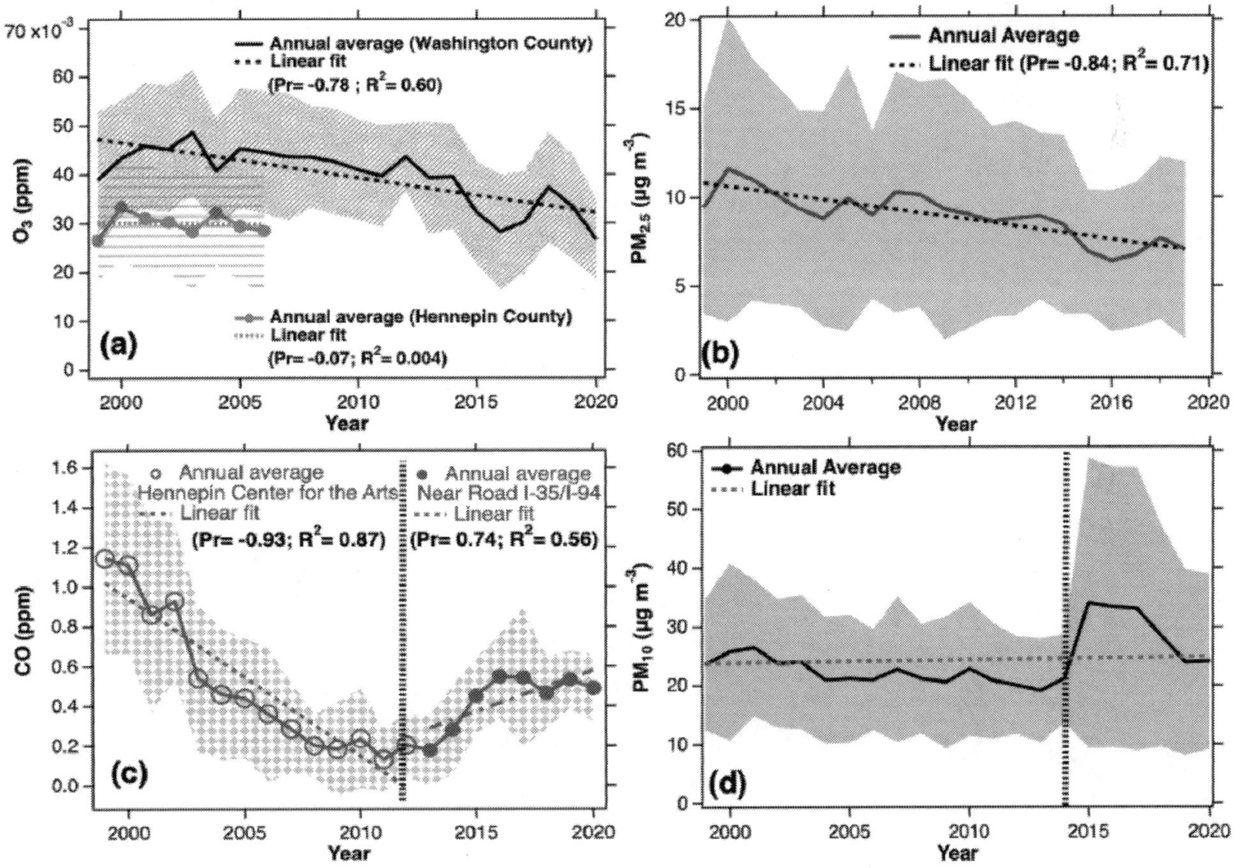

71h Development of Large-Scale Fire Whirls for Offshore Oil Spill Cleaning

Mitchell Huffman, Joseph Dowling, Bhushan Pawar, Wuquan Cui, Mohammadhadi Hajilou, Karen Stone, Elaine Oran, Michael J. Gollner, Qingsheng Wang

Fire whirls are a phenomenon that can intensify combustion, often formed in uncontrolled situations such as wildfires. Researchers have begun to explore the potential of harnessing this phenomenon in order to enhance combustion, increase burn efficiencies and reduce pollution. The fire whirl provides a method to promote oxygen entrainment though circulation to provide a more efficient burn, and increase heating rates to the fuel surface thereby removing more spilled fuel, resulting in less environmentally harmful burn products and residues. The potential for reduced combustion emissions as well as high efficiency burns creates a motivation to understand these dynamics at large-scale for the full implications of this phenomenon. Most knowledge of fire whirls has come from laboratory-scale experiments with a focus on scaling laws to predict their formation and structure, with some success in scaling up these approaches to large-scale observations. Only a limited subset of these experiments has captured emissions. However, these results have shown improved efficiencies suggesting the need for scale-up testing to verify this phenomenon. The current study will perform the largest fire whirl test to this point, utilizing a 1.2-meter diameter pool of HOOPS crude oil to establish the conditions necessary for a fire whirl to form outdoors at large-scale. Three sidewalls will be used to induce the circulation necessary through natural entrainment. Emissions will be measured using a UAV flying over the experiments as well as thermal, flow, and video measurements of the fire whirls. The successful execution of this experiment will provide the potential for cleaner burns in situations such as large oil spills on the ocean, releasing less harmful byproducts and less soot into the air.

Improved Approach for Benefit Analysis of Carbon Reduction Methods

Leisa Porter

In the world-wide efforts to reduce the emissions of carbon dioxide and other greenhouse gases, many methods and types of facilities have been proposed. The choice of technology is usually determined by government organizations, rather than corporations, because the investments can be extremely expensive and rarely is any immediate profit to be made by reducing greenhouse gases. Scientific comparison of alternatives would ideally be done by cost-benefit analysis, taking into account who would pay the costs and who would receive the benefits. But there are problems with cost-benefit analysis in this situation. Government projects have a history of overrun budgets; even the amount of overrun is difficult to estimate. It is impossible to agree on the economic value of benefits (actually disaster avoidance) for events such as the disappearance of endangered species.

Instead of cost-benefit analysis, benefit ranking provides a more useful comparison in this situation. In this study, the benefit is carbon dioxide reduction in terms of tons per year (or equivalent for other greenhouse gases). Because there are not enough resources nor enough time to implement all carbon reduction technologies at once, ranking is necessary to accomplish the most benefit over time, especially because the climate benefits are cumulative. The ranked benefits can be evaluated within engineering accuracy using the well-developed sciences of chemistry, physics, thermodynamics and statistical reliability, adjusted for the real-world effects of natural phenomena and human behavior.

For this paper, the benefit of electrical storage batteries is compared for two design applications. The first is storage batteries in connection with solar power generation facilities, to extend the facility production to cover the peak electrical demand that typically occurs in the evening. The second application is batteries to power vehicles that will be recharged from the electrical grid. Battery effectiveness is measured as avoided carbon emissions per kilowatt of battery capacity (realizing that batteries paired with solar plants will be different size than those in vehicles). Other technology applications could have been evaluated in a similar way, for instance batteries paired with wind generators, or used to produce hydrogen by electrolysis. The application of batteries to solar generators compared to electric vehicles was chosen for this paper because there is ample evidence of a world-wide shortage of the materials needed to produce efficient rechargeable batteries, and. because there are published data on hourly solar potential and peak demand. Calculations were done for different locations because data for sunlight power, peak load demands, average driving distances, line losses, and sources of grid supply power vary across the country. As an option to batteries, the paper investigated the minor effect of load-shedding of specific, customer-chosen nonessential loads during periods of peak demand.

The results showed that for both locations studied, the pairing of batteries with solar power facilities is orders-of-magnitude more effective at avoiding carbon-dioxide emissions than using the equivalent battery capacity in all-electric vehicles. This is true regardless of the electrical energy sources available, and is insensitive to uncertainties in future battery technology including the energy cost of producing and disposing or recycling of waste battery materials. The uncertainties and assumptions associated with the calculations are addressed in the Appendix section of the article.

80a The Feasibility of Palm-Based Biomass Co-Firing and Different Oxygen Carriers to Enhance the Performance of Coal-Based IGCC Power Plant

Nasir Al Lagtah

The constantly growing global energy demand for energy is still produced from fossil fuels with coal being the most abundant fossil fuel. Even with the rise in the global awareness about the negative effects non-renewable resources of energy, the abundance and the low cost of coal still ensures the existence of coal-fired power plants.

This research investigates the potential of retrofitting coal-based IGCC power plants with palm-based biomass co-firing and chemical looping combustion (CLC) carbon capture for near zero carbon emissions. The investigated biomass are palm coir, palm empty bunch and palm leaflets (agro-wastes in UAE). The effect of different biomass co-firing and CLC oxygen carriers on the techno-economic performance of coal-based IGCC power plants was modelled and simulated using Aspen Plus software.

The results showed that the use of iron oxide as an oxygen carrier resulted in better performance than the traditional oxygen carrier of nickel oxide in terms of total plant efficiency, total plant energy output and carbon specific emissions. However, to achieve near zero carbon emissions, the flow rate of the oxygen carriers was 15 times the flow rate of the feed.

Increasing the percentage of palm coir co-firing resulted in better performance regardless of the used oxygen carrier. However, the performance has diminished when the percentage of palm empty bunch and palm leaflets co-firing increased. One of the major drawbacks of biomass co-firing is the particulate matter that remains. The presence of particulate matter results in downstream equipment fouling and hence, lowering the overall power plant efficiency. This issue has been addressed by adding cyclones to the syngas cleaning section.

The use of CLC for carbon capture has increased the cost of power production but the increase depended on the type of oxygen carrier. The cost of power production has increased by 32% when iron oxide was used, whilst the cost has increased by 45% when nickel oxide was used. However, there was no clear trend to indicate if increasing biomass co-firing percentage will either increase or decrease the cost of power production. Unfortunately, the lowest cost was greater than actual market trading prices. Therefore, currently no clear economic drive for developing coal and biomass co-firing IGCC power plants with CLC carbon capture technology.

Even though, a satisfactory coal and biomass co-firing IGCC power plant with suitable decarbonisation scenario might be economically feasible, depending on a trade-off between the decarbonisation degree and CO_2 specific emissions. All that will be driven by the introduction of new national/international legislations regarding allowable carbon emissions from coal-fired power plants.

80b Process Modeling, Techno-Economic and Life-Cycle Assessments of Producing Potentially Carbon-Negative Building Material from CO_2 and Waste Lignin or Lignite

Yuan Jiang, Francesca Pierobon, Nicholas Nelson, Keerti Kappagantula, Satish Nune, David J. Heldebrant

Many technologies have been developed to capture CO_2 from point sources and convert it into fuels or chemicals, which typically emit CO_2 at the end of their life. However, there are limited studies of technologies that convert CO_2 into materials that can retain atmospheric carbon in a sequestered form for a long duration. To address this gap, the project team at Pacific Northwest National Laboratory developed an emerging two-step technology to produce a potentially carbon-negative building material from CO_2. In the first step, CO_2 captured from point sources is reacted with waste lignin or lignite to form CO_2-loaded lignin/lignite (CO_2LIG) material via carboxylation, forming durable carbon-carbon bonds. Then CO_2LIG filler is mixed with commodity polymers (i.e. high density polyethylene) to produce CO_2LIG-polymer composites via injection molding. A CO_2 loading up to 5 wt.% in CO_2LIG and a CO_2LIG loading up to 70-80% in the polymer composite have been achieved in the experimental study, which leads to a CO_2 loading up to 4 wt.% in the final product. In addition, water-lean solvents developed by PNNL will be used to capture CO_2 from point sources. Compared with the industrial benchmark, aqueous amines, the energy consumption, and carbon capture cost of water-lean solvent are roughly 22% and 15%, respectively. In this talk, we present process models developed in Aspen Plus based on lab-scale experimental data, techno-economic and cradle-to-gate life-cycle GHG emission assessments for both lignin and lignite cases and assess their viability as a carbon-negative approach. We also detail comparisons with commodity materials in the present market, and examine the sensitivity of how key economic and environmental measures are impacted by feedstock resources and logistics, reactor performance and mixing ratio between CO_2LIG and polymer.

80d Evaluating on-Board CO_2 Capturing Methods for the Ship with Outlook on Storage and Utilization Options

Mamoun Al-Rawashdeh, Jaafar Ballout, Dhabia Al-Mohannadi, Gareth Burton, Joseph Rousseau, Patrick Linke

Today there are around 60,000 large-scale ships that are moving in the oceans contributing to 2.9% of total global CO_2 emissions. They are faced with a significant challenge to meet the strict and ambitious upcoming regulations planned by the International Maritime Organization (IMO) to reduce carbon intensity by at least 40% by 2030 and aim towards 70% by 2050. CO_2 capturing technology options available from on-shore applications do not fit directly the ship requirements in terms of space limitation, limited resources, and sources of energy, and it is mobile and dynamic not like the typical stationary emission sources of power plants and those in the chemical industry.

This work aims first at (1) reviewing the published studies concerning onboard ship carbon capture, storage, and utilization and synthesizing gaps in existing knowledge to provide a focus for future studies. Then (2) it evaluates five captured CO2 technologies from the perspective of the shipping industry as absorption, adsorption, membrane, cryogenic, and chemical looping. Till now, a limited number of studies have been carried out for onboard CO2 capturing mainly focused on absorption with MEA as the most studied solvent. Techno-economic studies showed a wide range of achieved capture rates for onboard capture systems from 54% to 90% and wide costs ranging from 77.5 to 389 €/ton CO_2. In addition, seven CO_2 storage conditions were reported in the literature for onboard capture systems. Our work reviewed the potential of CO_2 capture technologies, highlighted the ship requirements, and proposed potential energy integrations between the ship energy system and CO_2 capture unit. A visualization was developed and demonstrated to quantify the emitted CO_2 per distance crosses for containerships. It is shown as the number of lost Twenty Equivalent Units (TEUs) on a containership per 1000 km crossed in a sailing mode. Reports for onboard CO_2 utilization work was limited to one study so far. We listed two high-level utilization routeways for captured CO_2 on a ship. Direct utilization works by converting CO_2 to a fuel or chemical while in-direct utilization integrates captured CO_2 with onshore applications.

In the second part, the integration of the above CO_2 capture technologies, compression, and storage as well as utilization options are explored in the context of the existing energy system onboard an LNG carrier ship. Close attention is paid to achieving efficient heat and power integration for overall energy efficiency as well as vessel-specific design constraints including space limitations and motion. The work culminates in the identification of the most suitable technologies and an outlook on required future development efforts.

80e A Cross-Sector Tool for Assessing and Strategizing Mitigation of Industrial Carbon Dioxide Emissions

Ajay Koushik V, Ram Kishore Sankaralingam, Peter Kamau Waiyaki, Achyuta Krishnan, Arun Muthukkumaran, Ramesh Murugan Natarajan, Gowri Shankar Navagana, Satyanarayanan Seshadri, Preeti Aghalayam, Raghunathan Rengaswamy

A flexible tool for the mitigation of CO_2 in various sectors

In recent decades, climate change has become a pressing issue that demands urgent attention from society. One of the primary contributors to this issue is increased greenhouse gas emissions from human activities. Carbon dioxide, the most prevalent greenhouse gas, accounts for approximately 65% of total greenhouse gas emissions. Since 1970, carbon dioxide emissions have risen by around 90%, primarily due to the combustion of fossil fuels and industrial processes. Therefore, it is crucial to monitor the amount of carbon dioxide that is being released into the atmosphere through the industrial, agricultural, and transport sectors.

Further, a deeper understanding of these processes would facilitate identifying critical parameters in each sub-sector responsible for CO_2 emissions. Identifying such parameters would pave the way for modifying existing technologies or implementing new technologies that would reduce CO_2 emissions. A comprehensive analysis on a pan-country/pan-sector, considering several interconnected sectors and sub-sectors could aid in identifying critical sectors and sub-sectors contributing to CO_2 emissions, and articulate the right way forward for meeting various country or sector-wide targets including net-zero carbon emissions.

To address this problem, a comprehensive tool that models sectors/sub-sectors using first principles (later validated with relevant data) accounting for emissions holistically is being developed. Implementing this tool can provide valuable insights into the carbon footprint of various sectors and help strategize efforts to reduce emissions. By comprehensively accounting for emissions, policymakers and industry stakeholders can identify the most significant sources of emissions and propose the most relevant and effective action plans to mitigate their impact.

Description of the tool and proposed methodology

The tool includes a fundamental primitive module that can model the emissions of any sector based on its input and output data. Further, the module can be enhanced with any phenomenological/data-driven model, whenever it is instantiated. Upstream and downstream processes involved in a particular sector, accounting for appropriate emissions from all associated sectors, can be incorporated using these fundamental modules. The first version of this tool will be validated using sector-wise data from India. This tool can be modified or extended to any country, provided necessary data is available.

The tool is proposed to be primarily a first-principles-based model, which would later be validated/improved using available data. Any sector, sub-sector or entity of a particular subsector could be considered the control volume on which the mass and energy conservation equations could be applied. For instance, the agrochemical sector, ammonia production industry and the reactor inside a particular ammonia production industry will all obey the mass and energy balance equations. Since CO_2 is a material stream leaving the process (in most cases), it can be quantified with respect to several factors such as feed quality, source of electricity and the efficiency of the process. This proposed tool development project aims to identify a set of parameters for each significant CO_2-emitting sector and sub-sector, with a clear definition of the dependency of CO_2 emissions on these parameters. These parameters can serve as baselines for policymakers to create new achievable policies and act as levers for sectorial users to decrease CO_2 emissions.

Uniqueness of the tool

Recently, several other tools that aid policymakers and the stakeholders have been proposed in the literature. Wang et al. [1] presented a spatial and temporal model for several decarbonization possibilities in China's power industries. Li et al. [2] modelled several scenarios where the emission footprint of China's light-duty trucks could be reduced by electrification. Tan et al. [3] proposed a leader-follower model for government and industry to decarbonize maritime emissions. Using this, the government could propose CO_2 mitigation techniques for emission reduction, and the industry could adopt some techniques with minimal investment and less penalty. The subsidy-penalty play between these two bodies would result in an optimal solution for decarbonization. Matamala et al. [4] studied the possibility of incorporating carbon capture and storage (CCS) in the power sector with the aid of negative carbon emission techniques such as biomass for the Latin American region. Hebada et al. [5] analyzed four different scenarios for decarbonizing Brazil's iron and steel industries towards net-zero emissions goal for 2050. Vernay et al. [6] reported a model for analyzing the impact on energy community businesses during an energy transition.

Despite the depth of study in all the above models, the proposed tool is unique in several ways. The reported models in the literature are restricted to a particular sector/sub-sector, whereas the proposed tool offers a generalized framework to integrate and model all sectors. Likewise, the reported models in the literature are confined to a particular geographic location relevant to a particular industry (for instance, Brazil was chosen for the decarbonization of Iron and Steel in ref [5]). However, in the proposed tool, this is expected as an outcome. In other words, if the proposed tool is used for the analysis of the iron and steel industry using Indian data, the results of the model would suggest CO_2 emissions corresponding to which region of India is more of a significant concern. This, in turn, can be used to take a closer look at that region and make some process/policy recommendations. Also, the proposed framework is rooted in first principles supported by data, whereas the reported models are purely data-driven with fundamental constraints. This feature makes the tool user-friendly, where the user can easily input the variables, parameters and the associated relations between them, in addition to providing significant flexibility and promoting wide-spread usage.

Features of the tool and the expected outcomes

1. The tool will have in-built 'boxes' that the user can consider as the system boundary/control volume.
2. The user can provide the variables and parameters associated with the box in a standard format.
3. By default, basic balance equations will be assembled.
4. The user can provide additional equations that capture various known phenomenology.
5. The tool would provide relevant parameters for some common CO_2 emission processes such as cement, iron and steel, and power plant. However, the user can customize/modify it as required.
6. The tool can be interrogated to generate policy decisions and their impact on decarbonization.

In general, the proposed tool is expected to be practical and easy to use for managing multiple CO_2-emitting systems of interest. By accurately identifying relevant parameters and defining system boundaries, this tool has the potential to offer valuable insights into identifying, reducing, and mitigating CO_2 emissions on a global scale.

References

[1] Y. Wang, Z. Zhao, W. Wang, D. Streimikiene, T. Balezentis, Interplay of multiple factors behind decarbonization of thermal electricity generation: A novel decomposition model, Technological Forecasting and Social Change. 189 (2023). https://doi.org/10.1016/j.techfore.2023.122368.

[2] J. Li, L. Wang, C. Ji, H. Liu, L. Lu, T. Zhang, M. Zhao, S. Xu, Assessing the decarbonization potential of China's light-duty truck fleet by electrification, Energy Reports. 9 (2023) 212–225. https://doi.org/10.1016/j.egyr.2023.03.018.

[3] R.R. Tan, I.H. V. Gue, J.F.D. Tapia, K.B. Aviso, Bilevel optimization model for maritime emissions reduction, Journal of Cleaner Production. 398 (2023). https://doi.org/10.1016/j.jclepro.2023.136589.

[4] Y. Matamala, F. Flores, A. Arriet, Z. Khan, F. Feijoo, Probabilistic feasibility assessment of sequestration reliance for climate targets, energy. 272 (2023) 127160. https://doi.org/10.1016/j.energy.2023.127160.

[5] O. Hebeda, B.S. Guimarães, G. Cretton-Souza, E.L. La Rovere, A.O. Pereira, Pathways for deep decarbonization of the Brazilian iron and steel industry, Journal of Cleaner Production. 401 (2023) 136675. https://doi.org/10.1016/j.jclepro.2023.136675.

[6] A.L. Vernay, C. Sebi, F. Arroyo, Energy community business models and their impact on the energy transition: Lessons learnt from France, Energy Policy. 175 (2023). https://doi.org/10.1016/j.enpol.2023.113473.

80f A Framework for the Combined Evaluation of Economic and CO_2 Fixation Feasibility of Carbon Capture and Utilization (CCU) Reaction Pathways

Gasim Ibrahim, Mohamed Sufiyan Challiwala, Mahmoud El-Halwagi, Nimir Elbashir

A global effort is underway to reduce GHG emissions to address the growing concerns of anthropogenic climate change. At 8.5 Gt_{CO2} in 2020, industrial greenhouse gas (GHG) emissions account for approximately 24% of total global anthropogenic emissions. Therefore, the industrial sector needs to reduce GHG emissions by 1.2% to 7.4 Gt_{CO2} by 2030 to meet sustainability and decarbonization targets. Carbon dioxide capture and utilization (CCU) has garnered significant attention as a means of reducing GHG emissions and for decarbonizing several industries, including the natural gas industry, the cement industry, and the steel manufacturing industry. One means of applying CCU to a process is through the chemical conversion of CO_2 into other products. Not all CCU processes may be economically feasible or environmentally sustainable in terms of CO2 fixation. Typically, to assess the CO2 fixation of a given CCU reaction, a life cycle assessment (LCA) is conducted. Additionally, a technoeconomic assessment (TEA) is carried out to evaluate the economic viability of the process. It should be noted that conducting these assessments requires significant effort and time, as well as a substantial amount of theoretical and empirical data. Therefore, it is recommended to perform these assessments at a later stage in the development and scale-up of a CCU technology.

In this work we develop a framework for the combined evaluation of economic and CO_2 fixation feasibility of CCU reaction pathways at early stages of the technology's development by integrating two metrics: the metric for inspecting sales and reactants (MISR) and the CO2Fix. The MISR is a metric used to assess the economic feasibility of any commercial reaction. The CO2Fix estimates the CO_2 fixation capacity for a given CCU reaction, accounting for Scope 1 and Scope 2 emissions, while providing an upper limit for CO_2 fixation. Both metrics use limited information obtained from the reaction stoichiometry and basic thermodynamic properties of the materials involved in the reaction. The framework categorizes a given CCU conversion technology into one of 4 types: Type 1 technologies that are both economically feasible (MISR>1) and lead to CO_2 fixation (CO2Fix>1), Type 2 technologies that lead to CO_2 fixation (CO2Fix>1) but are not economically feasible on their own (MISR<1), Type 3 technologies that are economically feasible (MISR>1) but do not lead to CO_2 fixation on their own (CO2Fix<1), and Type 4 technologies that are neither economically feasible (MISR<1) nor lead to CO_2 fixation (CO2Fix<1). Furthermore, the framework enables the estimation of various parameters necessary to determine the economic and CO_2 fixation viability of different CCU technologies. Specifically, (a) it allows for the estimation of the minimum subsidy required for a Type 2 CCU conversion technology to become economically feasible, (b) the minimum amount of CO_2 capture necessary for a Type 3 technology to become a CO_2 fixing technology, (c) the minimum CO_2 credit required for a Type 2 technology to become economically feasible, (d) and the maximum CO_2 capture cost required for a Type 3 technology to become CO_2 fixing while maintaining economic feasibility. By providing these estimates, the framework can assist decision-makers in understanding the potential economic and environmental implications of different CCU technologies, and help in determining the most appropriate course of action for the development and scaling up of these technologies.

We provide a demonstration use of this framework to assess two commonly discussed reactions in the realm of CCU: CO_2 hydrogenation to methanol and the dry reforming of methane (DRM). The DRM reaction was assessed to be a Type 3 technology, with a CO2Fix of 0.92 and an MISR of 1.6. Furthermore, it was estimated that for CO_2 capture and sequestration cost less than 0.0.08 \$/kg-CO_2, the technology can be considered a Type 1 technology. At a CO_2 capture and sequestration cost of 0.05 \$/kg-CO_2 the MISR was calculated to be 1.57 at a CO2Fix of 1. The CO_2 hydrogenation reaction was assessed to be a Type 2 technology with a CO2Fix of 2 and a MISR of 0.9. The developed framework estimates that the CO_2 hydrogenation reaction can be considered a Type 1 technology with a CO_2 utilization credit above 0.073 \$/kg-CO_2. Moreover, at a CO_2 utilization credit of 0.1 the reaction has an estimated MISR of 1.02.

The developed framework is a powerful means of quickly screening CCU conversion technologies at early stages of research and development, providing insights on the necessary conditions for the reaction to be economically viable and CO_2 fixing. It can also be used to generate early quantitative estimates of the economic and CO_2 fixation potential of such reactions for the purpose of comparison and optimization studies.

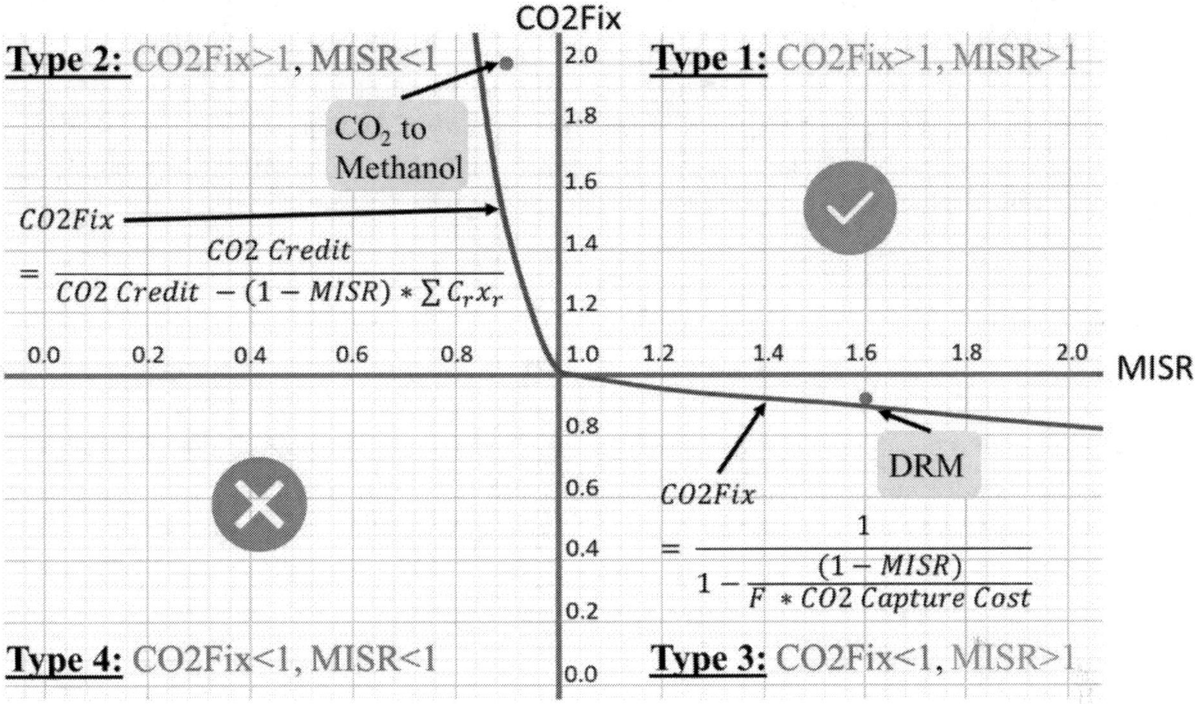

80g Determination of CO_2 Solubility in Brines and Hydrocarbons for Carbon Storage and Enhanced Oil Recovery

Sushobhan Pradhan, Rupom Bhattacherjee, Clint Aichele, Prem Bikkina

The release of CO_2, a major greenhouse gas (GHG), into the atmosphere is predominantly a consequence of excessive fossil fuel usage for industrial and transportation purposes. GHG emissions are known to cause environmental issues such as climate change. CO_2-based enhanced oil recovery (EOR) methods have been considered as a viable option to mitigate atmospheric CO_2 concentrations, due to their potential to permanently sequester a significant fraction of the injected CO_2, via solubility-, capillary-, and stratigraphic- trapping mechanisms, while meeting the growing demand for energy resources. During the CO_2 EOR, some of the injected CO_2 dissolves in the oil and brine phases and impacts the trapping mechanisms. Therefore, it is important to quantify the solubilities of CO_2 in the oil and brine phases.

In this research, we developed two novel experimental setups to measure CO_2 solubility in brines and hydrocarbons. The solubilities of CO_2 in brines of various salinities were measured as a function of pressure (150 psig to 1200 psig) at 65 °F (18 °C). The experimental results were compared with the CO_2 solubility predicted using Duan and Sun (2003) equation of state model. The experimental results are found to be in good agreement with the model predictions. The solubilities of CO_2 in *n*-hexane, *n*-decane, *n*-dodecane and a crude oil were measured at pressures ranging from 150 psig to 700 psig at 18 °C, using a swelling-based procedure. Then, the CO_2 solubilities were estimated using the swelling factors.

80h Research on the Construction of Advanced CO_2 Capture Absorbent and the Regeneration Enhancement Strategy

Weixin Kong

Amine-based adsorbents have the advantages of high absorption rate and high CO_2 selectivity in terms of industrial CO_2 capture performance. However, their poor desorption performance has reduced the recyclability of absorbents and increased the process cost. To solve the above problems, this research aims to develop a novel type of ionic liquid deep eutectic absorbents based on product-oriented regulation strategy. By strengthening the deep eutectic interaction between ionic liquids and solvents, the CO_2 absorption-desorption reaction path will be regulated and new products with low activation barriers can be generated, so as to reduce the heat duty of desorption reaction and overcome the thermodynamic bottleneck. By regulating the interaction between anions and cations, the stability and self-aggregation degree of the carbamate product will be weakened, so as to increase the desorption rate, and overcome the bottleneck of desorption kinetics. Through theoretical calculations to guide experimental synthesis, a functionalized ILs [DBNH][1,3-DMU]-EG deep eutectic absorbent was constructed, which can maintain ultra-high CO_2 loading (1.87 mol/mol) at low concentrations (0.50 mol/L), and the desorption efficiency up to 84.00%. The quantum chemical calculation results showed that the anions and cations of [DBNH][1,3-DMU] were dominated by weak van der Waals forces, and its intermolecular interactions were low. In addition, the ILs had high steric hindrance and electron-inductive properties, which could reduce the binding energy of amino groups to CO_2, thus reducing the stability of the product and making it easier to desorb CO_2.

18a Screening of Ionic Liquids for CO_2 Capture at Ambient Conditions

Rohit Chauhan, Rohan Sartape, Meenesh R Singh

The anthropogenic emission of CO_2 to the environment enhances its concentration in the atmosphere and becomes a significant source of global warming and the greenhouse effect. Therefore, via the Paris agreement, countries have decided to reduce their CO_2 emission drastically so that global temperature rises do not exceed 2 °C. The application of room-temperature ionic liquids (RTILs) is a practical approach presently being industrialized to capture CO_2 by chemisorption. Recently, it was observed that large amounts of CO_2 could be dissolved selectively and efficiently in a particular RTIL. Therefore, screening the RTILs is an essential parameter for the selective capture of CO_2. The simple measurement of the significant parameter of the CO_2 in RTILs, like diffusivity and solubility, is crucial for the progress of this research field. In this manner, an airtight glass vial was used to capture CO_2 into RTILs, and the equilibration of the absorption was analyzed by estimating Henry's constant at room temperature. In addition, time-dependent CO_2 absorption was also performed to screen the RTILs for capturing CO_2.

18b Controlled Morphology, Post-Synthetic Modification, and Scale-up of MOFs for Direct Air Capture of Carbon Dioxide

Xakin Ramirez Isunza, Brittany Bonnett, Connor Farrell, Amanda J. Morris, Stephen Martin

Net-zero emissions technologies have been identified by the IPCC as an important tool to achieve the goal of a less than 2 °C global average temperature rise over pre-industrial levels. Several companies are currently working to commercialize direct-air capture (DAC) technologies to remove CO_2 directly from the atmosphere, but significant strides need to be made in the development of sorbent materials to make this an economically viable approach. Sorbent materials need to exhibit high CO_2 capacity at low CO_2 concentrations (i.e. at atmospheric concentrations of ~400 ppm), rapid sorption and desorption kinetics to facilitate capture and regeneration cycles, low regeneration energy to minimize operating cost and carbon footprint, and low material capital costs. Sorbent materials also need to exhibit high stability over numerous sorption/regeneration cycles to increase the lifetime of the materials.

We have studied a series of metal organic framework (MOF) materials with the aim of optimizing and scaling them up as solid sorbents for direct air capture of CO_2. Specifically, we are interested in observing the effect of synthesis scale-up and particle morphology on packed bed CO_2 sorption in DAC-relevant conditions. Previously, a test series of controlled UiO-66 materials with well-characterized defect-levels and particles sizes was examined, and additional MOFs were selected based on high capacity (mmol/g) of CO_2 while also considering the isosteric heat of adsorption (Q_{st}). Herein we will present on DAC using **Mg-MOF-74** and **MIL-101-Cr** based sorbents.

Modulated syntheses of Mg-MOF-74 using different modulators (benzoic and salicylic acid) as well as a variety of reaction scales (100 mL, 250 mL, and 500 mL) resulted in variable and controllable particle morphologies, indicating that Mg-MOF-74 is highly favorable for morphological and size control studies relating to synthesis and gas sorption performance. MIL-101-Cr was synthesized in 25 mL, 265 mL and 600 mL reaction vessels using acetic acid as a modulator, resulting in negligible crystallographic or morphological difference between the samples. Additionally, a fraction of each synthesized MOF sample underwent post-synthetic ethylenediamine grafting to increase chemisorption site density. The resulting MOF samples were studied using isothermal gravimetric and volumetric sorption techniques. Based on these results, a subset of the MOFs was studied under multiple sorption/regeneration cycles in a packed bed sorption apparatus under DAC operating conditions.

18c Energy-Efficient and Stable Amine Structures Supported By Mesoporous Silica for Direct Air Capture

Joo-Youp Lee, Soumitra Payra

A large portion of CO_2 emissions in ~5.2 gigatons (Gt) is released in relatively small quantities from distributed sources emitted each year in the U.S. Therefore, for such emissions, point source CO_2 capture is not feasible and direct air capture (DAC) is an indispensable part of a diversified portfolio of technologies to mitigate U.S. greenhouse gas emissions. Successful DAC technologies require the separation of high purity CO_2 (e.g., >95%) with high selectivity toward CO_2, low energy requirements, minimum chemical and thermal degradations, reliability, long lifetime, etc.

In this regard, DAC technologies based on amine-based sorbents has emerged as one of the most promising approaches. However, amine-based sorbents are known for thermal and oxidative degradations and high desorption energy requirement for sorbent regeneration process due to high binding energy of amines toward CO_2. In this study, we examined various chemical modifications to amines for DAC and the impact of the modifications on CO_2 adsorption performances. *In-situ* FTIR spectroscopy has been implemented to probe the mechanism of the modifications for CO_2 adsorption and degradation on both unmodified and modified amine-based sorbents. The *in-situ* FT-IR results endorsed that unmodified amine sorbents exhibited strong chemisorption as well as high oxidative degradation compared to modified amine sorbents. Differential scanning calorimetry (DSC) results also supported that the heat of desorption requirement for modified amine-based sorbents are significantly lower than unmodified amine-based sorbents.

18d CO_2 Capture Using Phase-Changing Bis-Iminoguanidines (BIGs) with Amino Acids; Analysis of a Direct Air Capture Process

Abishek Kasturi, Gyoung Gug Jang, Diana Stamberga, Radu Custelcean, Sotira Yiacoumi, Costas Tsouris

CO_2 capture using phase-changing bis-iminoguanidines (BIGs), such as the methyl-glyoxal-bis-iminoguanidine (MGBIG), with aqueous amino acids (e.g., sarcosine) is investigated in this study. The objective is to provide a better understanding of the process and its scaleup for the direct air capture (DAC) of CO_2. DAC processes involving BIGs are of major interest due to their comparatively lower regeneration energy requirements and environmental risks. Unlike in monoethanolamine (MEA), which does not change phase, only the insoluble carbonate precipitate undergoes thermal regeneration instead of the bulk solution, leading to energy savings.

This work explores three main aspects of a DAC process using MGBIG and sarcosine: (i) CO_2 loading of the solvent, (ii) solid-liquid separation of the carbonate precipitate from the slurry through the use of press and vacuum filtration, and (iii) regeneration of the BIG precipitate using processes that employ temperature swing, vacuum-assisted temperature swing, and microwave exposure. A holistic technoeconomic analysis of the entire process is also developed to estimate the capital and operating costs and overall energy requirements for a scaled-up DAC plant capable of sequestering 1 million tonnes of CO_2 annually using MGBIG and sarcosine.

Calorimetric and spectroscopic studies indicate that the proposed DAC system can achieve CO_2 loadings of 0.6 mol CO_2 per mol solvent and will require approximately 8 GJ per tonne of CO_2 for solvent regeneration. Preliminary cost analyses peg the total cost of CO_2 sequestration of the proposed process to be approximately $210 per tonne of CO_2.

These results provide the fundamental basis for developing an effective carbon capture technology with phase-changing amino acid/guanidine solvents that can be used to effectively capture atmospheric CO_2 at scale.

18e Investigating Oxidative Degradation Products and Mechanisms of Aminopolymer Sorbents for Direct Air Capture (DAC)

Yoseph Guta, Tianchang (Athena) Xu, Juliana Carneiro, Giada Innocenti, Simon H. Pang, Miles Sakwa-Novak, Nga Lee Ng, Carsten Sievers, Christopher Jones

Aminopolymer based sorbents have shown promising results in direct air capture (DAC) of CO_2 from ambient air owing to their high CO_2 adsorption capacity and selectivity. While those adsorptive properties are important, the stability of a sorbent is important in creating high-performing, cost-effective and long-lasting sorbents. The high concentration of molecular oxygen (O_2) in ambient air (~21 %) and temperature swings used during sorbent regeneration cause oxidative degradation of these sorbents with time. Either via process upsets that could cause rapid oxidation, or slow oxidation under milder, controlled operating conditions, oxidative degradation has the potential to increase operation costs. This work focuses on identifying major oxidation products of aminopolymer sorbent (45 wt.% poly(ethylenimine) (PEI) on γ-Al_2O_3) to develop a deeper insight into the oxidative degradation mechanisms and pathways of aminopolymer sorbents under varied experimental conditions. We explore the effect of oxidative degradation on the CO_2 adsorption capacity using thermogravimetric analysis, *in situ* HATR-IR spectroscopy, High-Resolution Time-of-Flight Chemical Ionization Mass Spectrometer (HR-ToF-CIMS), ^{13}C NMR spectroscopy and elemental analysis (C/N ratio). Preliminary results show the formation of low molecular weight amine compounds formation as gas-phase products and the degradation mechanism seems to follow free radical initiated autoxidation.

18f Direct Air Capture with Amino Acid Solvent: Operational Optimization Using a Crossflow Air-Liquid Contactor

Keju An

One of the negative emission technologies, Direct Air Capture(DAC) is important technology for industrial decarbonization and CO_2 removal processes. Solvent-based DAC is a well-known process, but it requires significant initial costs and energy demand, particularly due to dilute CO_2 concentration in the air. In this study, the conventional packed bed and 3D printed Gyroid packing were utilized with 1M K-Sar (Potassium Hydroxide and Sarcosine) for passive, wind-driven, DAC. We used additive manufacturing for the prototype structured packing, Gyroid, in the vast design space accessible via advanced manufacturing and computer aided design. The effects of feed wind speed, solvent flowrate, contactor geometry and ambient temperature condition are tested to suggest optimized running condition. The tested wind speed range was 0 to 3.5 LPS and 0.06 to 0.16 GPM for the solvent flow rate. The ambient temperature and relative humidity condition were controlled by environmental chamber.

Three conditions of conventional packed bed: 360, 438 and 578 m^2/m^3 were tested and compared with advanced structured, Gyroid, packing: 579 m^2/m^3. The conventional packed bed (using Polypropylene balls) with the highest specific surface area, 578 m^2/m^3, showed nearly 50% CO_2 capture efficiency at optimum running condition. The optimum running condition was found at 1 LPS air flow rate and 0.1 GPM solvent flow rate. Compared to the lower specific surface area packed bed results: 25% for both 360 and 438 m^2/m^3, there is significant CO_2 capture efficiency shift from 578 m^2/m^3 packing. The Gyroid packing showed 4.5 times lower pressure drop while maintaining the same CO_2 capture efficiency, 50%, at optimum running condition.

We used Gyroid packing, 578 m^2/m^3 with the optimum running condition for the ambient temperature effect on CO_2 capture efficiency. The relative humidity conditions are fixed at 80% for all experiments for consistency. Three different ambient temperature conditions: 35, 65, 95 F are tested and showed 24%, 36% and 47% CO_2 capture efficiency as an average value for 1 hour test. This is because of temperature-dependent inherent solvent's reaction kinetics. DAC technology is sensitive to ambient temperature conditions meaning that DAC performance, CO_2 capture efficiency, varies under different climate conditions.

It is essential that we locate DAC system offer the best techno-economic performance given the high cost of the DAC technology, the need to secure public and private investment, and the pressing need to meet regional net-zero to negative emissions targets. Performance can change depending on the climate, which can affect DAC operational costs and, consequently, deployment potential. To reduce the overall capital and operating costs of the passive contacting DAC systems and allow for their scalability, the design of the contactor and the adsorption operating conditions must be optimized.

Amino Acid-Based Functionalized Deep Eutectic Solvents as a Promising Alternative for Efficient CO_2 Capture and Consideration of Its Application

Jaehan Jo[1], Minseok Park [1], Seokho Kwon[1] and Dongwoo Kang[2]
Chungbuk National University, Cheongju-si, Republic of Korea
(1)Chungbuk National Universitym Cheongju-si, Korea, Republic of (South), (2)Department of Engineering Chemistry, Chungbuk National University, Cheongju, Chungbuk, South Korea

Abstract

The development of effective carbon capture technologies is of great importance in mitigating the impacts of climate change. Absorption technologies that use conventional alkanolamine-based absorbents are currently the most widely used method for capturing CO_2, but they have several limitations, including high energy requirements, corrosion problems, and a low absorption capacity. Therefore, there is a need to explore alternative materials that can improve the efficiency of CO_2 capture.
Functionalized DESs are emerging as a promising alternative to conventional absorbents, and this study aimed to investigate the potential of amino acid-based HBAs as functional groups for improving their CO_2 absorption characteristics. The study employed two amino acids, glycine (GLY) and proline (PRO), as HBAs and alkanolamines, such as monoethanolamine (MEA), diethanolamine (DEA), and N-methyldiethanolamine (MDEA), as HBDs. The results demonstrated that the functionalized DESs exhibited a higher CO_2 absorption capacity and faster reaction rate than conventional absorbents.
The functionalization of DESs with amino acids offers several advantages. First, amino acids are inexpensive and readily available. Second, the reaction to convert amino acids into amino acid salts is simple and efficient. Third, the resulting amino acid-based HBAs provide additional sites for CO_2 absorption, leading to improved absorption capacity. Fourth, the enhanced intermolecular interactions resulting from the amino acid-based HBAs promote the formation of stable DESs, which can lead to a more efficient capture of CO_2. The results of this study are important for the development of more effective carbon capture technologies. By enhancing the absorption capacity and reaction rate of DESs, these functionalized materials can reduce the cost and energy required for capturing CO_2. Moreover, the low vapor pressure of DESs makes them a safer alternative to conventional absorbents, which can be highly volatile and flammable. Finally, the use of amino acids as HBAs provides an opportunity to design DESs with tailored absorption characteristics. In conclusion, this study demonstrates the potential of functionalized DESs for improving the efficiency of CO_2 capture. The use of amino acid-based HBAs provides an effective means of enhancing the absorption capacity and reaction rate of DESs. The results of this study suggest that these functionalized materials could be used as an alternative to conventional absorbents for capturing CO_2, potentially leading to more efficient and cost-effective carbon capture technologies.

Introduction

The continued use of fossil fuels has led to a rapid increase in global temperatures since industrialization, primarily due to the excessive emission of greenhouse gases. This has resulted in various issues, including abnormal climate patterns, rising sea levels, and elevated temperatures [1]. To combat global warming, it is crucial to develop technologies that can reduce CO_2 emissions. Carbon capture, utilization, and storage (CCUS) technology has been extensively studied, comprising two main branches: Carbon Capture and Storage (CCS) and Carbon Capture and Utilization (CCU) technologies. In CCS technology, CO_2 is captured and stored in deep-ocean crusts or underground reservoirs, while CCU technology involves capturing CO_2 and transforming it into useful materials. Although the treatment of CO_2 differs in these approaches, the primary focus is on effective CO_2 capture.

Among various CO_2 capture technologies used during fossil fuel combustion, the wet absorption method is considered the most efficient. It is currently nearing commercialization readiness due to its rapid CO_2 reaction rate and high capture capacity compared to alternatives such as dry absorption, membrane separation, and adsorption. Wet absorption primarily employs alkanolamines like monoethanolamine (MEA), diethanolamine (DEA), and N-methyldiethanolamine (MDEA) as absorbents [2]. These molecules contain amine groups (-NH), making them reactive to CO_2. The amine order, determined by the number of alkyl groups (-R) attached to the nitrogen atom, influences the chemical reaction behavior with CO_2.

MEA requires high regeneration energy because it reacts with CO_2 to form primary carbamate (RNHCOO-), which is thermodynamically stable. This implies that the process using MEA has a disadvantage in terms of the regeneration energy required for stripping, which is as high as approximately 4 GJ/ton CO_2 [3]. This incurs additional costs. Additionally, evaporation loss may occur during the process owing to the high volatility of the absorbent, and there is a disadvantage in the maintenance of the facility because it is corrosive [4]. Primary amines share disadvantages with secondary amines, and the reaction mechanism with CO_2 is similar for each. Nevertheless, studies on alkanolamine-based blended absorbents have been conducted because their use is appropriate for the treatment of a large amount of CO_2 at a high rate. Hence, the wet absorption process using liquid absorbents and alkanolamines is widely accepted in conventional CO_2 emission facilities. However, owing to the limitations of alkanolamine-based materials, research on the development of new absorbents is ongoing. Attempts have been made with alkanolamines, followed by ionic liquids (ILs), and finally with deep eutectic solvents (DESs), which are currently known as 3rd generation absorbents. A DES consists of a hydrogen bond donor (HBD) and hydrogen bond acceptor (HBA) and exists in a liquid state around 298.15 K because its eutectic point is lower than the melting points of its constituent HBA and HBD. DESs have attracted increasing attention as next-generation wet absorbents because they share advantages with ILs, such as high thermal stability and low vapor pressure [5]. Additionally, they are economical because they can be synthesized at a relatively low cost compared to ILs, owing to the less complex process and low cost of reactants.

A DES not only captures CO_2 physically in the voids between the HBA and HBD but can also selectively capture CO_2 chemically. This is achieved by introducing suitable functional groups, a process known as functionalization of DESs [6]. DESs can retain the characteristics of the functional groups offered by HBAs and HBDs, and it would be appropriate to select suitable HBAs and HBDs to improve the chemical absorption capacity. For this reason, compounds containing highly reactive amine groups can be considered for use as HBDs, and their physicochemical properties may change if converted to DESs [7]. Unlike previous studies, the possibility of using new types of HBAs was investigated in this study because the effects of HBDs on CO_2 capture have already been widely reported. It is important to use suitable HBAs to design molecular architectures that are appropriate for binding HBDs. DES production is assumed to be achievable when groups that can absorb and donate hydrogen bonds are utilized as HBA and HBD, respectively. However, it has been reported in the literature that DES formation is not always possible, even when suitable materials are mixed [8]. This implies that the interactions between HBAs and HBDs play an important role in the synthesis of DES complexes.

Hence, electrically charged materials were attempted to be used as HBAs for high static interactions. One such candidate is amino acids; however, they develop charges only when dissolved in water by forming zwitterions. Under conditions of low water content, they exist in the form of non-charged molecules. This can be resolved by converting them into amino acid salts, thereby enhancing the intermolecular interactions. Additionally, when metal hydroxides, such as potassium hydroxide, are added to amino acids, the protons attached to the carboxyl groups are detached, thus lowering the acidity of the components, and resulting in an increase in CO_2 absorption capacity. Amino acid salts are less toxic than alkanolamines, and their volatility is extremely low because of their low vapor pressure owing to their high interactions. Low volatility is advantageous in CO_2 absorption and desorption processes because it can reduce evaporation losses during the process [9]. Amino acid salts have an amine group that is not involved in the formation of DES and can absorb CO_2.

Herein, two types of amino acid salts that may have high CO_2 capture efficiencies were prepared and used as HBAs. Based on pre-screening, proline and glycine salt aqueous solutions showed excellent CO_2 absorption capacity [10-12]. Therefore, the potassium salts of proline and glycine [K-Prolinate and K-Glycinate, respectively] were selected as HBAs. By converting amino acids into their potassium salts, intermolecular interactions may be strengthened, and amine groups attached to them can facilitate CO_2 absorption. In addition, alkanolamines with different amine orders, that is, MEA, DEA, and MDEA, were selected for use as HBDs. Functionalization of DESs for CO_2 capture can be achieved when these HBDs are used in their synthesis. The CO_2 absorption characteristics of the DESs synthesized using these HBAs and HBDs were analyzed using Fourier transform infrared (FT-IR) spectroscopy, differential scanning calorimetry (DSC), and ^{13}C-nuclear magnetic resonance (NMR) spectroscopy. In addition, the physicochemical properties, including density, pH, and free volume, were measured to determine the relationship between the equilibrium absorption capacity and reaction rates with respect to the HBAs and HBDs.

Experimental

2.1 Materials and Method

Glycine (Sigma-Aldrich, CAS:56-40-6, 99.0%) and proline (Sigma-Aldrich, CAS:147-85-3, 99.0%) were used to prepare the HBAs. An aqueous solution of 8 N KOH (Acros Organics, CAS:1310-58-3) was used to prepare the potassium salts of the amino acids. Three types of alkanolamines were used as HBDs: MEA (Sigma-Aldrich, CAS:141-43-5, 99.0%), DEA (Sigma-Aldrich, CAS:111-42-2, 99.0%), and MDEA (Sigma-Aldrich, CAS:105-59-9, 99.0%). All the materials were used as received, without any treatment. Deionized water was used throughout the procedure.

2.2 Methods and detailed procedures

2.2.1 Synthesis of DESs

Two amino acid salts, potassium prolinate (K-PRO) and potassium glycinate (K-GLY), were prepared and used as HBAs. DES complexes were synthesized using MEA, DEA, and MDEA as HBDs. To convert the amino acids into their potassium salts, a stoichiometric ratio of the amino acids and KOH solution was mixed. The reaction was performed by a dropwise addition of 8 N KOH solution per second to the solution of amino acid using a burette. Water was evaporated using a rotary evaporator for 24 h at a temperature of 363.15 K to obtain the amino acid salt in the solid state (HBA) [13]. After preparing the HBAs, the DESs were synthesized. For this, the total mass of the DESs was set at 80 g, and the corresponding HBAs and HBDs with different molar ratios, that is, from 1:3 to 1:10, were prepared and mixed using a magnetic stirrer. The mixing process was continued for 24 h, and subsequently, 10 wt.% of water based on the total masses of the HBA and HBD was added to lower the viscosity of the DESs. Detailed data on DESs synthesis were suggested in the Table 1. To maintain a constant temperature, a double-jacket reactor was utilized, the temperature was maintained at 298.15 K, and the mixture was stirred at a speed of 300 rpm for a homogeneous reaction. During the DES synthesis, high-purity N_2 gas was injected into the reactor to prevent the absorption of CO_2 from the air. Using this process, the exact ratio of HBA and HBD at which a uniform liquid-phase product is obtained can be determined.

2.2.2 Investigation of DES formation by instrumental analysis

Instrumental analyses were performed to investigate the properties of the products in the uniform liquid state synthesized according to the previous procedure. Herein, the functional groups and hydrogen bond formation of DES were qualitatively analyzed by FT-IR [Cary 670 (Main Bench) + Cary 620 (Microscope)] spectroscopy. From the FT-IR spectra, the hydrogen bond formation was evaluated based on the peak broadness. Additionally, the thermal stability and uniformity of the final product were determined using DSC (DSC Q2000) by interpreting the thermal behavior when heat energy is supplied to the molecular complex.

2.2.3 Characteristics of CO_2 absorption and loading capacity

To measure the reaction efficiency, initial reaction rate, and CO_2 loading capacity for each DES, CO_2 absorption experiments were conducted. To simulate the flue gases produced from conventional coal-fired power plants, the experiment was conducted using a mixture of nitrogen (>99%) and carbon dioxide (>98%) gases at a ratio of 85:15 vol.%. A mass flow controller (MFC) was used to control the mass flow rate of each gas. A gas mixer was used to obtain a uniform mixture of simulated gas, and the temperature of the gas mixer was maintained at approximately 275.15 K to prevent the evaporation of water vapor from the gas mixer. Twenty milliliters of DES were placed into the reactor, and a magnetic stirrer was used to promote mass transfer. Because the chemical reaction between DES and CO_2 is exothermic, there is a possibility of a loss of water content from the water that was initially added to the DESs to lower the viscosity. Therefore, a condenser was installed at the top of the reactor, and a water circulator was connected. The circulator temperature was maintained at 275.15 K so that the vaporized water could be restored to the reactor. The mixed gas was injected underneath the gas-liquid interface, and CO_2 absorption commenced. The gas passing through the gas-liquid interface flows into the gas analyzer [Multi Gas Analyzer, Sensoronic Co., Ltd.] that records the in situ concentration of CO_2. Initially, CO_2 was selectively removed so that the gas analyzer could detect its concentration as 0 vol.%. However, the concentration gradually increased as the absorbent became saturated with CO_2. Subsequently, the experiment was terminated at 15 vol.% of CO_2, considering that CO_2 is no longer absorbed and discharged [12]. Using the data obtained from the absorption experiments, the CO_2 loading capacity for all the DESs was determined. To investigate the interaction between the amino groups and CO_2, FT-IR spectroscopy was used to analyze the DES-CO_2 intermediates and the corresponding chemical species in the system, including carbamate, carbonate, and bicarbonate. Finally, ^{13}C-NMR (AVANCE III 500 MHz spectrometer) was used to determine the structures of the DESs combined with CO_2 and to investigate the molecular behavior of the system.

Results and Discussion

3.1 Results of DES formation by instrumental analysis

When the potassium salts of amino acids (HBAs) and alkanolamines (HBDs) were mixed in different molar ratios of 1:3 to 1:10, a uniform and transparent mixture was obtained only at a ratio of 1:10, regardless of the type of HBDs used. Using FT-IR analysis, the reactive groups on HBA and HBD were investigated, and the results are shown in Fig. 4. The FT-IR spectra of K-GLY and K-PRO were in good agreement with those reported in the literature [13]. Therefore, it can be concluded that the synthesis of the potassium salts of the amino acids was successful. The broad peak in the 3200–3550 cm^{-1} region can be attributed to stretching vibrations of hydroxy groups in MEA, DEA, and MDEA. The formation of DES was confirmed from the 2800–3000 cm^{-1} region. N-H bending peaks were observed in the 1580–1650

cm^{-1} region in all the curves, confirming the presence of amino groups in HBA, HBD, and DES. In the FT-IR spectra of the synthesized DES, peaks reflecting the characteristics of HBD and HBA were observed. As shown in Fig. 1., in all the FT-IR spectra, the N-H bending peak in the range of 1580–1650 cm^{-1} is broadened after transformation into DESs compared to that in the FT-IR spectrum of HBA or HBD. The O-H stretching vibration in the 3200–3550 cm^{-1} range showed similar broadening. From DSC analysis, it was confirmed that all the DESs were in a stable liquid state in the temperature range from structural phase transition temperature (T_{st}) to 323.15 K. T_{st} is the structural phase transition temperature and can be detected at approximately 223.15 K for all the DESs. The reason for the occurrence of T_{st} is thought to be that the structure of DES transforms into another structural phase due to the rearrangement of hydrogen bonds or changes in molecular interactions. The values of T_{st} for all the DESs are shown in Fig. 2. When MEA was used as the HBD, T_{st} was the highest, regardless of HBA. In addition, T_{st} decreased when the order of the amines (HBD) increased. The molecular interaction caused by the molecule structure and subsequent hydrogen bonding may be the source of this shift in T_{st}. When MEA is used as the HBD, interactions can be induced by two functional groups: hydroxyl and amine groups. For DEA and MDEA, the interactions can be induced by three functional groups: two hydroxyl groups and one amine group. Hence, DESs, in which MEA was used as the HBD, can transform into another structure at relatively high temperatures when cooled. This is because the molecular interactions that affect the stability of the complex are maintained by fewer hydrogen bonds. For DESs with DEA and MDEA as the HBDs, the T_{st} for DESs with MDEA were lower than those for DESs with DEA. This may be because of the presence of an alkyl group with a high electron density. Typically, the alkyl group is classified as an electron-donating group (EDG). In MDEA, there is one methyl group attached to the N atom, which provides the N atom with a higher electron density compared to that in DEA. Hence, the N atom in MDEA may accept many more hydrogen bonds from the hydroxyl hydrogen of neighboring HBD molecules. In summary, DESs can maintain their complex structures based on the intensity of their molecular interactions.

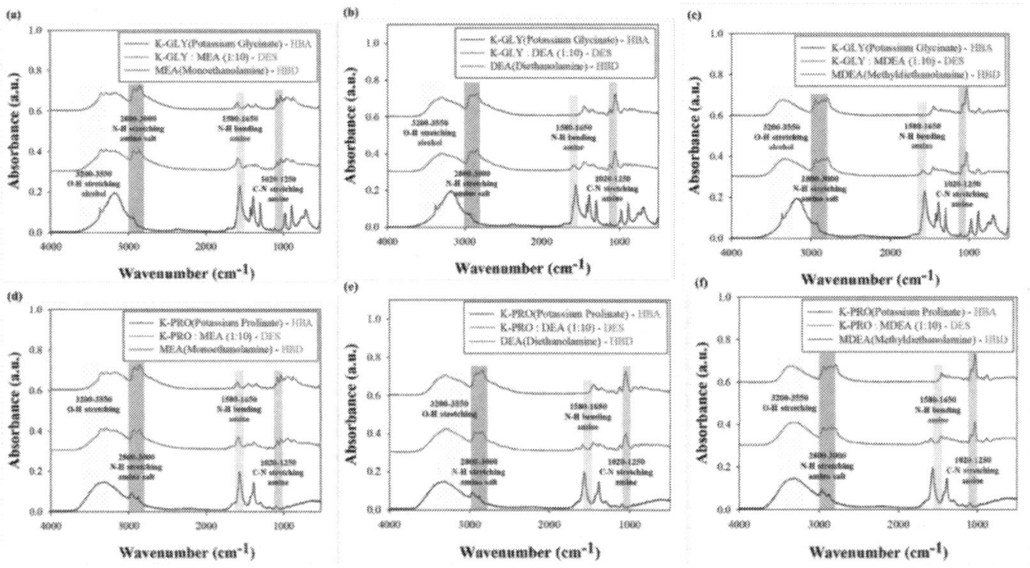

Fig. 1. FT-IR spectra of (a) K-GLY+MEA, (b) K-GLY+DEA, (c) K-GLY+MDEA, (d) K-PRO+MEA, (e)K-PRO+DEA, (f) K-PRO+MDEA.

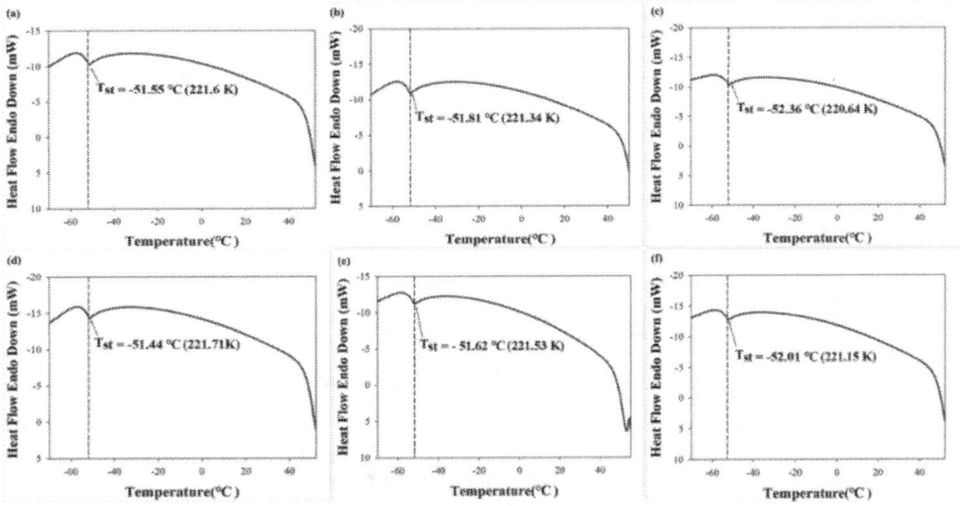

Fig. 2. DSC cooling curves of (a) K-GLY+MEA, (b) K-GLY+DEA. (c) K-GLY+MDEA, (d) K-PRO+MEA. (e) K-PRO+DEA, (f) K-PRO+MDEA.

3.2 Analysis of CO_2 absorption characteristics and loading capacity

After synthesizing the DESs, the CO_2 absorption capacity of each DES was measured, and the results are shown in Fig. 6. The gas analyzer measures the concentration of CO_2 in the gas stream passing through the absorbent. Using the CO_2 concentration data and the correlation suggested by Eq. (10), the CO_2 absorption capacity with respect to time was determined.

$$\frac{15-a}{15} * \frac{b}{1000} * \frac{273.15}{22.4(273.15+T_{rxn})} * 0.5 = \text{Amount of } CO_2 \text{ absorption (mol)} \quad (10)$$

where a and b denote the concentration of CO_2 in vol.% and the flow rate of CO_2 in cubic centimeters per minute, respectively, and T_{rxn} is the reaction temperature.

Fig. 3.(a) and Table 2 illustrate the CO_2 loading capacities of different DES combinations. The highest CO_2 loading capacity is observed with K-GLY+MEA, followed by K-PRO+MEA, K-GLY+DEA, K-PRO+DEA, K-GLY+MDEA, and K-PRO+MDEA. To express CO_2 loadings in mol CO_2/mol DES, we considered the combined molecular weights of HBA and HBD with a 1:10 stoichiometry. Regardless of the HBA type, DESs with MEA as the HBD showed high CO_2 absorption capacity, followed by DESs with DEA and MDEA. When compared with literature values [14-17], the DESs developed in this study demonstrated significantly improved performance. Moreover, for DESs with the same HBDs, those with K-GLY as the HBA exhibited higher absorption capacities than those with K-PRO as the HBA. Specifically, K-GLY+MEA and K-PRO+MEA displayed absorption capacities of 4.223 and 4.146 mol CO_2/mol DES, respectively, whereas conventional MEA exhibited an absorption capacity of only 0.5 mol CO_2/mol DES. In contrast, DESs with MDEA showed relatively lower CO_2 absorption capacity compared to other DESs. Although K-GLY+MDEA exhibited a 100% increase, K-PRO+MDEA showed only a slight improvement compared to pure MDEA. This may be attributed to MDEA forming sterically hindered intermediates (R_3NCOO-) when reacting with CO_2, particularly pronounced in K-PRO+MDEA, where both proline rings and MDEA contribute to steric hindrance. The influence of the donor on CO_2 absorption capacity is emphasized when expressed as g CO_2/g DES. Notably, when K-GLY served as the HBA, it consistently outperformed K-PRO, irrespective of the HBD. This is attributed to the steric hindrance associated with the cyclic structure of K-PRO, potentially interfering with mass transfer, and forming unstable intermediates during absorption processes [18].

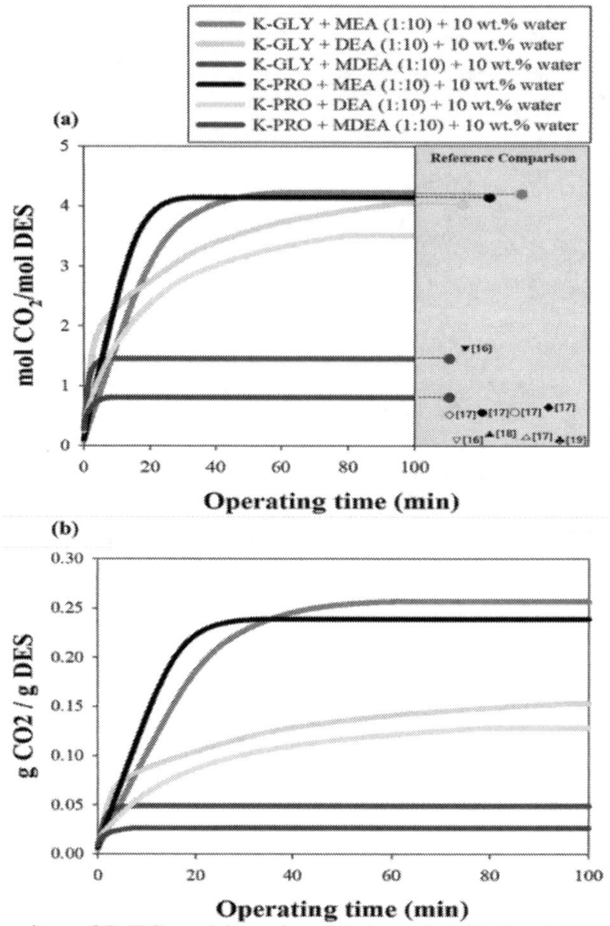

Fig. 3. CO_2 loading capacity of DESs with unit of (a) mol CO_2/mol DES and (b) g CO_2/g DES.

Moreover, the CO_2 loading capacity appeared to be affected by the pH of the DESs. CO_2 combines with the amine group of the DES to form COO^-. The dissolved CO_2 combines with amine molecules or water and releases H^+, which is acidic. In other words, it can be assumed that the higher the pH of the DES, the easier it is to capture CO_2; therefore, DES with a high pH will be more advantageous for capturing CO_2. The results shown in Table 2 are in accordance with previously reported results [19].

Table 1 Mass ratios of the materials used in DES synthesis.

DES	Material		Mass ratio (g)		Water added (g)
	HBA	HBD	HBA	HBD	
K-GLY+MEA (1:10)	K-GLY	MEA	12.504	67.496	8.0
K-GLY+DEA (1:10)	K-GLY	DEA	7.774	72.226	8.0
K-GLY+MDEA (1:10)	K-GLY	MDEA	6.938	73.062	8.0
K-PRO+MEA (1:10)	K-PRO	MEA	16.044	63.956	8.0
K-PRO+DEA (1:10)	K-PRO	DEA	10.176	69.824	8.0
K-PRO+MDEA (1:10)	K-PRO	MDEA	9.114	70.886	8.0

Table 2 CO_2 absorption capacity and pH of each DES.

Types of DES	CO_2 loading		pH
	g CO_2 / g DES	mol CO_2 / mol DES	
K-GLY+MEA	0.2567	4.2231	14.26
K-GLY+DEA	0.1530	4.0495	12.83
K-GLY+MDEA	0.0491	1.4546	12.50
K-PRO+MEA	0.2388	4.1458	13.58
K-PRO+DEA	0.1283	3.5118	12.27
K-PRO+MDEA	0.0263	0.8049	11.85

FT-IR analysis was performed to investigate the changes in functional groups and new chemical bonds after CO_2 uptake, and the results are shown in Fig. 4. For all the samples after CO_2 uptake, the peak at 3340 cm^{-1} indicates strong hydrogen bonding. The peaks from 1327–1571 cm^{-1} correspond to $-COO^-$ (carboxylate), confirming that DES has absorbed CO_2 chemically in the form of $-COO^-$ [20]. In the FT-IR spectra of the sample with MDEA as the HBD, only a slight change in the peaks was observed before and after CO_2 uptake, unlike that for the other DES. These results agree with the theoretical expression are shown in Fig. 4. [21].

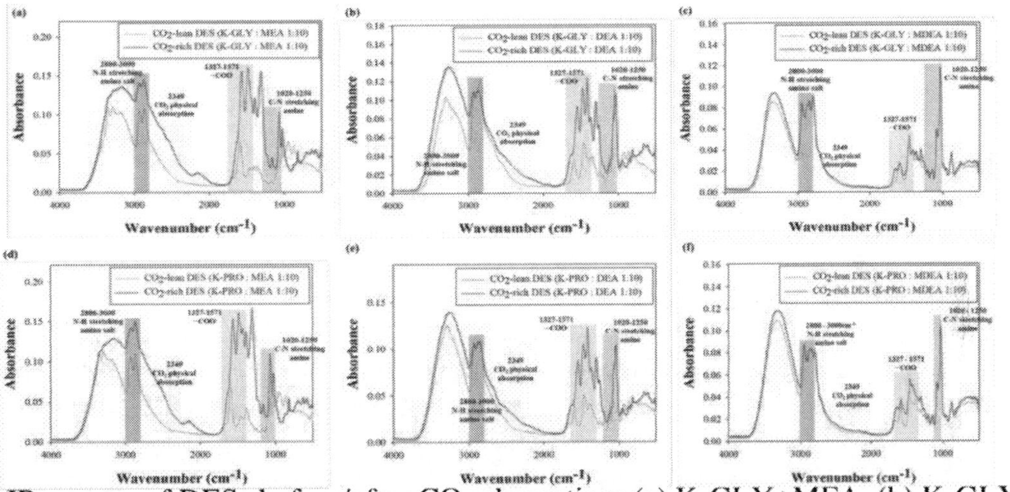

Fig. 4. FT-IR spectra of DESs before/after CO_2 absorption: (a) K-GLY+MEA, (b) K-GLY+DEA, (c) K-GLY+MDEA, (d) K-PRO+MEA, (e) K-PRO+DEA, and (f) K-PRO+MDEA.

^{13}C-NMR analysis was performed to evaluate the structures of the DES complexes and the chemical bonding before and after CO_2 uptake, and the results are shown in Fig. 5. The main NMR peaks are summarized below [22].

(a) R-COOR: 165–175 ppm, (b) COO^-: 165 ppm, (c) C-O: 50–80 ppm, (d) C-N: 40–60 ppm, (e) R-CH_2-R: 15–50 ppm, and (f) R-CH_3: 8–35 ppm. The characteristic peaks of the HBAs and HBDs were confirmed from the NMR data. The peaks in the ranges of 40–60 ppm and 50–80 ppm, corresponding to alkylamine and alcohol groups, respectively, appear strongly because of the presence of several functional groups in MEA, DEA, and MDEA. The peaks corresponding to R-COO-R' in the range of 165–175 ppm indicate that two moles of amine groups were involved in capturing one mole of CO_2 during the absorption process [23]. Additionally, a peak at approximately 165 ppm was observed in the ^{13}C-NMR spectra of all the DESs, which indicates the chemical absorption of CO_2 in the form of COO^- by all the DESs [24].

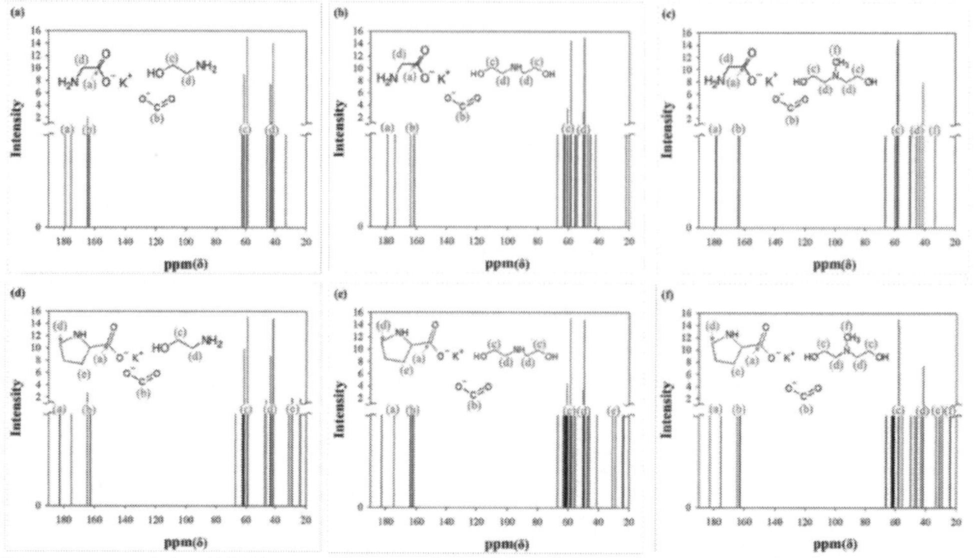

Fig. 5. ^{13}C NMR spectra of the DES complexes after CO_2 absorption: (a)K-GLY+MEA, (b) K-GLY+DEA, (c) K-GLY+MDEA, (d) K-PRO+MEA, (e) K-PRO+DEA, (f) K-PRO+MDEA.

Conclusions

In this study, two types of amino acid salts, K-PRO and K-GLY, were prepared and used as HBAs for DES synthesis, and MEA, DEA, and MDEA, which are alkanolamine-based solvents, were used as HBDs. The DESs were synthesized using HBAs and HBDs at a molar ratio of 1:10, and 10 wt.% water was added to reduce the viscosity after CO_2 uptake [25]. Their thermal stabilities and the structural phase transitions in the around 298.15 K range were confirmed by DSC analysis. As no structural phase transitions were observed beyond T_{st}, that is, approximately 223.15 K, it was interpreted that the DESs exist in a stable liquid phase at ambient temperature and are thermodynamically stable. Moreover, all the DESs showed a significantly higher absorption performance towards CO_2 compared to pure alkanolamines [26]. The highest absorption capacity was found to be 4.223 mol CO_2/mol DES for the DES with K-GLY as the HBA and MEA as the HBD. This is approximately 8 times more absorption capacity compared to that of pure MEA. FT-IR analysis confirmed the successful synthesis of the DESs and CO_2 absorption. After DES formation, FT-IR spectra for each constituent HBA and HBD were obtained. Additionally, peaks before and after CO_2 absorption were compared, and a characteristic peak of COO^- in the range of 1327–1571 cm^{-1} and a peak at 2349 cm^{-1} corresponding to the region of physical absorption of CO_2 were observed [27]. Thus, it was concluded that both physical and chemical absorption are possible using the DESs synthesized in this study. The behavior of the absorbent was confirmed by ^{13}C-NMR analysis, and the chemical absorption of CO_2 in the form of COO^- was confirmed from the peak at 165 ppm.

From the results of this study, it can be concluded that the CO_2 absorption characteristics of DESs are affected by the type of HBA and HBD [28]. In particular, the absorption capacity and rate were affected by the HBA and HBD, respectively. As a result, it can be suggested that DESs with high CO_2 loading capacities and absorption rates can be prepared by controlling the properties of HBA and HBD. If these types of DES are commercialized, CO_2 emissions can be significantly reduced [29].

References

1. A. Olabi, M.A. Abdelkareem, "Renewable energy and climate change," Renewable and Sustainable Energy Reviews 158 (2022) 112111.
2. A. Aroonwilas, A. Veawab, "Integration of CO2 capture unit using single-and blended-amines into supercritical coal-fired power plants: Implications for emission and energy management," International Journal of Greenhouse Gas Control 1(2) (2007) 143-150.
3. R. Sakwattanapong, A. Aroonwilas, A. Veawab, "Behavior of reboiler heat duty for CO2 capture plants using regenerable single and blended alkanolamines," Industrial & engineering chemistry research 44(12) (2005) 4465-4473.
4. T.R. Sekharan, R.M. Chandira, S. Tamilvanan, S. Rajesh, B. Venkateswarlu, "Deep eutectic solvents as an alternate to other harmful solvents," Biointerface Res. Appl. Chem 12 (2022) 847-860.
5. M.S. Jagirani, M. Soylak, "Deep eutectic solvents-based adsorbents in environmental analysis," TrAC Trends in Analytical Chemistry (2022) 116762.
6. F.P. Pelaquim, A.M. Barbosa Neto, I.A.L. Dalmolin, M.C.a.o.d. Costa, "Gas solubility using deep eutectic solvents: review and analysis," Industrial & Engineering Chemistry Research 60(24) (2021) 8607-8620.
7. A. Paiva, R. Craveiro, I. Aroso, M. Martins, R.L. Reis, A.R.C. Duarte, "Natural deep eutectic solvents–solvents for the 21st century," ACS Sustainable Chemistry & Engineering 2(5) (2014) 1063-1071.
8. Z. Yuan, H. Liu, W.F. Yong, Q. She, J. Esteban, "Status and advances of deep eutectic solvents for metal separation and recovery," Green Chemistry (2022).
9. D. Kang, S. Park, H. Jo, J. Min, J. Park, "Solubility of CO2 in amino-acid-based solutions of (potassium sarcosinate),(potassium alaninate+ piperazine), and (potassium serinate+ piperazine)," Journal of Chemical & Engineering Data 58(6) (2013) 1787-1791.
10. R. Haghbakhsh, S. Raeissi, "Deep eutectic solvents for CO2 capture from natural gas by energy and exergy analyses," Journal of Environmental Chemical Engineering 7(6) (2019) 103411.
11. Z. Bao, Q. Li, N.G. Akhmedov, B.A. Li, M. Xing, J. Wang, B.I. Morsi, B. Li, "Innovative cycling reaction mechanisms of CO2 absorption in amino acid salt solvents," Chemical Engineering Journal Advances 10 (2022) 100250.
12. S.D. Nielsen, L.M. Jakobsen, N.R. Geiker, H.C. Bertram, "Chemically acidified, live and heat-inactivated fermented dairy yoghurt show distinct bioactive peptides, free amino acids and small compounds profiles," Food Chemistry 376 (2022) 131919.
13. K. Kim, H.S. Park, H. Lim, J.H. Kang, J. Park, H. Song, "Sulfur dioxide absorption characteristics of aqueous amino acid solutions," Journal of Industrial and Engineering Chemistry 105 (2022) 491-501.
14. Y. Yoo, D. Lee, J. Park, "Characteristics and configurations of task-specific deep eutectic solvents with CO2-philic functional groups," Journal of Environmental Chemical Engineering 10(3) (2022) 108034.
15. E. Ali, M.K. Hadj-Kali, S. Mulyono, I. Alnashef, A. Fakeeha, F. Mjalli, A. Hayyan, "Solubility of CO2 in deep eutectic solvents: Experiments and modelling using the Peng–Robinson equation of state," Chemical Engineering Research and Design 92(10) (2014) 1898-1906.
16. X. Liu, B. Gao, Y. Jiang, N. Ai, D. Deng, "Solubilities and thermodynamic properties of carbon dioxide in guaiacol-based deep eutectic solvents," Journal of Chemical & Engineering Data 62(4) (2017) 1448-1455.

17. M. Lu, G. Han, Y. Jiang, X. Zhang, D. Deng, N. Ai, "Solubilities of carbon dioxide in the eutectic mixture of levulinic acid (or furfuryl alcohol) and choline chloride," The Journal of Chemical Thermodynamics 88 (2015) 72-77.
18. D. Kang, Y. Yoo, J. Park, "Accelerated chemical conversion of metal cations dissolved in seawater-based reject brine solution for desalination and CO2 utilization," Desalination 473 (2020) 114147.
19. B. Lv, B. Guo, Z. Zhou, G. Jing, "Mechanisms of CO2 capture into monoethanolamine solution with different CO2 loading during the absorption/desorption processes," Environmental science & technology 49(17) (2015) 10728-10735.
20. Z. Zhang, Y. Li, W. Zhang, J. Wang, M.R. Soltanian, A.G. Olabi, "Effectiveness of amino acid salt solutions in capturing CO2: A review," Renewable and Sustainable Energy Reviews 98 (2018) 179-188.
21. D. Kang, H. Jo, M.-G. Lee, J. Park, "Carbon dioxide utilization using a pretreated brine solution at normal temperature and pressure," Chemical Engineering Journal 284 (2016) 1270-1278.
22. K. Zhang, Y. Hou, Y. Wang, K. Wang, S. Ren, W. Wu, "Efficient and reversible absorption of CO2 by functional deep eutectic solvents," Energy & Fuels 32(7) (2018) 7727-7733.
23. Z. Li, L. Wang, C. Li, Y. Cui, S. Li, G. Yang, Y. Shen, "Absorption of carbon dioxide using ethanolamine-based deep eutectic solvents," ACS Sustainable Chemistry & Engineering 7(12) (2019) 10403-10414.
24. D. Xiong, Q. Zhang, W. Ma, Y. Wang, Y. Shi, J. Wang, "Temperature-switchable deep eutectic solvents for selective separation of aromatic amino acids in water," Separation and Purification Technology 265 (2021) 118479.
25. Y.E. Kim, J.A. Lim, S.K. Jeong, Y.I. Yoon, S.T. Bae, S.C. Nam, "Comparison of carbon dioxide absorption in aqueous MEA, DEA, TEA, and AMP solutions," Bulletin of the Korean Chemical Society 34(3) (2013) 783-787.
26. F. Barzagli, F. Mani, M. Peruzzini, "A comparative study of the CO2 absorption in some solvent-free alkanolamines and in aqueous monoethanolamine (MEA)," Environmental science & technology 50(13) (2016) 7239-7246.
27. Y. Gu, Y. Hou, S. Ren, Y. Sun, W. Wu, "Hydrophobic functional deep eutectic solvents used for efficient and reversible capture of CO2," ACS omega 5(12) (2020) 6809-6816.
28. J. Wang, H. Cheng, Z. Song, L. Chen, L. Deng, Z. Qi, "Carbon dioxide solubility in phosphonium-based deep eutectic solvents: an experimental and molecular dynamics study," Industrial & Engineering Chemistry Research 58(37) (2019) 17514-17523.
29. L.F. Zubeir, D.J. Van Osch, M.A. Rocha, F. Banat, M.C. Kroon, "Carbon dioxide solubilities in decanoic acid-based hydrophobic deep eutectic solvents," Journal of Chemical & Engineering Data 63(4) (2018) 913-919.

18h Ultra-Functional CO_2 Capture Sorbent Development Via One-Pot Deep Eutectic Solvent Based Thermochemical Conversion

Al Ibtida Sultana, Md Tahmid Islam, Josh Calhoun, Robert Cheatham, Toufiq Reza

Rising global emission of carbon dioxide (CO_2), a greenhouse gas, mostly as a result of the frightening amount of fossil fuel combustion has urged the research community towards CO_2 capture and utilization. Increased attention in deriving highly functionalized and simultaneously inexpensive CO_2 capture sorbents is therefore the primary challenge which can be circumvented via utilization of forest residues like loblolly pine (LP) and environmentally friendly solvents like deep eutectic solvents (DES). This study, therefore, investigates one-pot functionalization during the step of thermochemical conversion of LP, by using DES as the medium of reaction, where the resulting hydrochars were further chemically activated using melamine to improve the N-containing functionalities in the activated hydrochars employed for CO_2 capture. Ultra-functional LP hydrochars were derived utilizing choline chloride-urea, methyl triphenyl phosphonium bromide-ethylene glycol. The mixture of LP and DES were hydrothermally carbonized at 170, 200, and 260°C followed by chemical activation at 800°C and functionalization at 500°C using activation agents of either KOH, melamine, or a combination of KOH and Melamine. The samples' surface area and porosity were then quantified through the utilization N_2 adsorption–desorption. X-Ray powder diffraction and thermographic analysis were respectfully employed to observe the crystallinity and thermal stability of the samples. Lastly CO_2 capture was assessed for the sorbents in atmospheric condition where results suggest exceptional multifold increase of CO_2 adsorption due to the one-pot functionalization, resulting from the enrichment of surface N-functionalities, in combination with ultra-high porosity of the adsorbents.

130a Carbon Capture Enabled By High-Performance Membranes: Process Modelling and Economic Assessment

Marina Micari, Kumar Varoon Agrawal

Membrane processes are emerging as an alternative to absorption for carbon capture. Membranes based on porous graphene selective layer have shown high promise in this regard with extremely large CO_2 permeance (10000 GPU) and attractive CO_2/N_2 selectivity (30) [1,2]. Further developments of graphene membranes in our laboratory have significantly increased the CO_2/N_2 selectivity up to 80, while still maintaining a large CO_2 permeance of 10000 GPU[3].

In addition to that, important advancements in the scale-up process have allowed to realize for the first time large membrane coupons with area of 50 cm^2. Graphene membranes will be evaluated in a pilot plant for post-combustion carbon capture that is going to be installed in Aigle (Switzerland) with a capture rate of 10 kg/day[4].

In order to design the process and to assess the economic feasibility at the large scale, we developed a techno-economic model for multi-stage membrane process based on graphene membranes.

Taking into account the performances of graphene membranes measured experimentally, we optimized a double-stage membrane process to target 90% CO_2 recovery and 95% CO_2 purity in the permeate stream. We assumed a specific membrane cost of 500 $/m^2 and an electricity cost of 0.05 $/kWh.

We compared two different pressure configurations for the first membrane stage: either working with compression of the feed and permeate under vacuum (FC+VP) or with ambient pressure in the feed and permeate under vacuum (VP)). Conversely, the second membrane stage is supposed to work always with the feed at ambient pressure and permeate under vacuum (VP). Compressing the feed of the second membrane stage is not needed, because the CO_2 concentration in the feed (that corresponds to the permeate of the first membrane stage) is around 50%, thus the driving force is much higher than in the first membrane stage.

When a post-combustion flue gas containing 13.5% CO_2 and saturated with water vapor at 50 °C is fed to the capture system, the optimized values of capture penalty range from 34.0 $/ton to 28.9 $/ton when selectivity increases from 30 to 80, with a double-stage membrane system with FC+VP in the first stage and VP in the second stage. Conversely, the capture penalty decreases from 34.3 $/ton to 28.5 $/ton when selectivity increases from 30 to 80, with a double-stage membrane system with VP in both stages.

It is worth noting that the optimized capture penalties of the two membrane systems are very similar: this is because the higher energy consumption of the process with FC+VP in the first stage is counterbalanced by the lower membrane area with respect to the VP-VP process.

The increase of selectivity from 30 to 80 brings to a significant decrease of specific energy consumption, that reduces to 0.97 MJ/kg in the case of FC+VP (1st stage) and VP (2nd stage) and to 0.74 MJ/kg in the case of VP in both stages.

In all cases, thanks to the high permeance, the specific area is quite limited, as it ranges between 1200 m^2/(kg/s) for the system with FC+VP in the 1st stage and VP in the 2nd stage and 2200 m^2/(kg/s) for the system with VP in both stages.

Furthermore, with a prospective membrane cost of 100 $/m^2, the capture penalty reduces to a minimum of 18.7 $/ton for the system with VP in both stages (see Figure). This value is lower than the capture penalties reported in the literature for the same targets with polymeric membranes (from 25 to 45 \$/ton), despite the higher specific membrane cost (assumed cost of polymeric membranes between 30 and 50 \$/m^2).

Finally, the graphene membranes developed in our laboratory have showed extremely large CO_2/N_2 selectivity values up to 1000 at low CO_2 concentration (0.5-2%), thus these could be particularly attractive for carbon capture from diluted sources [3]. Estimated values of capture penalty from a diluted stream with CO_2 concentration of 1% range from 40 to 50 \$/ton and are very promising for Direct Air Capture integrated multi-stage strategies.

Overall, these results show how optimized membrane processes based on graphene membranes are highly promising and competitive for carbon capture from various sources.

(1) Huang, S.; Li, S.; Villalobos, L. F.; Dakhchoune, M.; Micari, M.; Babu, D. J.; Vahdat, M. T.; Mensi, M.; Oveisi, E.; Agrawal, K. V. Millisecond Lattice Gasification for High-Density CO2- And O2-Sieving Nanopores in Single-Layer Graphene. *Sci. Adv.* **2021**, *7* (9), 1–13.

(2) Hsu, K. J.; Villalobos, L. F.; Huang, S.; Chi, H. Y.; Dakhchoune, M.; Lee, W. C.; He, G.; Mensi, M.; Agrawal, K. V. Multipulsed Millisecond Ozone Gasification for Predictable Tuning of Nucleation and Nucleation-Decoupled Nanopore Expansion in Graphene for Carbon Capture. *ACS Nano* **2021**, *15*, 13230–13239.

(3) Hsu, K.-J.; Li, S.; Micari, M.; Chi, H.-Y. C.; Zhong, L.; Villalobos, L. F.; Huang, S.; Duan, X.; Züttel, A.; Agrawal, K. V. Pyridinic Nitrogen Substituted Two-Dimensional Pores for Rapid and Selective CO2 Transport. *Under Rev.*

(4) Agrawal, K. V.; Bautz, R. Capture Du Carbone Par Membranes En Graphène à Nanopores. *Aqua gas* **2022**.

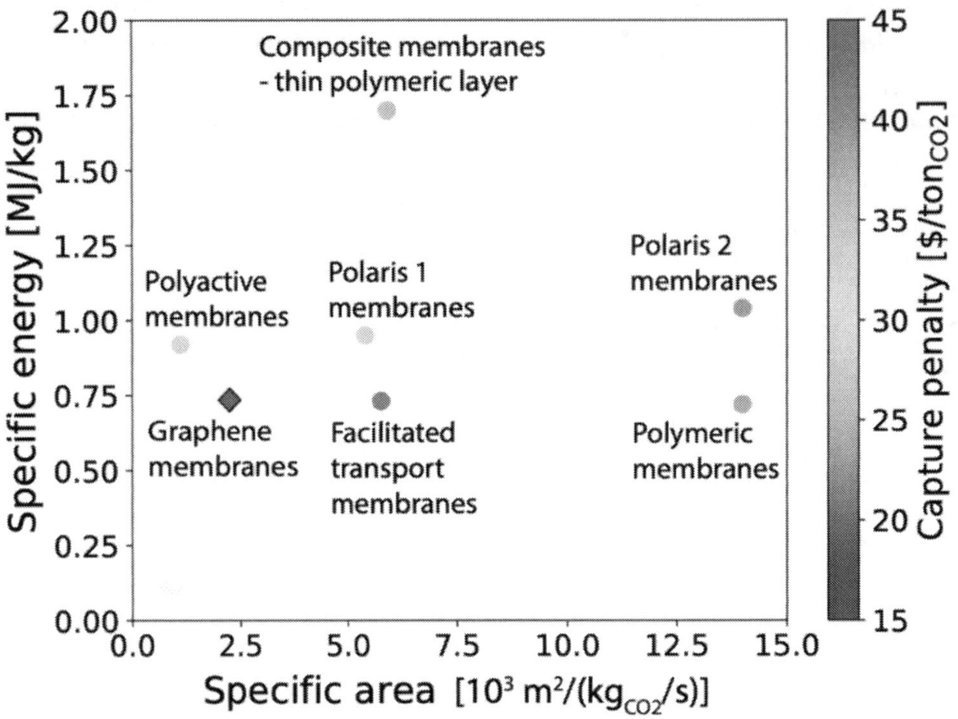

Specific energy [MJ/kg] vs. specific area [$m^2/(kg/s)$] with graphene membranes (optimized VP-VP process) and with membranes from the literature. All cases refer to an inlet CO_2 concentration of 13.5%, CO_2 recovery of 90% and purity of 95%. The colors correspond to the capture penalty [$/ton] in the colorbar.

130b Design Space Analysis: Flexibility Quantification for Carbon Capture Adsorbent Screening

Haditya K. Purwanto, Steven Sachio, Adam Ward, Ronny Pini, Maria Papathanasiou

Pressure-vacuum swing adsorption (PVSA) is an established technology used to capture carbon dioxide (CO_2) from post-combustion flue gas, representing a key step towards achieving a sustainable energy system on a global scale. The status-quo for PVSA design involves solving a constrained multi-objective optimization problem [1], which seeks to maximize productivity while minimizing energy consumption subject to purity and recovery constraints mandated by regulatory bodies [2]. Similarly, for adsorbent screening , the emphasis is typically placed on minimizing the parasitic energy of the adsorbent [3]. While this approach leads to highly efficient designs, they may be constrained to operate within a narrow parameter space that ensures optimal performance. Therefore, assessing process flexibility is essential to ensure robustness and controllability of the process, especially when regulatory purity requirements are stringent.

In this study, we propose a model-based framework to facilitate an integrated assessment of process flexibility and performance. Specifically, we deploy the proposed framework to screen adsorbent materials suitable for a PVSA process. Contrary to previous studies, flexibility is added as the metric to evaluate process controllability. The framework consists of three main steps: (1) model development and problem formulation, (2) design space identification, and (3) design space analysis. Firstly, a mathematical model of the process is developed and experimentally validated, providing the basis for subsequent analysis. The problem is then defined by selecting the design inputs (DIs), key performance indicators (KPIs), and process constraints. Quasi-random Sobol sampling is employed to explore the process efficiently with low discrepancy. Based on the sampled data set, an artificial neural network is trained and used to enhance the design space identification using alpha shapes. Once the design space boundary is quantified, design space analysis is performed to identify acceptable ranges and KPI distributions in an integrated manner.

The proposed framework is implemented to screen adsorbents for post-combustion carbon capture via PVSA. The model consists of partial differential equations and has been experimentally validated as presented by Ward and Pini [4]. A techno-economic assessment framework is coupled to the process model to enable the quantification of capture cost. The assessment of the adsorbent is integrated within the evaluation of process flexibility. The design inputs of the problem are the process intermediate pressure, high pressure, and feed velocity. While the KPIs recorded are purity, recovery, energy consumption, productivity, and capture cost. Constraints with respect to regulations are implemented [1-3] and design spaces are identified for all adsorbents. Integrated quantitative analysis between adsorbents and design cases are carried out. The most flexible designs are identified, acceptable parameter ranges are quantified, and KPI distributions are examined for each adsorbent.

We demonstrate how the proposed framework can be utilized for adsorbent screening of a PVSA process. In contrary to the status-quo approach for PVSA design and adsorbent screening, our methodology is capable of performing integrated assessment between performance and flexibility of the adsorbent-design combination with respect to industrially relevant constraints.

References

[1] R. Haghpanah *et al.*, 'Multiobjective Optimization of a Four-Step Adsorption Process for Postcombustion CO_2 Capture Via Finite Volume Simulation,' *Industrial & Engineering Chemistry*

Research, vol. 52, no. 11, pp. 4249-4265, 2013, doi: 10.1021/ie302658y.

[2] R. T. Maruyama, K. N. Pai, S. G. Subraveti, and A. Rajendran, 'Improving the performance of vacuum swing adsorption based CO_2 capture under reduced recovery requirements,' *International Journal of Greenhouse Gas Control,* vol. 93, 2020, doi: 10.1016/j.ijggc.2019.102902.

[3] V. Subramanian Balashankar and A. Rajendran, 'Process Optimization-Based Screening of Zeolites for Post-Combustion CO_2 Capture by Vacuum Swing Adsorption,' *ACS Sustainable Chemistry & Engineering,* vol. 7, no. 21, pp. 17747-17755, 2019, doi: 10.1021/acssuschemeng.9b04124.

[4] A. Ward and R. Pini, 'Efficient Bayesian Optimization of Industrial-Scale Pressure-Vacuum Swing Adsorption Processes for CO_2 Capture,' *Industrial & Engineering Chemistry Research,* vol. 61, no. 36, pp. 13650-13668, 2022, doi: 10.1021/acs.iecr.2c02313.

130d Retrofitting Direct Air Capture Systems to Natural Gas Combined Cycle and Nuclear Power Plants

Mahpara Habib, Iza Lantgios, Katherine Hornbostel

As CO_2 concentrations continue to rise rapidly due to increased use of fossil fuels resulting from industrialization, it is necessary to combat the effects of global warming by not only curbing the amount of CO_2 being released, but also by re-capturing the CO_2 already in the atmosphere. This process, called direct air capture, or DAC, is challenging because the concentration of CO_2 in ambient air is far more dilute than in flue gas, thus making it harder to perform DAC than post-combustion capture. It requires a sorbent that can adsorb CO_2 at low partial pressures, as well as an abundant source of power to run the DAC system. In this work, we have designed and optimized a direct air capture system that can be retrofitted to a natural gas combined cycle (NGCC) power plant. Tetraamine-appended Mg_2(dobpdc)(3-4-3) metal oxide framework (MOF) is chosen as the solid sorbent material because it has demonstrated that it can effectively capture CO_2 at both ambient air and flue gas concentration. These MOF sorbents are packed into a series of fixed bed reactors through which CO_2-rich gas is flown and adsorbed by the sorbents. The beds regenerate via temperature swing adsorption assisted by steam drawn from the NGCC power plant. The benefit of retrofitting the DAC system to an NGCC plant lies in the fact that it can power the DAC system, while the DAC system in combination with a membrane CO_2 capture system will scrub CO_2 from both the flue gas exhaust from the plant, as well as performing direct air capture. Running in two different operational modes, the DAC system has the two-pronged advantage of cleaning up the NGCC plant's tailpipe (Mode 1), as well as performing DAC (Mode 2) during the off-peak operation hours of the plant, where electricity produced during this period is used to run the DAC system. This can lead to overall negative emission of CO_2 from the power plant. However, the results from this study show that the solid sorbent system, when combined with a membrane capture system, can achieve 98.7% and 99.2% capture of CO_2 during Modes 1 and 2 of operation respectively, thus achieving near-net zero carbon footprint from the NGCC plant instead.

In order to achieve net-negative emissions from a power plant, we propose that the DAC system be retrofitted to a nuclear power plant, since the operation of a nuclear plant does not produce CO_2, thus allowing for negative net CO_2 emissions. The DAC system does not need to be in constant operation, but rather can be operated when it is economically advantageous, likely at off-peak times when it is more profitable to capture CO_2 than to supply electricity to the grid. When the DAC system is in operation, steam and electrical energy from the nuclear plant will be used to power the DAC system. Extracting steam from the nuclear plant will decrease the efficiency of the nuclear plant, thus making the design and optimization of this system a complex challenge. This nuclear-powered DAC system has not yet been designed, and there are many factors at play. This work will discuss the factors that influence the design and optimization of a nuclear-powered DAC system along with the modeling efforts underway to address this challenge. Aspects of the DAC system, such as sorbent material and bed design, should be chosen to optimize performance under the constraints of the nuclear plant steam conditions. There are also a number of factors that must be taken into consideration pertaining to the integration of the DAC system and the nuclear plant, including the type of nuclear reactor (e.g. large light water reactor, small modular reactor, etc.), the system integration (e.g. turbine tapping, turbine bypass, waste heat, etc.) and the system control (e.g. grid demand vs economic demand). Modeling efforts are underway to better define this system with the goal of identifying important system parameters such as the amount of CO_2 that can be captured, the impact on nuclear plant efficiency, and an operating pattern that is profitable.

Flow Diagram of DAC System Model Fitted to an NGCC Power Plant

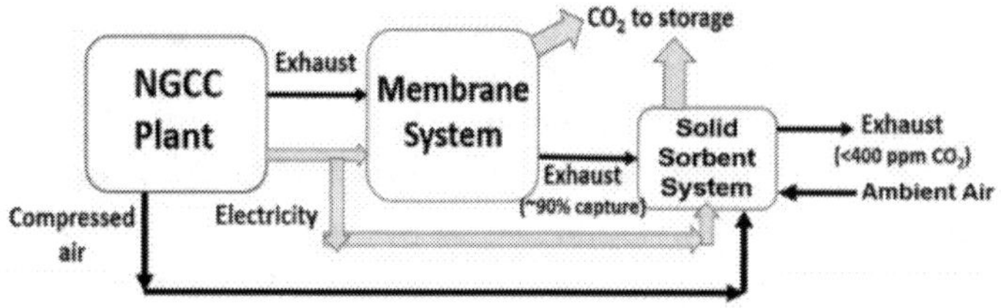

Flow Diagram of DAC System Model Fitted to a Nuclear Power Plant

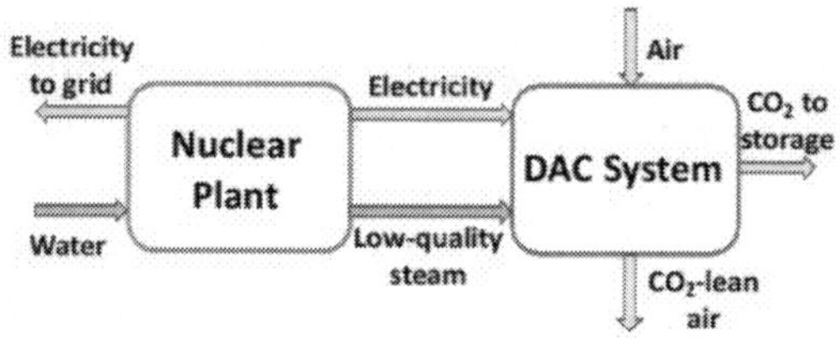

130f CO_2 Adsorption Kinetic Model for Direct Air Capture

James Akinjide, Aashish Priye, Joo-Youp Lee

To successfully gear the planet towards a net-zero emissions scenario by 2050, direct air capture (DAC) is particularly crucial among the diverse collection of technologies currently employed in sequestering CO_2 as it could potentially resolve the emissions from small distributed sources which account for ~20% of total CO_2 emissions. In addition, the inherent benefits of DAC such as low requirements of land and water utilization and flexible location constraints, make it a promising candidate for DAC. Adsorption kinetic model plays a vital role in determining the performance of DAC sorbent for scale-up. In this work, a comprehensive model combining CO_2 adsorption reaction kinetics with external and internal pore diffusion under a fluid flow was studied. Both external mass transfer in the sorbent bed and intraparticle mass transfer within the sorbent particles are considered in the model. Equilibrium adsorption isotherms were determined from lab-scale experimental data under sub-ambient to ambient temperatures and relative humidity conditions. These isotherm data were used to determine adsorption and desorption kinetic constants at different temperatures. The figures below show breakthrough curves for CO_2 adsorption with respect to different inlet CO_2 concentrations of <400 ppm and the accompanying adsorption isotherm collected at 22 °C.

The design of our DAC system includes a sorbent-washcoated monolith approach. The adsorption kinetic expression determined for a powdered form of a sorbent is applied to a sorbent-washcoated monolith block being operated under different flow conditions to validate a feasibility of the conceptual approach. Important adsorption isotherm and kinetic parameters determined from this model will enable us to scale up the DAC system by investigating the overall CO_2 capture efficiency, volumetric productivity, and throughput for a successful development of our DAC system.

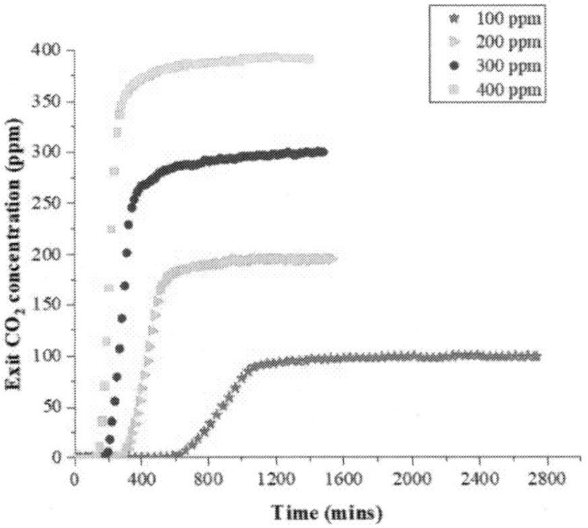

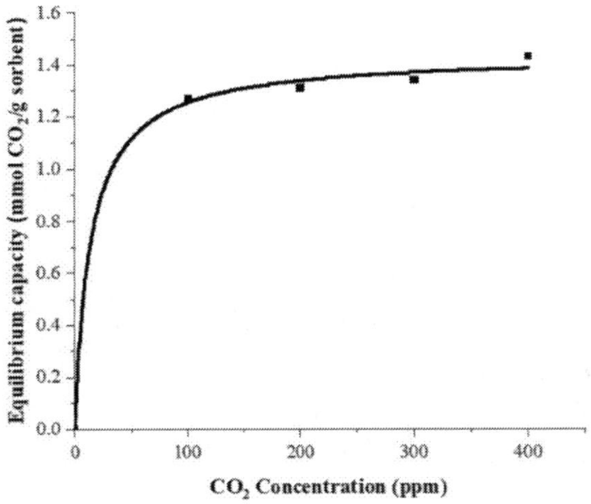

130g Direct Ocean Carbon Capture Using Hollow Fiber Membrane Contactors

Joanna Rivero, Katherine Hornbostel, Tagbo Niepa, PhD, Austin Lieber, Christopher Snodgrass, William Gamble, Donna Hildebrandt, Zoe Neil

Although much research has been done on direct air capture (DAC), far less has been conducted on direct ocean capture (DOC) to-date. Whereas DAC suffers from large system sizes and high energy costs to blow large volumes of air through the system, DOC could be more compact, require less energy due to the higher CO_2 concentration per unit volume, and (at a large-scale), it could have the added benefit of controlling local ocean pH in a high-risk marine environment (i.e., coral reef or shellfishery). We propose a novel approach to removing CO_2 from sweater using a hollow fiber membrane contactor (HFMC) that flows seawater on one side, and an aqueous solvent (NaOH) on the other side. HFMCs have been investigated thoroughly for traditional carbon capture and are a promising fit for DOC due to low energy costs, compact/efficient separation, and the benign aqueous solvents. This work will present lab-scale experimental results that validate a developed 1D models. The model applies the resistance-in-series model for mass transport, and detailed reaction chemistry on both seawater and solvent sides. Then, parametric studies will be shown to demonstrate design and operating parameters for HFMCs for DOC. This work concludes that CO_2 flux increases in a HFMC as the seawater is flown on the shell side of the membrane and that flow direction, co-current or counter-current, has minimal effects on performance. Other parameters such as fiber diameter, temperature, and solvent loading times are also studied. Finally, a techno-economic assessment (TEA) was performed on a theoretical system similar to a scaled-up DAC system, as this is the first of it's kind. The system size was matched to Carbon Engineering's proposed DAC system, which captures 980,000 CO_2 tonnes/year. The TEA here considers both capital and operating costs for all components needed for absorption, regeneration and CO2 compression for a greenfield installation. The results found that lowering the local seawater pH to around 5-6 will be necessary to make this approach economically competitive with state-of-the-art DAC systems. Next steps involve working with experimental collaborators at Arizona State University and UC Irvine to develop a hybrid DOC/desalination membrane that locally lowers pH on the seawater side to improve CO_2 flux and therefore lower capture cost.

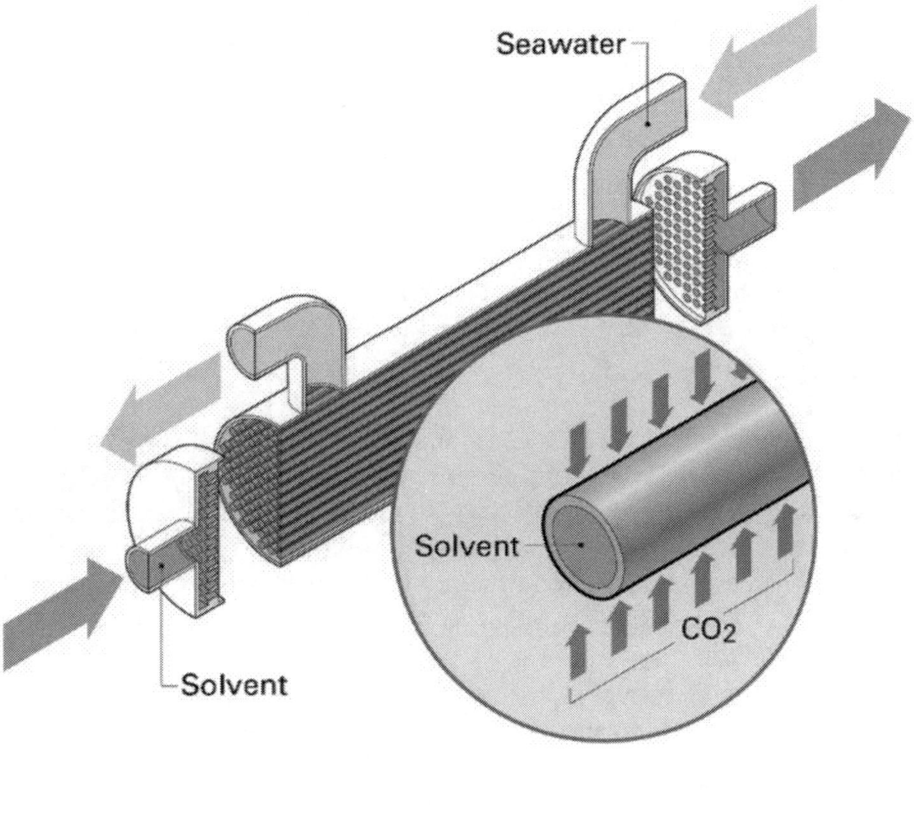
Seawater
Solvent
CO2
Solvent

130h Designing Algal Biodiesel Production Networks for Maximum Carbon Sequestration Using Hybrid Mechanistic Machine Learning Approach

Abhimanyu Raj Shekhar, Shweta Singh

The pursuit of sustainable and carbon-neutral energy sources is crucial in addressing climate change and fostering energy security. Algal biodiesel has gained significant attention as a viable renewable biofuel, due to its high lipid content and rapid rate of utilization as Sustainable Aviation Fuel (SAF).[1] However, the optimization of the algal biodiesel production network comprising of several interconnected industrial systems to achieve maximal decarbonization remains a complex challenge. This research introduces an innovative approach that utilizes surrogate modeling and optimization of the algal biodiesel production network via the Sparse Identification of Nonlinear Dynamics (SINDy) algorithm to enhance the decarbonization rate.[2]

The research methodology encompasses the development of data-driven surrogate models for the algal biodiesel production network, leveraging the ML-based SINDy algorithm. By exploiting sparse regression techniques, the SINDy algorithm effectively identifies generalizable parsimonious models that captures the nonlinear dynamics and interactions within the production network. These surrogate models substantially reduce computational complexity while preserving the essential characteristics of the underlying system, providing a robust foundation for subsequent optimization.[3,4]

With the established surrogate models, a sophisticated telecoupled integrative framework is constructed to maximize the decarbonization rate of the algal biodiesel production network. This framework employs the use of numerically integrated values of coupled surrogate models to concurrently optimize economic and environmental objectives of the research, resulting in a sustainable and balanced network. The optimization process pinpoints optimal operating conditions and strategic interventions, including resource allocation and process improvements, to elevate the decarbonization rate.

The outcomes of this study demonstrate the robustness of the SINDy-based surrogate modeling and optimization approach in significantly enhancing the decarbonization rate of the algal biodiesel production network. By incorporating the identified strategies, the algal biodiesel industry can contribute to climate change mitigation, support sustainable energy production, and curtail reliance on fossil fuels.[5] Thus, this research advances the scientific understanding of sustainable biofuel production and has potential applications in other renewable energy systems.

References:

1. Kandaramath Hari, T., Yaakob, Z. & Binitha, N. N. Aviation biofuel from renewable resources: Routes, opportunities and challenges. *Renew. Sustain. Energy Rev.* **42**, 1234–1244 (2015).
2. Brunton, S. L., Proctor, J. L. & Kutz, J. N. Discovering governing equations from data by sparse identification of nonlinear dynamical systems. *Proc. Natl. Acad. Sci.* **113**, 3932–3937 (2016).
3. Kaptanoglu, A. *et al.* PySINDy: A comprehensive Python package for robust sparse system identification. *J. Open Source Softw.* **7**, 3994 (2022).
4. Kukreja, S. L., Löfberg, J. & Brenner, M. J. A LEAST ABSOLUTE SHRINKAGE AND SELECTION OPERATOR (LASSO) FOR NONLINEAR SYSTEM IDENTIFICATION. *IFAC Proc. Vol.* **39**, 814–819 (2006).
5. Leite, G. B., Abdelaziz, A. E. M. & Hallenbeck, P. C. Algal biofuels: Challenges and opportunities. *Bioresour. Technol.* **145**, 134–141 (2013).

48a Systematic Comparison of CO_2 Capture Towards the Renewable Production of Methanol and Methane: Spain Case

Guillermo Galán Iglesias, Mariano Martin, Ignacio Grossmann

The high increase in CO_2 emissions from the development of transport, industry, and other activities and their accumulation in the atmosphere, with levels above 32Gt CO_2 yearly (EPA, 2022), require to remove it, either by storing it in geological deposits and formations, or by using as a reagent to synthesize chemicals, i.e. methanol (Martin & Grossmann, 2017) or methane (David & Martin, 2014). Using renewable CO_2 and H_2 to produce methanol and methane is a potential and attractive option since they constitute the first C1 in the chain of organic chemicals, being valuable products in the market. Methanol constitutes one of the seven fundamental primary chemical products in the chemical industry (IEA, 2022). Methane can also be utilized in boilers, heating systems, generation energy in cogeneration and thermal power plants, and for domestic use, substituting natural gas. To produce both chemicals, CO_2 and renewable H_2 are needed. On the one hand, the capture of CO_2 can be carried out by employing solutions based on monoethylamine (MEA), which captures the CO_2 directly from concentrated gaseous effluents from thermal power plants, cement plants, etc. Alternatively, the technology based on direct air capture (DAC) using the conventional process based on basic solutions also constitutes a complementary option to remove CO_2 from the atmosphere compared to biomass (Galán et al., 2023). On the other hand, H_2 is produced by water electrolysis through the utilization of PV panels or wind turbines as a renewable source of energy (Martin & Grossmann, 2016). O_2 is generated as a by-product, which may be used in secondary processes or to obtain additional credit. Solar and wind can supply the energy needed by the water electrolysis and provide some or a total of the energy requirements for all the processes. This aspect is strongly influenced by the climatic conditions present in the location.

A facility location problem is formulated, which allows deciding which chemical to produce, the source of CO_2, the source of renewable energy, and the location of the capture and production facilities aiming at avoiding a certain level of emissions. To achieve this goal, the development of a multiscale optimization model based on the work by Heras & Martín (2021) is carried out. At the process scale, a techno-economic analysis is developed to determine the yield and the investment and product cost of MEA, conventional DAC, renewable H_2, methane, and methanol production. All the processes are modeled employing an equation-based approach. The conventional DAC process employs alkaline solutions based on KOH. The CO_2 is hydrogenized in each of the products. Water electrolysis, powered by renewable energy from PV panels or wind turbines, is considered the best alternative to produce renewable H_2. This is purified from traces of O_2 and H_2O. The O_2 stream could supply an extra chemical to employ in other processes or obtain earnings for the complete system. Methanation of CO_2 with renewable H_2 allows producing synthetic methane. Water is also generated in the process, and recycled to the electrolyzer, reducing freshwater makeup. Alternatively, CO_2 can be hydrogenized to methanol. The process is controlled by equilibria that require unreacted H_2 and CO_2 to be recycled. Referring to the macro scale, the KPIs of the different processes are used to formulate a network optimization problem to decide on installing the different technologies across the land/country for a given budget. A multi-period, multi-objective MINLP problem, is formulated, including social issues related to generating wealth and jobs in the region where the facilities are installed. A study of the required supply chain is carried out to determine the most appropriate area in which to locate the CO_2 capture, water electrolysis, and methane/methanol synthesis plants, as well as the type of technology necessary for the generation of solar energy and/or wind and the area or region and ground required in which to locate the technology. At the same time, the social and economic aspects of the use of CO_2 capture plants, the generation of methane and methanol and its distribution, and the use of PV panels and wind turbines are considered to mitigate the impact of climate change on the areas to be treated and developed the local and national economy. A sensitivity analysis is performed considering target prices and different budget allocations to assess the infrastructure to be installed. The years 2022, 2030, and 2050 are considered to evaluate the capture of some and/or the total CO_2 produced. The year 2022 is taken as a reference, achieving around 223.8 Mtons of CO_2 emitted.

As a case study, Spain is considered, except for the autonomous cities of Ceuta and Melilla, the Balearic and Canary Islands, and other Spanish enclaves in the Mediterranean Sea due to their small area. The target process for methanol and methane corresponds to market prices of 0.480 €/$kg_{Methane}$ (Eurostat, 2022) and 0.400 €/$kg_{Methanol}$ (Methanol Institute, 2022). The evaluation assumes 1%, 1.5%, and 2% of the total surface of Spain to install the facility processes, PV panels, and wind turbines, obtaining maximum volumes of CO_2 captured of 51.76%, 74.72%, and 96.57%, requiring investments above 800,000 MM€-1,600,000 MM€, and power supplies between 140 GW-280 GW in 2022. In the year 2030, there is a predicted reduction of 55% (European Commission, 2020) in GHG compared to 1990 in the European Union. In the year 2050, there is a predicted net zero emissions, considering a reduction of 90% of emissions, and the remaining 10% will be absorbed naturally (MITECO, 2020). Generally, methanol production is chosen over methane due to the more competitive production and investment costs, allowing obtain sell prices of 0.400 €/$kg_{Methanol}$, 0.390 €/$kg_{Methanol}$, 185 €/$kg_{Methanol}$ in 2022, 2030, and 2050 (Lloyd's Register-UMAS, 2019). The same evaluation is carried out over methane production to determine the competitive production costs to produce it, resulting in 0.944 €/$kg_{Methane}$, 0.871 €/$kg_{Methane}$, and 0.424 €/$kg_{Methane}$. As is shown, these values can be reached through state subsidies and grants. Moreover, indirect aid will help to facilitate reaching these prices. The production of solar energy is placed in the central-southern peninsular locations since they have higher rates of solar radiation and the wind turbines in the coastal area, overall the northern third of the peninsula, some Mediterranean places, as provinces of Gerona, Alicante, and Murcia, and the Strait of Gibraltar, which have higher wind velocities. In the case of methane, regasification plants, and the pre-existing transportation network provide an adequate infrastructure for the national and international transportation levels.

References:

Davis W, Martín M. Optimal year-round operation for methane production from CO_2 and water using wind and/or solar energy. J Clean Prod. 2014;80:252-261.

European Commission. 2030 Climate Target Plan. Communication From the Commission to the European Parlament, the Council, the European Economic and Social Committee and the Committee of the Regions. https://eur-lex.europa.eu/legal-content/EN/TXT/PDF/?uri=CELEX:52020DC0562&from=EN; 2020.

Eurostat. Eurostat Statistics Explained. Natural Gas Price Statistics. https://ec.europa.eu/eurostat. https://ec.europa.eu/eurostat/statistics-explained/; 2022.

Galán G, Martín M, Grossmann IE. Systematic comparison of natural and engineering methods of capturing CO_2 from the air and its utilization. Sustainable Production and Consumption. 2023;37:78-95.

Heras J, Martin M. Multiscale analysis for power-to-gas-to-power facilities based on energy storage. Computers and Chemical Engineering. 2021;144:107147.

48b Amine Vapor Emissions Control Technology for Water-Lean Solvents for Point Source CO_2 Capture

Jacob Nelson, Lucas Cody, Hannah Honeycutt, Addison Lane, Daniel Mogollon, Moumita Bhattacharya, Paul Mobley, Jak Tanthana, Marty Lail

Water-lean solvent (WLS) systems are an emerging technology for point source CO_2 capture. Although the technology shows a decrease in energy consumption over aqueous solvents, an increase in amine emissions presents new environmental and economic challenges for the technology. The development of economical and effective amine emissions control technologies (ECT) is critical for the advancement and commercialization of WLS systems. One ECT in development is an advanced water or organic wash system, which is capable of reducing amine content in flue gas to < 1 ppm leaving the CO_2 capture system. The novel wash system includes a sorbent bed that recovers the amine from the wash liquid and returns the amine to the CO_2 capture system, resulting in reduced fugitive emissions and improved process economics.

Common adsorbents were tested in a lab-scale adsorbent bed. Amine was adsorbed from flow of amine solutions in water and an organic solvent through the bed. Adsorbents were regenerated from countercurrent flow of steam through the bed. Amine content of the bed effluent was monitored periodically during adsorption and regeneration. Adsorbent performance was evaluated based on the amine adsorption capacity and cost. The material down-selection and results obtained from this lab-scale setup will be further tested at a larger-scale adsorbent bed in future development.

48c Modeling CO2 Adsorption in an Euler-Lagrange Framework Accounting for Pseudo-Turbulence and Intraparticle Diffusion

Rebecca Grawe, Jesse Capecelatro

As global temperatures rise, there is an ever-increasing need for improved carbon capture technology. CO_2 adsorption has emerged as a promising method of CO_2 removal for both post-combustion capture and direct air capture applications. Amine-based sorbents are a promising choice for both applications due to their ability to capture a relatively high amount of CO_2 with relatively low energy costs, even at low concentrations of CO_2. This technology is currently not yet mature enough to be implemented on a large scale, in part due to challenges in scaling up. The flow physics and chemistry taking place at the microscale (i.e., at the scale of individual sorbents) control macroscale behavior in the reactor. Because such effects are not well understood, they are frequently excluded from numerical simulations. The aim of this work is to model CO_2 adsorption using the most state-of-the-art models available to account for microscale thermo-chemistry and flow physics. Modeling is performed in NGA2, an open source multi-phase flow solver that solves the volume-filtered Eulerian-Lagrangian equations. Subgrid-scale models for pseudo-turbulent velocity fluctuations and pseudo-turbulent heat/mass fluxes are employed. These terms account for turbulent mixing induced by particles, which is unresolved in Eulerian-Lagrangian simulations. Additionally, unlike many simulations that model chemical kinetics using a simple linear driving force, intrinsic kinetics are employed and corrected using an effectiveness factor to account for intraparticle diffusion limitations within the particles. Including this effectiveness factor allows transport inside the particle to be accounted for in a computationally efficient manner. This work will look at flow through random arrangements of solid sorbents for both fixed and fluidized systems. The relative importance of the pseudo-turbulent terms and intraparticle diffusion will be discussed.

"Optimizing CO_2 Capture through Solid Adsorption: Developing Optimal Strategies"

Ilse María Hernández-Romero [a] & Antonio Flores Tlacuahuac [a]

[a] *Institute of Advanced Materials for Sustainable Manufacturing, Tecnológico de Monterrey, Monterrey, N.L., 64849, Mexico*

Abstract

Greenhouse gas levels have increased over the last 70 years, mainly CO_2, from 300 to 420 ppm. The high concentration of CO_2 in the atmosphere is a major cause of climate change. Thus, carbon capture and sequestration are promising solutions to mitigate greenhouse gas emissions. In this sense, Pressure Swing Adsorption technology (PSA) is promising for direct CO_2 capture. However, its capture efficiency is highly dependent on the diffusion behavior of CO_2 in porous media. Therefore, this paper presents a simulation and optimization approach to incorporate the time-dependent operation aspects of the PSA column packed with 13X zeolite as adsorbent material to maximize the capture of atmospheric CO_2. A two-bed, four-step PSA system is considered. The four steps are pressurization, adsorption, depressurization, and regeneration. At the same time, the beds correspond to the adsorber and the regeneration columns.

In addition, this proposal aims to develop a rigorous mathematical model for CO_2 diffusion in porous media, considering three coupled modeling scales: macro, meso, and micro. Furthermore, the CO_2 diffusion behavior at each scale will be analyzed, and the results will be compared to investigate if considering such fine details presents significant changes in CO_2 capture. It is expected that these considerations will improve the CO_2 capture process. In addition, Neumann and Dirichlet boundary conditions define the problem and establish the link between the PSA operation cycles and the three modeling levels. Finally, the numerical formulation, the finite difference method, was used to discretize the spatial domain during an orthogonal fitting for the spatial domain.

Keywords: dynamic optimization, pressure swing adsorption, CO_2 capture, adsorbents, partial differential equations (PDEs).

1. Introduction

Energy production, transport, and industry activities have been mentioned as the main reasons for the increased greenhouse gases in the atmosphere. However, sustainable CO_2 capture from the atmosphere is a difficult problem for several reasons. One of the main problems is the relatively low and highly diluted CO_2 concentration (420 ppm) and the high energy consumption (Madejski et al., 2022). Therefore, carbon dioxide capture technologies, as well as the selection of porous materials, have been proposed as a promising alternative to reducing CO_2 emissions from the atmosphere. In this sense, Pressure Swing Adsorption (PSA) technology represents an important gas separation process (Wiheeb et al., 2016). Moreover, this technology has reached a state of maturity, making it one of the best alternatives for dealing with such pollution problems and providing one of the best economical alternatives (Haines et al., 2014). On the other hand, porous adsorbent materials, such as zeolites, activated carbon, or hybrid materials (MOFs), have been proposed due to their high adsorption capacity and selectivity towards CO_2 compared to other gases present in the atmosphere (Ghanbari et al., 2020).

It is important to note that in the adsorption process, gas or liquid molecules are concentrated on a solid's surface. On an industrial scale, the process is carried out in adsorption beds, where the structure of the solid is selected to maximize the capture of molecules of a compound of interest, known as selectivity. For example, in the purification of industrial gases, activated carbon or zeolites are used as adsorbent solids since they have demonstrated good efficiency in adsorption. This material type is characterized by porous structures that favor the adsorption process.

On the other hand, the modeling of PSA technology represents great challenges since the models for PSA processes are multiple instances of partial differential equations (PDEs) in time and space with periodic boundary conditions linking the processing steps (Zhang et al., 2022). Furthermore, the solution of this coupled PDE system is highly time-dependent; therefore, the optimization of these systems represents a great computational challenge (Hao et al., 2021). In addition, the real-time analysis of the PSA system is highly dependent on the choice of key decision variables and operational considerations, such as processing steps and column pressure-time profiles, making its design and operation another challenge (Subraveti et al., 2019).

One of the major areas of innovation is developing advanced computer models that can predict and optimize the performance of CO_2 capture systems (Nikolaidis et al., 2018 & Subraveti et al., 2019). These models help researchers and engineers design more efficient and cost-effective systems (Haines et al., 2014). Some of the latest modeling techniques include process simulation and process optimization. These tools are essential for pushing the boundaries of CO_2 capture technology beyond the current state of the art. However, the ultimate success of CO_2 capture and utilization technologies will depend on their practicality, cost-effectiveness, and ability to compete with other greenhouse gas reduction strategies in the energy market. Therefore, in this work, an optimization framework is developed to incorporate the most relevant aspects of the operation of the PSA column. Specifically, the following two aspects are considered: I) The simulation of the CO_2 adsorption and desorption process and II) The dynamic optimization of the process, which involves the determination of the optimal time-dependent values of the main decision variables (composition, temperature, and pressure). For this, Neumann and Dirichlet boundary conditions are considered as a restriction of the problem. In addition, for the numerical formulation, the finite difference method was used to discretize the spatial domain during orthogonal collocation for the spatial domain.

2. Overview of the PSA technology

Pressure Swing Adsorption (PSA) is a technique for separating gases in which an adsorbent material is regenerated by lowering the partial pressure of the absorbed component. A quick reduction in partial pressure is achieved by lowering the total column pressure through purging (Abdeljaoued et al., 2018). This process typically follows a four-step cycle as shown in Figure 1:

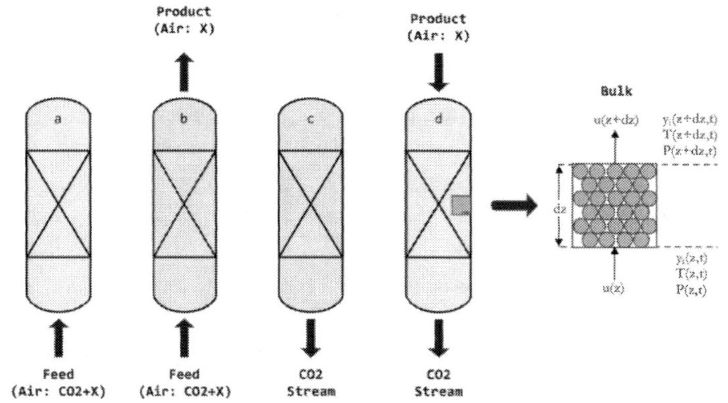

Figure 1. Operational steps of the pressure swing adsorption:
a) Pressurization, b) Adsorption, c) Depressurization and d) Desorption

In the first stage, a mixture of gases is fed into the column until the desired pressure is reached. In the Adsorption stage, the purified product stream is obtained. When the bed becomes saturated, the feed flow to the column is stopped, and the column is purged to atmospheric pressure in the Depressurization stage. Finally, a pure product stream is passed through the column in the Regeneration stage, replacing the remaining molecules in the bed and restoring the initial conditions for the next cycle.

3. Problem definition

The problem addressed in this work can be stated as follows:

- A dilute CO_2 stream of specified composition, pressure, and temperature is available.
- Detailed distributed dynamic mathematical model capable of describing the process behavior (adsorption rate, mass, energy balances, temperature, and pressure profiles) with sufficient accuracy for the proposed objectives.
- Information on physical, thermodynamic, and equilibrium properties.

A generalized PSA mathematical model is proposed that aims to evaluate the performance of different adsorbents (including MOFs), based on critical factors like capture efficiency, cost, and environmental impact. The model inputs the physical, chemical, and thermodynamic properties of the adsorbents and provides a basis for comparison. Zeolite 13x was selected as a sample adsorbent to validate the model, as extensive experimental information on it is available in the literature. In addition, it is possible to determine optimal operation strategies subject to the gas inlet conditions through the optimal adsorption process.

Finally, the following assumptions are made to develop the modeling: (a) gases behave as ideal gases, (b) the variation of temperature, pressure, and concentration is negligible in the radial direction, (c) the surface velocity of the gas is constant in adsorption, and desorption stages, (d) piston flow with axial dispersion and non-steady state conditions assumed, (e) non-isothermal conditions, (f) physical properties of compounds are temperature-independent, (g) Langmuir Isotherm is used as the equilibrium model for gas mixtures, (h) mass transfer rate evaluated using the Linear Driving Force (LDF) model, (i) pressure drop across the packing calculated using the Ergun equation.

4. Mathematical formulation

This section summarizes the proposed mathematical model based on Figure 1. The formulation mainly involves the mass and energy balances, the adsorption kinetics, and the estimation of the pressure in the cycle, among others, as well as their respective boundary conditions. It is important to note that the equations are indexed and *i* represents the component (CO_2 and N_2).

4.1. Mass balance.

A mass balance is carried out for component *i* (*i*= CO_2, N_2) over a differential control volume with respect to the position (z) and time (t). It considers the mole fraction of each component *i* in the gas phase (y), the axial dispersion coefficient (D_L), the flow velocity (v), the packed bed density and porosity (ρ_{bulk} and ε), and the concentration of component *i* in the solid (q).

$$-D_L\left[\frac{\partial^2 y_i}{\partial z^2}-2\left(\frac{1}{T}\right)\left(\frac{\partial y_i}{\partial z}\right)\left(\frac{\partial T}{\partial z}\right)+2\left(\frac{1}{P}\right)\left(\frac{\partial y_i}{\partial z}\right)\left(\frac{\partial P}{\partial z}\right)\right]+\frac{\partial y_i}{\partial t}+v\frac{\partial y_i}{\partial z}+\frac{RT}{P}\frac{\rho_{bulk}}{\varepsilon}\left(\frac{\partial q_i}{\partial t}-y_i\sum_i\frac{\partial q_i}{\partial t}\right)=0 \quad (1)$$

4.2. Energy balance.

The energy balance can be expressed as follows:

$$\left(\rho_g Cp_g+\frac{\rho_{bulk}}{\varepsilon}Cp_s\right)\frac{\partial T}{\partial t}+\rho_g Cp_g v\frac{\partial T}{\partial z}-\frac{K_L}{\varepsilon}\frac{\partial^2 T}{\partial z^2}-\frac{\rho_{bulk}}{\varepsilon}\sum_i \Delta H_i\frac{\partial q_i}{\partial t}+\frac{K_{eq}}{\varepsilon}(T-T_W)=0 \quad (2)$$

where, ρ_g is the density of the gas mixture, Cp_g and Cp_s are the heat capacity of the gas and adsorbent respectively. In addition, the axial and radial conductivity coefficient are represented by K_L and K_{eq}. ΔH_i is the heat of adsorption of component *i*, T is the bed temperature and T_W is the temperature at the column surface.

4.3. Mass transfer rate estimation.

In this case, the mass transfer rate $\left(\frac{\partial q_i}{\partial t}\right)$, depends on the concentration of component *i* in the solid phase in the equilibrium. In other words, the mass transfer associated with the amount of adsorption at equilibrium is evaluated. And is represented using the LDF model as follows.

(3)

$$\frac{\partial q_i}{\partial t} = k_i\left(q_i^{eq} - q_i\right)$$

Where the adsorption equilibrium q_i^{eq} is explained by the Langmuir isotherm:

$$q_i^{eq} = \frac{a_i b_i P_i}{1 + \sum_j b_j P_j} \tag{4}$$

The Langmuir parameters $\boldsymbol{a}$ and b are a function of temperature.

4.4. Pressure drops.

Pressure drops (P) along the length of the packed bed, given some fluid velocity (v), are estimated with the Ergun equation (Wilkes & Brown, 2022). It also tells us that the pressure drop depends on the packing size, length of the bed, viscosity, and density of the fluid. Where μ and ρ_g are the viscosity and density of the gas stream.

$$-\frac{\partial P}{\partial z} = \frac{150\mu}{4R_p^2}\frac{(1-\varepsilon)^2}{\varepsilon^3}v + \frac{1.75\rho_g}{2R_p}\left(\frac{1-\varepsilon}{\varepsilon^3}\right)v|v| \tag{5}$$

4.5. Optimization approach

This paper does not yet address this approach in detail; however, it is noteworthy that the work is still in progress. However, the objective function used in the preliminary results is presented:

$$Max\ \mathrm{CO_2Capture\ [\%]} = \frac{\text{Total moles of } \mathrm{CO_2} \text{ in extract product}}{\text{Total moles of } \mathrm{CO_2} \text{ fed into the PSA column}} x100 \tag{6}$$

5. Case study

The following case study is addressed, which consists of a dynamic PSA column operating in a four-step cycle with a length of 1 m for atmospheric CO_2 capture using Zeolite 13X as adsorbent material (Regufe et al., 2020). Table 1 shows the physical properties of zeolite 13X and air, respectively.

Table 1. Physical properties of gas and the material adsorbent.

Zeolite 13X	Value	Air	CO_2	N_2
Particle radius (m)	$1x10^{-3}$	Molecular weight	44	28
Particle tortuosity	4.5	Diffusion volume	26.9	18.5
Pore diameter (m)	$1x10^{-9}$	Mass axial dispersion coefficient (m^2/s)	$5.68x10^{-4}$	
Bed Porosity	0.348	Specific heat of gas (J/Kg-K)	994.2	
Bulk density (Kg/m^3)	1060	Effective axial thermal conductivity (J/s-m-K)	1.7948	
Specific heat of bed (J/Kg-K)	504	Effective radial thermal conductivity (J/s-m^3-K)	6614.68	

The column is fed with a mixture of CO_2 and N_2 (420 ppm CO_2) at a pressure of 4 atm, with a velocity of 1 m/s and a temperature of 305 K. On the other hand, regeneration is carried out with a pure stream of N_2 at a pressure of 2.5 atm and a temperature of 305 K. In addition, in Table 2, the boundary conditions and the initial conditions used in the adsorption/desorption processes are presented.

Table 2. Main boundary conditions.

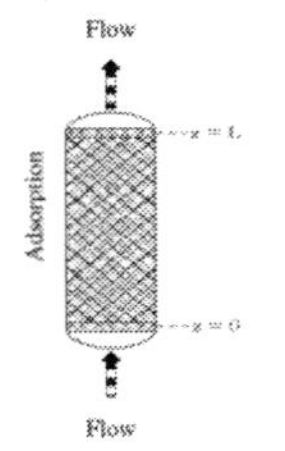	Boundary conditions	Initial conditions
	$D_L \frac{\partial C_i}{\partial z}\Big\|_{z=0} = -v(C_i^-\|_{z=0} - C_i^+\|_{z=0})$	Adsorption
	$\frac{\partial C_i}{\partial z}\Big\|_{z=L} = 0$	$C_i = C_{ads,i}$ $T = T_{ads}$
	$K_L \frac{\partial T}{\partial z}\Big\|_{z=0} = -\rho_g C p_g v(T^-\|_{z=0} - T^+\|_{z=0})$	Desorption
	$\frac{\partial T}{\partial z}\Big\|_{z=L} = 0$	$C_i = C_{des,i}$ $T = T_{des}$

6. Results

The NLP model was implemented in JuMP embedded in Julia Language (Dunning et al., 2017). The results presented below correspond to the simulation phase of the PSA column. In addition, the four-step process is being considered. The idea is to obtain a rigorous model based on the considerations made and validate it from experimental data (already reported in the literature) to evaluate and properly estimate all the PSA operating variables (Daeho et al., 2003 & Uebbing et al., 2021). In this way, Figures 2-4 show the concentration, temperature, and pressure profiles throughout the PSA operation. In Figure 2, the pressure profile shows the change in pressure with respect to time. The pressurization time is 5 min (300 s), reaching a maximum pressure of 4 atm. The concentration profile is not altered during this time because the pressurization time is not long enough for significant adsorption. However, it can be observed in the temperature profile that the temperature of the column rises drastically during this period of time due to the increase of the pressure inside the column.

On the other hand, the pressurization stage lasts 50 min (3000 s), and the pressure is kept constant at 4 atm. In this period, the adsorbent solid captures CO_2 until it is saturated. At the entrance of the column (0 m), the solid is quickly saturated, while at the middle of the column (0.5 m), it takes more than 30 min to reach saturation. At the end of the column, a CO_2-free current is obtained for almost 40 min. After this time, when the concentration begins to increase, the column begins to be completely saturated, and the pressure change must be made to give way to the regeneration of the bed. Finally, it can be observed that in Figure 3, during the adsorption time, the column temperature remains close to the feed temperature, even though the adsorption process is exothermic. The temperature remains constant because the rate of energy transfer to the environment is higher than the rate at which energy is released due to the adsorption phenomenon. It can be observed in the first minutes when the adsorption process starts.

The depressurization process lasts 5 min (300 s), during which time the feed is interrupted, and the column is purged from the same side where it was previously fed (0 m). In this stage, the column reaches a minimum pressure of 1 atm at one end, while at the other end, it reaches a pressure of 2.5 atm, as shown in Figure 2 in the pressure profile. This is a result from a pressure drop inside the column due to the packed bed, which is approximately 1.5 atm.

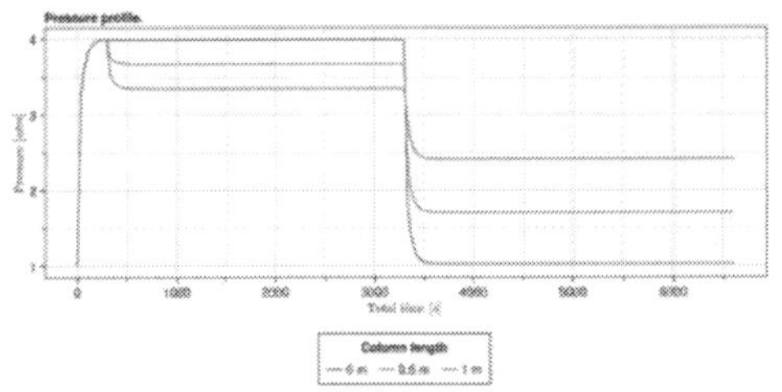

Figure 2. Pressure variation in the column

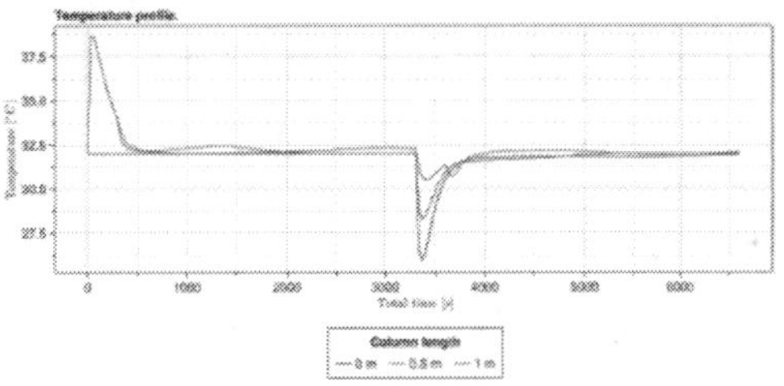

Figure 3. Temperature profile in the column

On the other hand, Figures 4 and 5 show the concentration profile, where a sudden rise in the concentration inside the column can be seen. It occurs because when the pressure changes, the adsorption equilibrium is modified, allowing the release of a large amount of CO_2 inside the adsorbent solid. In addition, at this stage, when there is a significant change in pressure, there is also a change in temperature inside the column, as shown in Figure 3, where the temperature reaches a minimum of 26 °C. During the regeneration stage, the bed is refreshed using a free N_2 stream, which enters at the opposite end of the feed stream at a pressure of 2.5 atm. Moreover, in this step, the CO_2 concentration inside the column decreases with time until the initial conditions of the column are reached again. Finally, a CO_2-rich current is obtained at one end of the column, as depicted in Figure 4.

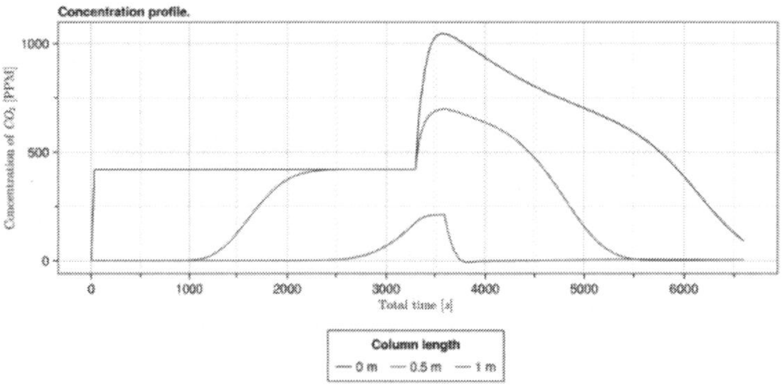

Figure 4. CO_2 concentration variation throughout the PSA column

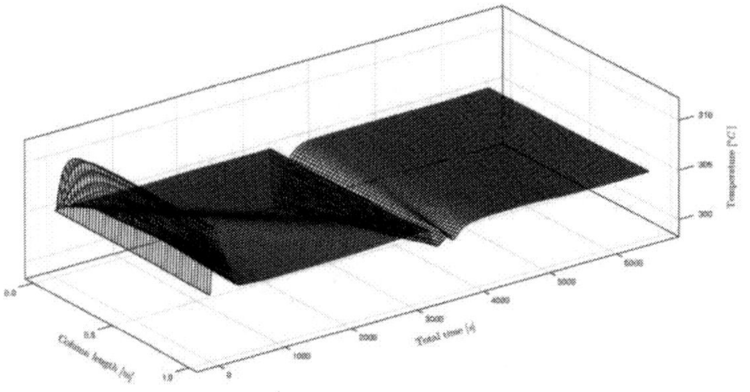

Figure 5. CO_2 concentration profile throughout the PSA operation

7. Conclusions

Since the cost of adsorbents in CO_2 capture processes is usually high, it is of the utmost importance that the design and operation of adsorption equipment be carried out, minimizing the total annual costs. The reality is that, regardless of its relevance or interest, the industrial use of virtually any technology requires economic profitability. In the present work, our main objective is to develop adsorption processes for CO_2 capture and operate this type of process in conditions of economic profitability. Moreover, in this work, the presented results show only the simulation of the dynamic PSA operation considering all the operation cycle, with the four steps. The optimization details will be considered later, in which the economic profitability of the process and the design are addressed. The basic process of CO_2 capture using adsorption towers is not new. However,

Ongoing work

This work aims to go deep into the optimization approach focusing on energy consumption, column sizing, and economic analysis. Specifically, we aim to estimate the power needed for pressurizing/depressurizing the column and gas/airflow. The column sizing will be optimized for optimal bed height and diameter based on feed gas flow rate, CO_2 concentration, and operating pressure. Finally, the economic assessment will include estimating capital and operating costs for the PSA system, including adsorbent cost, energy consumption, and maintenance.

References

A. Abdeljaoued, F. Relvas, A. Mendes & M. H. Chahbani, 2018, Simulation and experimental results of a PSA process for production of hydrogen used in fuel cells. Journal of Environmental Chemical Engineering, 6(1), 338-355.

A.D. Wiheeb, Z. Helwani, J. Kim, & M.R. Othman, 2016, Pressure swing adsorption technologies for carbon dioxide capture, Separation & Purification Reviews, 45(2), 108-121.

D. Ko, R. Siriwardane, L.T. Biegler, 2003, Optimization of a pressure-swing adsorption process using zeolite 13X for CO2 sequestration. Industrial & engineering chemistry research, 42(2), 339-348.

G. N. Nikolaidis, E. S. Kikkinides & M. C. Georgiadis, 2018, A model-based approach for the evaluation of new zeolite 13X-based adsorbents for the efficient post-combustion CO2 capture using P/VSA processes. Chemical Engineering Research and Design, 131, 362-374.

I. Dunning, J. Huchette & M. Lubin, 2017, JuMP: A modeling language for mathematical optimization. SIAM review, 59(2), 295-320.

J. Uebbing, L.T. Biegler, L. Rihko-Struckmann, S. Sager & K. Sundmacher, 2021, Optimization of pressure swing adsorption via a trust-region filter algorithm and equilibrium theory. Computers & Chemical Engineering, 151, 107340.

M. D. Wilkes & S. Brown, 2022, Flexible CO2 capture for open-cycle gas turbines via vacuum-pressure swing adsorption: A model-based assessment. Energy, 250, 123805.

M. Haines, J.Kemper, J. Davison, J. Gale, P. Singh & S. Santos, 2014, Assessment of emerging CO2 capture technologies and their potential to reduce costs. International Energy Agency (IEA): Paris, France.

M. J. Regufe, A. F. Ferreira, J. M. Loureiro, A. Rodrigues & A.M. Ribeiro, 2020, Development of hybrid materials with activated carbon and zeolite 13X for CO2 capture from flue gases by electric swing adsorption. Industrial & Engineering Chemistry Research, 59(26), 12197-12211.

P. Madejski, K. Chmiel, N. Subramanian & T. Kuś, 2022, Methods and techniques for CO2 capture: Review of potential solutions and applications in modern energy technologies, Energies, 15(3), 887.

R. Zhang, Y. Shen, Z. Tang, W. Li, & D. Zhang, D, 2022, A review of numerical research on the pressure swing adsorption process. Processes, 10(5), 812.

S. G. Subraveti, K. N Pai, A. K. Rajagopalan, N. S. Wilkins, A. Rajendran, A. Jayaraman & G. Alptekin, 2019, Cycle design and optimization of pressure swing adsorption cycles for pre-combustion CO2 capture. Applied energy, 254, 113624.

T. Ghanbari, F. Abnisa & W.M.A. Daud, 2020, A review on production of metal organic frameworks (MOF) for CO2 adsorption, Science of The Total Environment, 707, 135090.

Z. Hao, A. Caspari, A.M. Schweidtmann, Y.Vaupel, A.A. Lapkin & A. Mhamdi, 2021, Efficient hybrid multiobjective optimization of pressure swing adsorption. Chemical Engineering Journal, 423, 130248.

48f Grey and Blue Hydrogen Production Costs in Hydrogen Plants: A Comparative Analysis

Mary Katebah, Mamoun Al-Rawashdeh, Patrick Linke

Global hydrogen demand has tripled since the 1970s [1], reaching 94 million tonnes in 2021, surpassing its historical maximum of 91 million tonnes in 2019 [2]. Its emergence as the fuel of the future is credited to its clean burning, long-term storage capacity, and high energy density [3]. Several pathways exist for hydrogen production, including "black/grey hydrogen" from fossil fuels, where the generated CO_2 emissions are released to the environment, "blue hydrogen" where the produced CO_2 is captured and stored by an industrial carbon capture and storage (CCS) process, and "green hydrogen" by electrolysis using renewable energy [4]. To date, most of the hydrogen produced is by natural gas (NG) reforming, accounting for over 75% of the annual global hydrogen production [1]. There are three main routes for hydrogen production from NG: steam methane reforming (SMR), auto-thermal reforming (ATR), and partial oxidation (POX). These processes are associated with significant amounts of CO_2, ranging from 7.5 [5] to 13 tons of CO_2 [6] per ton of produced hydrogen. Producing low-carbon hydrogen is key to future energy systems as it can play a major role in the clean energy transition. Current methods being explored for low-carbon hydrogen generation include blue hydrogen from natural gas, and electrolysis [7], with blue-hydrogen being the most promising short-term solution for CO_2 mitigation from hydrogen processes [8].

Numerous studies in the literature have focused on the techno-economic analysis of blue-hydrogen production by SMR, with very limited analyses on ATR and POX. The literature is lacking a detailed integration, analysis and comparison of blue hydrogen production from NG for increasing emission levels following the three prominent routes. This study addresses the aforementioned gaps by presenting a comparative techno-economic assessment on hydrogen production by SMR, ATR and POX. The hydrogen plants are assumed to be standalone, and each process is energy-integrated with CO_2 capture and compression units to obtain a relationship between hydrogen production costs and CO_2 emissions' mitigation for a wide range of CO_2 removal rates (0-90%). Different design regimes and energy system designs are developed for heat and power integration across the hydrogen plants and CO_2 capture and compression for grey to blue hydrogen production. Impact of natural gas price and plant capacities are also investigated on the hydrogen costs. The analysis is extended to assess the impact of life-cycle emissions outside the boundary of the hydrogen plants, and the results are compared with state-of-the-art electrolysis technology. Results show that hydrogen costs are lower than those reported in the literature, highlighting the importance of energy integration on overall production costs reduction.

Acknowledgments. This work was made possible by funding from Qatar National Research Fund (QNRF) project number NPRP12S-0304-190222 and co-funding by Qatar Shell Research and Technology Center (QSRTC). The statements made herein are solely the responsibility of the author(s).

References:

[1] IEA, "The Future of Hydrogen," 2019. [Online]. Available: https://www.iea.org/reports/the-future-of-hydrogen

[2] IEA, "Hydrogen," 2022.

[3] N. Liu, "Increasing blue hydrogen production affordability," *Hydrocarbon Processing*, 2021. [Online]. Available: https://www.hydrocarbonprocessing.com/magazine/2021/june-2021/special-focus-process-optimization/increasing-blue-hydrogen-production-affordability

[4] M. Katebah, M. Al-Rawashdeh, and P. Linke, "Analysis of hydrogen production costs in Steam-Methane Reforming considering integration with electrolysis and CO2 capture," *Clean Eng Technol*, 2022.

[5] G. Collodi, "Hydrogen production via steam reforming with CO2 capture," *Chem Eng Trans*, vol. 19, pp. 37–42, 2010, doi: 10.3303/CET1019007.

[6] F. O. Ayodele, N. Mohammad, S. I. Mustapa, and B. V. Ayodele, "An overview of economic analysis and environmental impacts of natural gas conversion technologies," *Sustainability (Switzerland)*, vol. 12, no. 23, pp. 1–18, 2020, doi: 10.3390/su122310148.

[7] Hydrogen Council, "Path to hydrogen competitiveness A cost perspective," 2020.

[8] N. Muradov, "Low to near-zero CO2 production of hydrogen from fossil fuels: Status and perspectives," *Int J Hydrogen Energy*, vol. 42, no. 20, pp. 14058–14088, 2017, doi: 10.1016/j.ijhydene.2017.04.101.

226a Techno-Economic Optimization of a Compressed Air Energy Storage System Integrated with a Natural Gas Combined Cycle Plant Considering Time-Varying Electricity Price

Pavitra Senthamilselvan Sengalani, Md Emdadul Haque, Manali S. Zantye, Akhilesh Gandhi, Mengdi Li, Faruque Hasan, Debangsu Bhattacharyya

Global electricity demand is increasing. However, it is desired that the greenhouse gas emissions due to electricity production be reduced. These challenges are being addressed by increasing use of renewable energy sources [1,2]. However, renewable energy production is intermittent affected by diurnal and seasonal variability that can adversely affect the reliability and resiliency of the electric grid [2]. One potential way to facilitate higher penetration of renewables without compromising the reliability and resiliency of the electric grid is to develop energy storage systems [3-5]. When electricity is in excess, energy can be stored, while stored energy can be used when there is high demand. While various energy storage technologies are being developed and investigated [1], compressed air energy system (CAES) is one of the technologies that is matured, has fast transients, and can enable large amount of energy storage especially when an appropriate natural formation such as cavern is readily available. While there are several works in the literature investigating the design and optimization of the CAES technology, those studies have been conducted for stand-alone storage systems. In this work, we have investigated CAES integrated with a natural gas combined cycle (NGCC) power plant. For storage, air can be extracted from the gas turbine (GT) compressor in an NGCC plant while injecting the air back to the GT combustor through expanders when needed. This approach utilizes the existing infrastructure in the NGCC plant, especially the GT compressor and turbine, thus reducing the capital costs and improving efficiency due to the use of the highly efficient components of the GT. For optimal design and operation of such systems, variabilities in the electricity demand supply over a long time horizon needs to be considered thus optimal operation of the power plant also needs to be considered simultaneously. Due to the transient characteristics of the storage system as well as that of the NGCC plant, the underlying optimization problem for optimal design and operation of the CAES system and optimal operation of the retrofitted power plant leads to a challenging large-scale dynamic optimization problem.

A dynamic model of the CAES is developed and validated with the literature data [3]. Storage in caverns and above-ground vessels is considered to be alternative options. As the first-principles dynamic model of the NGCC plant is highly nonlinear and computationally expensive, it is difficult to use that model for dynamic optimization. A linear reduced order model of the NGCC plant is developed with air extraction/injection included by linearizing the nonlinear model around steady-state conditions and with further reduction in orders through balanced truncation.

Maximization of the net present value (NPV) is considered to be the objective function for the optimization problem. Optimizations are done in the Python/Pyomo platform considering the whole year locational marginal price (LMP) of electricity for 14 regions [6] with varying carbon taxes. The integrated CAES system is found to be economically superior compared to the standalone NGCC system for some regions. For those regions where the integrated system is not superior, a study is done to evaluate the impact of the reduction in the capital and operating costs of the CAES technology and to understand what minimum reduction of costs will be needed for a specific region for the CAES technology to be economically viable. Levelized cost of storage for cavern and vessel storage is computed for all regions. The study shows interesting optimal dynamic profiles of air injection and extraction that are not only affected by the instantaneous LMP but also the LMP profile before and after of a given time instant showing that consideration of long time windows will be critical for optimal design of the energy storage systems.

Bibliography

1. Heidari, M.; Parra, D.; Patel, M.K. Physical design , techno-economic analysis and optimization of distributed compressed air energy storage for renewable energy integration. *Energy Storage*, **2021**, 35, 102268.
2. Zhou, Q.; Du, D.; Lu, C.; He, Q.; Liu, W. A review of thermal energy storage in compressed air energy storage system. *Energy*, **2019**, 188, 115993.
3. Raju, M.; Khaitan, S.K. Modeling and simulation of compressed air storage in caverns: A case study of the Huntorf plant. *Applied Energy*, **2012**, 89, 474–481.
4. Jeong, J.H.; Yi, J.H.; Kim, T.S. Analysis of options in combining compressed air energy storage with a natural gas combined cycle. *Mech. Sci. Tech*, **2018**, 32(7), 3453–3464.
5. Sadhukhan, J.; Sen, S.; Randriamahefasoa, T.M.S.; Gadkari, S. Energy System Optimzation for Net-Zero Electricity. *Chem. Eng.*, **2022**, 3, 100026.
6. Sun, Y.; Wachche, S.; Mills, A.; Ma, O.; Meshek, M.; Buchanan, S.; Hicks, A.; Roberts, B. 2018 Renewable Energy Grid Integration Data Book. *NREL*, **2020**, Golden, CO.

226b Optimal Design of Liquid Oxygen Storage System for Flexible Operation of a Direct-Fired Supercritical CO2 Power Cycle

Radhakrishna Tumbalam Gooty, Anderson Soares Chinen, Sandeep Pidaparti, Jaffer Ghouse, Eric A. Liese

Low-carbon generators, such as fossil generators equipped with carbon capture systems, are needed to reduce the cost of decarbonization. Such generators need to be flexible to respond to grid conditions, especially in a grid containing a high percentage of variable renewable energy. The direct natural gas-fired supercritical CO2 (sCO2) power cycle is a promising technology for cost-effectiveness, highly efficient power generation with inherent carbon capture that captures more than 97% of the CO2. Nevertheless, slow dynamics and a long startup time associated with the operation of the air separation unit (ASU) required to produce oxygen needed for combustion make the overall operation less flexible. One proposed solution to improve operational flexibility is to include an on-site liquefaction and liquid-oxygen (LOx) storage system, which helps in shifting the load depending on the grid requirements. Therefore, in this work, the economic benefits of installing a LOx storage system are assessed along with whether the additional capital expenditure (CAPEX) needed for the liquefaction and storage system is justified. Three systems are considered: (1) an sCO2 power cycle without a LOx storage system to serve as a reference, (2) an sCO2 power cycle with a partial LOx withdrawal from the ASU, and (3) an sCO2 power cycle with a dedicated liquefaction unit and a LOx storage system. Following the workflow in [1], a multiperiod price-taker problem is formulated that determines the optimal design of the ASU and liquefaction and storage systems as well as the optimal hourly operating decisions, which maximize the net present value for a given electricity market. Detailed models with off-design performance relations for the subprocesses are developed in process simulators. Using the data obtained from detailed models, simple linear surrogate models are constructed and used in the multiperiod optimization problem for computational tractability. The resulting mixed-integer linear program is implemented in the IDAES integrated platform [2] and solved to global optimality using Gurobi. With this formulation, the overall economics of the system across multiple electricity markets is investigated for different carbon pricing scenarios. Finally, a sensitivity analysis over a few important parameters is performed (i.e., CAPEX), and target values, which make the overall economics favorable, are identified.

Acknowledgments:

We gratefully acknowledge support from the U.S. DOE's Office of Fossil Energy and Carbon Management through the Turbines Program. We thank Gurobi Optimization for providing a free research license.

Disclaimer

necessarily state or reflect those of the United States Government or any agency thereof.

References

1. R. Tumbalam Gooty, J. Ghouse, Q. Le, B. Thitakamol, S. Rezaei, D. Obiang, R. Gupta, S. J. Zhou, D. Bhattacharyya, D. Miller. 2023. Incorporation of Market Signals for the Optimal Design of Post Combustion Carbon Capture Systems. Applied Energy, 337, pp.120880.

2. A. Lee, J. Ghouse, J. Eslick, C. Laird, J. Siirola, M. Zamarripa, D. Gunter, J. Shinn, A. Dowling, D. Bhattacharyya, L. Biegler, A. Burgard, D. Miller. 2021. The IDAES process modeling framework and model library – Flexibility for process simulation and optimization. Journal of Advanced Manufacturing and Processing, 3(3), pp.e10095.

226c Identifying Opportunities and Recommendations for the Integration of Nuclear Integrated Energy Systems with Industrial Heat Users

Rami Saeed, Byung-Hee Choi

Idaho National Laboratory is developing a roadmap to identify and assist industrial heat and electricity users with decarbonization by integrating with nuclear power plants (NPPs) to provide clean, abundant, and dispatchable energy. This roadmap has a library of documents and models to guide various industries toward understanding nuclear technologies based on their needs. Considerations include specific industrial hazards which impact siting of an NPP, heat transport requirements and associated technologies, and implementation feasibility based on site-specific demand profiles. Facility process models are based on real data obtained from a survey of baseline requirements and process information (e.g., methods, quality, quantity of required heat inputs) of industrial facilities in the United States.

The industrial processes are prioritized by: operational heat characteristics that can be provided by nuclear systems, sufficient energy requirements to merit nuclear plant construction, and environmental benefit of replacing existing energy production with carbon-free nuclear power. Other decarbonization opportunities considered include the addition of nuclear-powered electrolysis processes for hydrogen-intense industries or high-temperature electric heating where thermal requirements exceed nuclear generation conditions. Through the selection of a few technologies, INL will focus on documenting and characterizing selected technologies to develop preliminary but detailed information of what energy streams within the industrial process may be replaced by a nuclear source. To promote integrated energy system deployments, the identification of generalized requirements within specific industries will support preliminary designs for thermal and electrical integration of nuclear power with industrial applications.

226d Nuclear Power-Driven Utilization of Coal As a Raw Material for Chemical Products, Including CO_2 Byproduct Utilization

Byung-Hee Choi, Elizabeth K. Worsham, Maria A. Herrera Diaz, Levi Larsen, Eliezer A. Reyes Molina, Joseph E. Oncken, L. Todd Knighton, Aaron S. Epiney, Richard Boardman

Coal is a globally abundant resource, but its utilization range is limited. More than 90% of the coal produced in the United States is utilized for electric power generation as a fuel through pulverized coal combustion and integrated gasification combined cycle plants. While some plants are fitted with carbon dioxide (CO_2) capture and storage, the sequestration of carbon has associated costs and environmental concerns. Alternatively, CO_2 can be combined with hydrogen to make many high-value chemical products such as formaldehyde, carboxylic acids, and fertilizers. Utilizing coal and CO_2 as carbon feedstocks beyond electric power generation can broaden coal industries while limiting carbon emissions. Besides being utilized as or reprocessed into fuels for power generation, coal can also be converted into chemical products. As most carbon will be captured and stored in the product material, this process has less concern of CO_2 emissions than that of coal for energy. This study presents the possibility of the utilizing high-quality coal as a carbon source to produce valuable chemical products via methanol pathways with aid of nuclear power. All heat, steam, and electricity power required to produce chemicals from coal can be supplied from carbon-free nuclear power instead of coal combustion. Multiple pathways for CO_2 utilization, including urea and formic acid, are considered to target and supply local agricultural markets.

The "carbon refinery" design converts coal to solid char and syngas using pyrolysis and hydrothermal gasification and incorporates methanol synthesis. The Rectisol process controls the amount of CO_2 and removes sulfur. Nuclear-driven hydrogen production from high-temperature steam electrolysis (HTSE) replenishes hydrogen to obtain the proper syngas composition for methanol production, avoiding the addition of a water-gas shift reactor. The design follows a "self-sustaining" approach to minimize waste streams and utilize coproducts throughout the refinery, such as converting char to an activated carbon product for flue gas mercury removal and using chilled methanol for CO_2 removal in the Rectisol process. Nuclear-driven hydrogen production from high-temperature steam electrolysis facilitates various chemical reactions within the refinery as well as CO_2 hydrogenation to produce carbon utilization products. This cost and material sharing approach can improve the economics of hydrogen production and de-carbonization compared to the traditional methods of producing these chemicals. This study will consider the integration of the carbon refinery with advanced light-water reactor (LWR) and high-temperature gas reactor (HTGR) technologies. Even with heat from a high-temperature gas reactor, nuclear-driven energy cannot reach the high temperatures obtained from coal combustion. We modeled chemical processes in Aspen Plus with consideration of the lower temperature heat source, with nuclear heat introduced at the highest temperature in the hydrothermal gasification process (up to 750°C) and recovered by the pyrolysis process (500°C) and coal drying (250°C).

This study investigates the technical feasibility of this process and through an performs economic evaluation using Aspen Plus and Aspen HYSYS along with Aspen Process Economic Analyzer and the Framework for the Optimization of ResourCes and Economics (FORCE). The refinery feedstock demands, product outputs, and equipment costs will be scaled based on the increase in coal input, but the chemical process operations are steady state. Optimization studies using FORCE will provide descriptive economics of the entire design, including the cost of the nuclear reactor. Figures of merit for these models include the total net present value of the refinery and the cost of carbon avoided compared to the incumbent production methods of each refinery product.

This project will provide an illustration of the technical and economic feasibility of nuclear-driven chemical production using coal as a feedstock with either CO_2 sequestration or CO_2 utilization. An analysis of will inform opportunities for CO_2 utilization at operating coal plants. The resulting economic estimates at both a product subsystem and refinery level will guide decisions for the optimal processing of coal to produce chemical products.

226f Design and Operation of a Renewable Polygeneration Energy System with Unit Commitment-Based Power Planning: A Deterministic Modelling Approach

Tuhin Poddar, Mohammed Alkatheri, Ali Almansoori, Peter L. Douglas, Ali Elkamel

With the energy transition underway, there is a consensus effort from both worldwide governments and industry, to facilitate the integration of more renewable energy into power grids. However, the main challenge with integrating renewables like wind and solar, is the intermittent nature of these sources that result in inefficiencies and a lack of reliability. One possible energy system that can allow for an effective integration with intermittent sources is a polygeneration energy system (PES). The concept of a PES has been proposed in academic literature in recent times, where a typical pathway is the use of excess power from a fossil power plant being directed towards the production of valuable liquid fuels, in addition to electricity to the grid. In this study, the concept of a polygeneration system has been modified and extended to include renewable energy generation from wind, and a chemical production pathway of methanol to act as a form of long duration storage for times when the wind generation is much higher for any load demand and therefore increasing the flexibility. The model takes advantage of both a deterministic and stochastic approach to better reflect the real-world phenomena of wind power that is intermittent.

The approach taken in this model development paves a different path away from past polygeneration modelling studies, by using a network constrained Unit Commitment (UC) model that is well established in the domain of power systems engineering, to optimally schedule the power planning and power flow. First, the design and operation of a power generation planning model is developed to showcase how the power system responds to the intermittency of wind in the form of wind scenarios. This model was extended to show a storage mechanism in the form of a typical hydrogen electrolysis system and fuel cell. Mixed-Integer Linear Programming (MILP) models are then developed for the chemical production of methanol and integrated with the power planning model as a multi-scale (design and operation) model of a renewable polygeneration energy system (RPES) with chemical storage. To showcase the design and operation of the proposed RPES, the model is solved in a deterministic manner, to capture the integrated model as a snapshot. The total RPES system cost, based on real world wind power data and load demand data, was found to be USD 2317.93 million. The chemical production block had a cost of USD 138.51 million when integrated as a part of the RPES and the power generation planning block had a cost of USD 2179.42 million. The integrated model resulted in costs for the chemical production block that were much lower than the stand-alone plant while the RPES model also showed how excess intermittent wind power could be used for driving the chemical production. A key contribution to this work is also the implementation of machine learning methods, like K-Means clustering to help with the model's solution tractability and representation of a full year's hourly wind data and load demand. The MILP models have been developed using the General Algebraic Modeling System (GAMS) software and solved using state of the art optimization solvers BARON and CPLEX. Future work is planned in this topic that takes advantage of stochastic optimization with recourse to further study the RPES's flexibility under increased uncertainty.

226g Robust Optimization of Integrated Biodiesel and Petroleum Supply Chain Considering Seasonality and Risk Management Under Carbon Policy

Wenhui Zhang, Yiqing Luo, Xigang Yuan

With the continuous development of human society, the demand for energy keeps growing. For the sustainable development of human living environment, it is necessary to reduce the use of fossil fuel, and consider biofuel with low carbon dioxide emission instead. This study proposes a mixed integer model for optimizing an integrated biodiesel and petroleum supply chain in conjunction with carbon policies to control the amount of carbon emitted along supply chain, from both production and transportation activities, and consider both economic and environmental prospective. To deal with the complexity and uncertainty of the actual biodiesel supply chain, the supply variance of biomass is subdivided into 12-month subintervals, and a robust optimization method is adopted considering the uncertainty of demand. To improve the anti-risk ability of the supply chain network, the risk management of the warehousing link is considered. The goal is to minimize annual operating expenses while keeping carbon emissions as low as possible. A real case located in Illinois, USA is considered. The optimal operating conditions for the supply chain are found. The results show this model can reduce the annual cost while ensuring the least carbon emissions.

226h Techno-Economic Analysis of IGCC for the High Ash Indian Coal Using Bubbling Fluidized Bed Gasifier

Narendra Yadav, Srinivas Seethamraju, Sanjay M. Mahajani

Energy is the basis for running the economy. Every decade energy demand is increasing. In the last decade, it has grown by 21.54% [1]. Among the various energy sources, coal contributes 26 % of the world's energy consumption which is significant after oil. However, coal consumption leads to negative environmental impacts such as carbon emissions during burning and soil & air pollution during mining, transport and storage. The world is moving towards renewable energy sources that are cleaner than carbon; however, current renewable technology is expensive and yet to overcome the problem of intermittency. Thus, coal is still the only option to meet the significant energy demand, especially in countries like India where there are large reserves of coal. It also provides energy security. The total amount of coal in the world is 1033.71 billion tons, and 36% of the world's electricity was produced by coal in 2021 [2]. India has around 11% of the world's coal reserves [3]. Therefore, a substantial research effort is going on to make coal technology cleaner. However, the challenges include reduction of 9.7 Gt of carbon emitted from the coal-fired power plant in 2021[4]. To overcome some of these challenges, developing countries should adopt clean coal technologies such as integrated gasification combined cycle (IG-CC), chemical looping combustion power generation (CLC-CC) or underground coal gasification coupled with combined cycle (UCG-CC). Among all these options, IGCC not only reduces carbon emission but also increases plant efficiency, and is a more mature technology than UCG.

IGCC incorporates different units such as air separation, gasification, syngas cleaning and power island. This work focuses on gasification which can be performed in a fixed bed, fluidized bed, or entrained flow gasifier. The entrained flow gasifier has limitation in the formation of ash agglomeration for high ash coal because it works above the ash-fusion temperature. Among the other two gasifier types, the fluidized bed gasifier is more efficient. It decreases char production by properly mixing coal particles and gasifying medium, and it works below ash-fusion temperature.

India has a vast amount of Bituminous low-rank coal, which typically has low carbon (60%-80%) and high ash (20%-45%) content [5]. Due to technological constraints and experimental repeatability issues, India still lacks a commercial gasification plant for high-ash coal. The development of fluidized bed gasification technology for high ash coals could provide a cleaner technology from an Indian perspective. Therefore, this study aims to investigate the feasibility of IGCC for high-ash Indian coals using a fluidized bed gasifier. An integrated gasification combined cycle (IGCC) model is developed in Aspen plus for the high ash coal and validated with the experimental data. The validation is done for the following parameters: steam/coal ratio, air (O_2)/coal ratio, and operating temperature. The models are based on the equilibrium approach where Gibb's free energy minimization has been performed to find the syngas composition. In addition, the temperature approach is used for the water gas shift and methane-steam reforming reactions.

The results suggest an increase in equivalence ratio (ER) from 0.19 to 0.28, increases the carbon conversion efficiency (CC) from 39.42% to 61.04% and cold gas efficiency (CGE) from 20.11% to 26.11%. Further, during air-steam gasification, keeping ER the same and increasing steam/coal ratio from 0.06 to 0.26 results in an increase in CC from 58.06% to 72% and CGE% from 29.03% to 46.18%. The model gave a syngas heating value between 1-2 MJ/m^3 during air-assisted gasification and ~3 MJ/m^3 during air-steam gasification.

In continuation of previous work, the study also focuses on (i) syngas cleaning and electricity production and (ii) economic analysis of the IGCC plant. However, this work is in progress. The process methodology for the syngas cleaning contains a single-stage Rectisol method with methanol taken as the physical solvent [6]. The low temperature (- 40) operation helps to increase the absorption of corrosive gases like carbon dioxide and hydrogen sulfide gases in the methanol. The model includes a corrosive gas absorber and solvent recovery column. Carbon capture is not considered in the methodology. The power island for the electricity production uses a combined cycle operation. Finally, the economic analysis of IGCC will be done to find the levelized cost of electricity (LCOE). The LCOE will contain the sensitivity analysis with respect to the discount rate, raw material and capital cost.

Overall, the study investigates fluidized bed gasification as cleaner technology for high-ash coal which suits the Indian scenario. In addition, a methodology has been developed to evaluate power production and economic analysis of the IGCC.

Keywords: Efficiency (Energy), Fossil fuels, IGCC, Fluidised bed gasifier

References:

[1] Electricity consumption, Web Link: https://www.iea.org/reports/electricity-information-overview/electricity-consumption, Accessed on 03-April-2023

[2] Proven coal reserves worldwide in 2020, by country Web link: https://www.statista.com/statistics/237096/proven-coal-reserves-of-the-top-ten-countries, Accessed on 03-April-2023

[3] Statistical Review of World Energy (70th edition), 2021, bp,Web link: https://www.bp.com/content/dam/bp/business-sites/en/global/corporate/pdfs/energy-economics/statistical-review/bp-stats-review-2021-full-report.pdf

[4] How much coal is left, Web link: How much coal is left - U.S. Energy Information Administration (EIA), Accessed on 03-April-2023

[5] Power Sector at a Glance – All India, Web link: https://powermin.gov.in/en/content/power-sector-glance-all-india, Accessed on 03-April-2023

[6] Sun, L. and Smith, R., 2013. Rectisol wash process simulation and analysis. Journal of Cleaner Production, 39, pp.321-328.

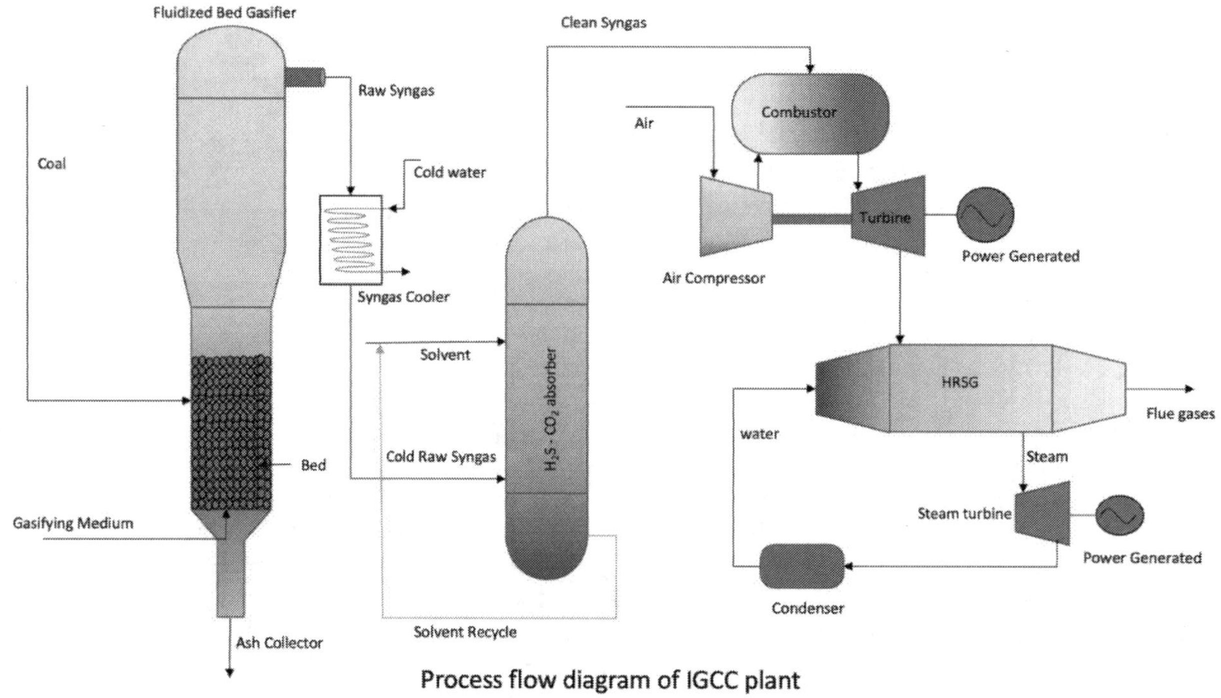

Process flow diagram of IGCC plant

8a Lowering the Barriers to Clean Water through Catalysis

Michael Wong

Catalysis science and engineering provide a powerful means to upgrade contaminated water to a quality level that matches its intended use, if operational constraints (ambient temperature, atmospheric pressure, variable water quality) can be overcome with materials with improved properties. I will describe two catalytic systems from my laboratory: nitrate hydrogenation to dinitrogen using In-supported Pd heterogeneous catalyst will be discussed, as a means to remove nitrate from water destructively (https://pubs.acs.org/doi/10.1021/acscatal.7b01371). I will show how the reaction selectivity (*e.g.*, N_2, NH_3, and N_2H_4) can be controlled by reaction conditions and metal composition (https://pubs.acs.org/doi/10.1021/acscatal.9b03239). I introduce a second catalytic system (boron nitride powders, BN) that address per/polyfluoroalkyl substances (PFAS), the so-called forever chemicals. Under commercially available UV-C light and even sunlight, BN composites can photo-oxidize perfluorooctanoic acid and other PFAS compounds (https://doi.org/10.1016/j.cej.2022.137735). These reaction systems exemplify the challenges and successes of materials synthesis, modeling, structure-property analysis, and systems-level design towards catalytic converters for water.

Electrochemical Wastewater Refining: Reactive Separations to Convert Pollutants into Products

William Tarpeh

Wastewater is an underleveraged resource; it contains pollutants that can be transformed into valuable high-purity products. This presentation will describe recent efforts to realize electrochemical wastewater refining—the use of electrochemical processes to tune and recover specific products from wastewaters— to valorize aqueous pollutants. We focus in particular on reactive separations, which combine selective separations and electrocatalysis to concentrate pollutants and transform them to high-value products. Specific case studies include electrochemical nitrate reduction to ammonia and other products; selective lithium recovery via electrodialysis; and selective nutrient recovery via electrified ion exchange.

175b Mars Regolith- Hydrochar Complex Synthesis with Water Retention Application

Robert Cheatham, Al Ibtida Sultana, Toufiq Reza

With the advancement of technology and the determination of humanity to inhabit Mars, the necessity of finding sustainable ways to cultivate and produce food, in deep space, has risen at an alarming rate. With the steep cost of transportation between Earth and Mars, NASA has emphasized the importance of utilizing the natural resources found on Mars. Research has found that hydrochars, biomass that has undergone the application of hydrothermal carbonization (HTC), is key in improving the water retention capacity of natural resources. The purpose of this study was to further the initiative through the development of Mars Regolith-Porous Carbon Complex (MRPCC) in order to enhance the water retention properties of Mars regolith via the utilization of hydrochars. Loblolly pine was selected as the biomass and was combined with Mars regolith (0, 5, 10, 25, 50 %w/w) before being chemically activated, with KOH at a 4:1 ratio respectively, at 800 °C for 2 hours to synthesize MRPCC. The resulting MRPCC was then characterized using N_2 adsorption–desorption for surface porosity quantification, scanning electron microscopy for morphology appearance analysis, X-ray powder diffraction for crystallinity analysis, proximate and ultimate analysis for detailed chemical composition analysis, and ion chromatography to determine the nutrient adsorption capacity. Additionally, the MRPCC's ability to hold water was tested by first saturating it with water and then slowly stirring it until excess water was visible. Results show that MRPCC's water retention increased by 2% when, compared to that of the pure Mars regolith, suggests that the addition of porous, organic material has furthered the initiative. Preliminary nutrients adsorption results have indicated that MRPCC has significant adsorption qualities, especially regarding ammonium, at 3000ppm and is expected to supersede these results at greater concentrations. With these considerations in mind, MRPCC is favorable when considering applications such as an agricultural aid, on Mars.

175c Adsorption of Harmful Algal Bloom Toxins Onto Commercially Available Biochar: An Exploration on the Mechanisms and Factors Influencing Adsorption

Cadianne Chambers, Toufiq Reza

Harmful Algal Bloom (HAB) events involve the growth of microalgae which release lethal toxins into fresh and brackish water bodies due to climatic and anthropogenic activities. Some common HAB examples are red tides, blue-green algae, and cyanobacteria, which are known to lyze neurotoxins and hepatotoxins during blooms causing massive fish kills, and serious threat to human life. These outbreaks have terrorized Great lakes and coastal waters such as Lake Erie, Lake Okeechobee, Florida's Coasts and Gulf of Maine, negatively impacting marine ecosystems, fisheries, and the economy. To mitigate the release of these toxins, physical water treatment such as adsorption with the use of biochar provides a low-cost solution with high removal efficiency. This study investigates the effectiveness of commercially available pine derived biochar for adsorption of HAB toxins: *Brevetoxin (PbTx-3)*, *Microcystin-LR (MCLR)* and *Saxitoxin (STX)*. Material characterization of the biochar presented positively charged with high carbon content, high porosity (S_{BET} = 261.06 m^2/g) with substantial oxygen functionality (941.33 μmol/g). Dosage, equilibrium and kinetic study were performed on *PbTx-3*, *MCLR* and *STX* under ambient conditions with commercial biochar. Enzyme-linked Immunosorbent Assay (ELISA) analysis and high -performance liquid chromatography (HPLC) was used for adsorption analysis of the investigated toxins. From the dosage reports all three toxins experienced high percent removal (> 90 %) at a dosage rate of 0.4 g/L. High adsorption capacities were observed for the target toxins which fitted well with the combined Langmuir-Freundlich adsorption model and Pseudo-Second Order and Elovich kinetic model. The results support possible adsorption mechanisms such as electrostatic attraction, pi-pi stacking, ion exchange, hydrophobic interactions, and more which is explored in detail. Thus, commercial provides as a high removal efficiency solution for the mitigation HAB toxins.

175d Low-Level Radionuclide Removal from Water Sources Using Waste-Derived Biochars Followed By Adsorbent Combustion

Shermal Fernando, Catherine Brewer

Low levels of heavy metals and radionuclides, such as uranium, can be the difference between a usable and an unusable water resource. Adsorption-based removal methods have the potential to be long-lasting for low-level metal concentrations over large volumes of water. The adsorption and removal of radionuclides from water using biochars derived from abundant and low-cost wastes has been seen as a promising alternative. In this study, the feedstocks (pecan wood, cattle manure, and hemp residue) were converted into biochars in a 5-L lab pyrolyzer at 400–500°C under nitrogen flow. Batch uranium adsorption studies were conducted for the effects of initial concentration, solution pH, contact time, and competing ions. Adsorption isotherms and kinetic models constructed from concentration measurements before and after adsorbate contact. After adsorption, loaded biochars were combusted in a tube furance and the fractionation of radionuclides tracked between the gas/vapor and ash fractions. Data on radionuclide fractionation during combustion wil be used to understand the potential of radionuclide adsorption and radioactive waste volume reduction without needing to regenerate the adsorbents.

175e Impact of Pyrolysis Conditions of CO_2 Adsorption of Biochar

Wesley Wong, Amanda Simson

As of 2022, the atmospheric CO_2 level has reached an all-time peak of over 420 ppm. In addition to rapid decarbonization, several methods of CO_2 removal such as direct air capture (DAC), afforestation, ocean fertilization, and soil carbon sequestration are needed to mitigate the harmful effects of climate change and decrease the level of CO_2 in the atmosphere. Biochar, made from pyrolyzing biowaste, has been cited for its potential in carbon sequestration due to its resistance to microbial degradation. Biochar has also been considered as an adsorbent for DAC. Previous work has found that biochar had CO_2 adsorption capacities up to 73.55 mg CO_2/g biochar, compared to industrial DAC sorbents. This study investigated how pyrolysis conditions affect the CO_2 adsorption capacity of biochar. Sawdust, coconut shell, corncob and walnut shell feedstocks were tested to see if nitrogen flow rate during pyrolysis impacted adsorption capacity. It was found that for walnut shell, maintaining high flow rate was important to capacity; when flow rate was dropped below 150 mL/min, a decrease of up to 16% was measured whereas above this flow rate threshold, no significant increase in adsorption capacity was observed. Biochar pyrolyzed from sawdust and coconut feedstock did not show such a strong correlation between nitrogen flow rate and adsorption capacity, indicating that this dependence is feedstock specific. In addition to flow conditions, the impact of pre-drying was studied. Adsorption capacity of the pre-dried walnut was approximately 40% less than the non-dried batch for pyrolysis temperatures between 500-800°C. However, this impact was found to be feedstock dependent as well, as sawdust and coconut biochar resulted in smaller differences. Our results indicate that changes in process conditions can greatly impact the chemical and physical properties of biochar and thus CO_2 adsorption.

91a Integration of the Carboxylate Platform with Mcl-PHA Fermentation to Enable a Universal Route from Waste to Bioplastics

Jorge Arreola Vargas, Bing Xu, Zach Foudeh, Cheng Hu, Cesar Granda, Susie Dai, Joshua Yuan

Medium chain length-polyhydroxyalkanoates (mcl-PHA) have unique advantages to replace petroleum-based plastics due to their biodegradability, excellent physicochemical properties, and capacity to be produced from organic wastes as raw materials. However, broader application of mcl-PHA is hindered by the high process cost and low yields. Therefore, we have developed efficient *P. putida* strains to overcome these issues using organic waste as raw material. To this end, we have engineered *P. putida* based on systems biology and evaluated the metabolic capacity of the strains to utilize organic acids as the main mcl-PHA precursors. More important, we have integrated the carboxylate platform with mcl-PHA fermentation to determine the potential application of this technology using real organic waste as feedstock. The carboxylate platform is a variant of the anaerobic digestion process that can transform organic matter into short chain length organic acids (scl-OA) by suppressing the methanogenic archaea. These scl-OA (C2-C5) can also be converted into more valuable medium chain length organic acids (mcl-OA) containing 6 to 8 carbons by modifying the carboxylate platform conditions to favor chain elongation. This is a technology that our commercial partners have scaled-up to demonstration plants but requires efficient methods for the extraction of organic acids. We therefore propose to bio-upgrade the carboxylate fermentation broth containing unextracted scl-OA and mcl-OA by using the *P. putida* metabolic versatility for mcl-PHA production, providing an attractive alternative to waste valorization and plastic sustainability. Our results show that our engineered *P. putida* strains containing overexpression of genes *phaC1* and *phaJ4* involved in mcl-PHA formation, as well as suppression of gene *phaZ* required for mcl-PHA depolymerization, are efficient to increase yields and titers during fermentation of model organic acids. Overall, mcl-PHA productions were higher when longer organic acids are present in the medium. More importantly, waste-derived organic acids –obtained as a mixed broth after the carboxylate platform–, reached similar mcl-PHA titers and conversion efficiencies compared to synthetic media containing model organic acids at same concentrations. Bioplastic characterization and further mcl-PHA fermentation improvement was carried out (> 5 g mcl-PHA/L and 70% mcl-PHA content) and will be discussed during the talk. Overall, our work demonstrates for the very first time the potential of these two biochemical platforms to integrate a universal route from waste to elastomer bioplastics. Further development of this technology may also help abating atmospheric CO_2 by sequestration using the Wood–Ljungdahl pathway during a modified carboxylate platform combined with mcl-PHA fermentation.

91b Technoeconomic Assessment of Emerging Technologies for the Enhancement of Anaerobic Digestion Processes at Municipal Wastewater Treatment Facilities

Peter Valdez, Lesley J. Snowden-Swan, Timothy E. Seiple, Phillip Teller, Richard Garrison, Fuad Ale Enriquez, Nalok Dutta, Anthony Giduthuri, Birgitte Ahring

Treatment and disposal of municipal wastewater sludges is an expensive activity. Within the United States, 35 billion gallons of water is treated daily producing over 40,000 tons of sludge solids per day [1], costing over $50 billion for wastewater management and treatment [2]. We investigated the economic benefits of add-on technologies that can boost biogas yields from anaerobic digestion, upgrade biogas carbon dioxide to methane, and reduce the net solids to be disposed. The technology pathway was developed by Clean-Vantage, LLC and Washington State University. The investigated technologies include advanced wet oxidation steam explosion (AWOEx), carbon dioxide biomethantion, and thermophilic anaerobic digestion. The AWOEx process uses dissolved oxygen and thermal hydrolysis to transform the recalcitrant materials in biomass, boosting the net yield of methane during anaerobic digestion [3]. Methanogensis or biomethantion of the carbon dioxide in the biogas with hydrogenotrophic microorganisms can increase the net upgrading of carbon [4]. Furthermore, thermophilic anaerobic digestion can reduce the net solids while boosting biogas yield.

In this work, we modelled the installation of each aforementioned technology at a treatment facility, with existing infrastructure for mesophilic anaerobic digestion, processing five million gallons of municipal wastewater per day. The process and economic models evaluate each technology individually and as integrated systems to maximize value and minimize capital investment, operational expenses, and net greenhouse gas emissions. Our findings show that installation of all three processes results in a 59% reduction in solids to be disposed and a 352% increase in methane production. Operational costs are 40% higher with the new processes, but the revenue from the sale of renewable natural gas and the value of the associated renewable energy credits can offset the costs and generate revenue. In this presentation we will summarize the modeling approach and report the economic and environmental impacts and detail the key process sensitivities.

[1] Seiple et al., Journal Environmental Management. 197 (2017) 673-680.

[2] http://www.usmayors.org/wp-content/uploads/2018/10/2016-Spend-Article.pdf, Accessed March 2023.

[3] Dutta et al., Waste Management, 154 (2022) 105-112.

[4] Khan et al., Renewable Sustainable Energy Reviews 149 (2021) 111343.

91c Hydrothermal Liquefaction of Waste Activated Sludge: Initial Composition Effects on Product Distribution and Nutrient Recovery

João Poli, Susan Williams, Belinda S.M. Sturm

Municipal sludge is a widely available feedstock that is often discarded through composting, landfilling, or agricultural land application, without energy recovery. Recent concern about the toxicity of land applying biosolids has driven research into new technologies for valorizing municipal sludge. Hydrothermal liquefaction (HTL) is a thermochemical process that occurs in the presence of an aqueous environment that can be used to recovery energy and nutrients from municipal sludge. The process conditions for HTL include temperatures ranging from 200 °C to 450 °C and pressures ranging from 4 MPa to 35 MPa. Under these conditions, four product streams are obtained including biocrude, biochar, gas (primarily CO_2), and an aqueous stream. To gain a better understanding of the viability of implementing HTL conversion of municipal sludge at wastewater treatment facilities, it is necessary to understand the impact of the wastewater treatment process on the sludge composition and ultimately the product properties. Specifically, the ability of the aqueous stream to be recycled back to the mainstream wastewater process is critical to the implementation of HTL at a wastewater treatment facility. Furthermore, the solids produced from HTL must have reduced toxicity compared to the biosolids and be suitable for land application.

This study aims to identify the effects of initial sludge composition on the HTL product distribution and product composition. Specifically, waste-activated sludge (WAS) was sampled from two different municipal facilities: one with and one without enhanced biological phosphorous removal (EBPR). In addition, the WAS had different calcium concentrations, due to differences in lime addition at the two wastewater treatment facilities. HTL was performed on both feedstocks and the HTL products, including the potential for nutrient recovery, are presented. Preliminary results show that WAS from EBPR processes have higher biocrude yields compared to the process without nutrient removal. Furthermore, The Ca content of the municipal sludge can change the biochar yield significantly and promotes a different product distribution in terms of elements such as P and Mg. WAS from the EBPR processes has a high feedstock P content, and > 94% of the P was recovered in the biochar after HTL. The storage of the sludge also plays an important role in nutrient distribution in the products. Sludge stored aerobically showed a lower recovery of P in the biochar phase, whereas anaerobically stored sludge showed a higher recovery of P in the biochar. The results indicate that P recovery is efficient for all WAS feedstocks, but the calcium content of the WAS affects the total mass of biochar produced as well as the P-recovery efficiency.

91e Process Modeling, Techno-Economic Analysis and Life-Cycle Assessment of Intensified Biogas to Liquid (IBGTL) Process.

Rarosue Amaraibi, John Kuhn, Babu Joseph

Landfill gas (LFG) is a natural byproduct of the decomposition of organic waste in landfills, consisting primarily of methane and carbon dioxide [1]. Despite its potential as a source of renewable energy, the low value of end products such as electricity and heat has limited its economic viability [2-3]. Recent developments in gas-to-liquid (GTL) processes, specifically Fischer-Tropsch (F-T) synthesis, have shown promise in converting biogas into high-value liquid transportation fuels [2-3].

The Intensified Biogas to Liquid (IBGTL) process, developed at the University of South Florida, aims to overcome the economy of scale barrier in the production of liquid fuels from biogas. This process utilizes a hybrid catalyst system capable of combining endothermic bi-reforming with exothermic oxidation of methane and Fischer-Tropsch reactions into a single intensified process.

In this study, we conducted a techno-economic analysis (TEA) and life cycle assessment (LCA) of liquid fuel production from LFG via the IBGTL process. We compared the IBGTL process to the conventional biogas to liquid (BGTL) process and other classes of drop-in replacement fuels. A schematic comparison of the two processes is shown in Figure 1. The IBGTL reactor was modeled using Aspen Plus for a facility generating 2500 scfm of LFG, with CAPCOST employed for equipment costing and net present value and levelized cost of energy used for the economic analysis. SimaPro was used to conduct the LCA, with the system boundary and functional unit shown in Figure 2.

Our preliminary results show that the IBGTL process has a minimum fuel selling price (MFSP) of $6/GGE for Scenario 1 and $4/GGE for Scenario 2, which includes material recycle of excess fuel gas to the IBGTL reactor to increase liquid fuel yield. The TEA and LCA results provide insights into the economic and environmental feasibility of the IBGTL process, which can help in making informed decisions regarding its commercialization. The IBGTL process offers several advantages over the conventional BGTL process, such as reduced energy and material input, leading to reduced capital investments.

Further analysis is needed to fully evaluate the potential of the IBGTL process for large-scale commercialization. Additional scenarios and the carbon intensity of the various routes analyzed will be presented. Overall, the IBGTL process shows promise in converting LFG into high-value liquid transportation fuels, which can help decarbonize the transportation sector and contribute to the sustainable management of organic waste in landfills.

References

[1] IEA, Outlook for biogas and biomethane: Prospects for organic growth. IEA, Paris, 2020.

[2] Naqi, A., J.N. Kuhn, and B. Joseph, Techno-economic analysis of producing liquid fuels from biomass via anaerobic digestion and thermochemical conversion. Biomass and Bioenergy, 2019. 130 DOI: 10.1016/j.biombioe.2019.105395.

[3] Zhao, X., et al., Conversion of landfill gas to liquid fuels through a TriFTS (tri-reforming and Fischer–Tropsch synthesis) process: a feasibility study. Sustainable Energy & Fuels, 2019. 3(2): p. 539-549 DOI: 10.1039/c8se00344k.

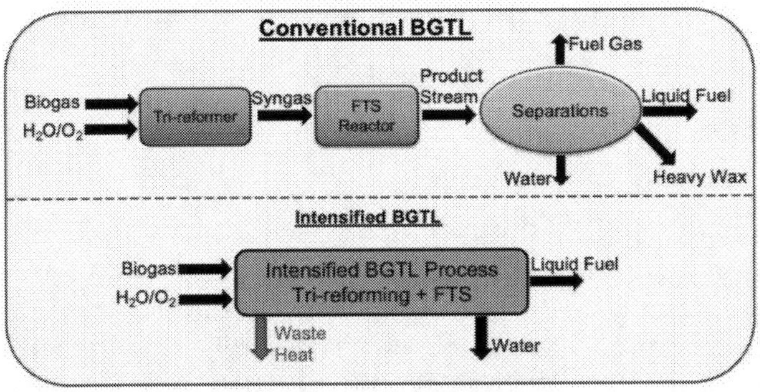

Figure 1: Schematic of the Conventional Biogas to Liquid (BGTL) Process and the Intensified Biogas to Liquid (IBGTL) Process.

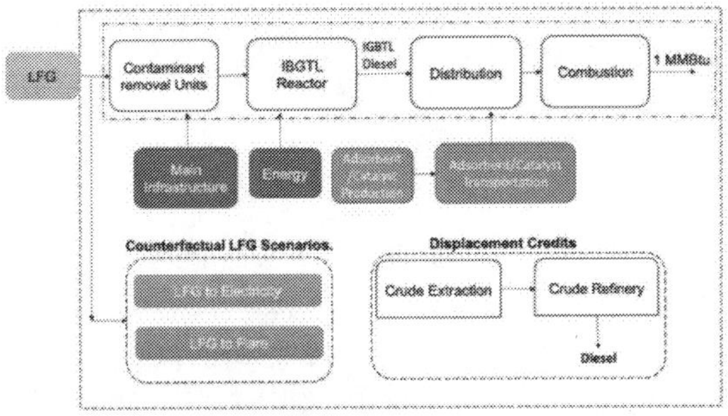

Figure 2: System Boundary and Functional Unit

91f Elucidation of Change in Oxidation State for Metal Recovery from WTE Ash in Reactive Environments

Janhvi Trivedi, Marco J. Castaldi

Nearly half of the 2800 million tons per year of municipal solid waste (MSW) that is produced in globally is sent to landfills. Waste-to-Energy (WtE) facilities are the only viable alternative to this that matches the scale of waste produced, producing electricity and reducing the waste volume. The declining cost of electricity due to increased natural gas and renewable electricity generation means the WtE facilities need to go beyond energy generation. MSW contains valuable materials that must be recovered and incorporated back into the economy.

This work focuses on treatment of WtE ash with other waste streams like gypsum wallboard and spent Fluidized Catalytic Cracking catalyst (FCC) in high temperature and oxidizing environments to understand the change in the oxidation state of metals and their formation into more extractible forms. Currently spent FCC catalysts are classified as non-hazardous and amount to nearly 400,000 tons annually which are currently sent to landfills. Gypsum waste is estimated at 13 million tons annually of waste produced annually with only 2% recycled.

Therefore, the aim is to combine these waste streams in different proportions with the MSW generated in the WtE plants and treat it thermally and process it to recover valuable materials which are landfilled right now. The type of reactions that takes place are solid-solid and gas-solid between the ash and the additives at high temperatures forming metal oxides and different metal complexes. This leads to changes in the oxidation states of the metals and depending on the oxidation state, metal can be easy or difficult to extract.

The addition of gypsum into MSW ash has been shown to convert metals like Al, Mn, Mg, and Fe into more extractible phases. The phase change of Aluminum from crystalline to the amorphous, reduces the evolution of hydrogen gas from the ash-additive mixtures in an inert environment. This helps to reduce the deleterious effects of hydrogen gas caused in the landfills. Addition of spent FCC catalyst to the ash can lead to formation of various alumino-silicates compounds causing phase change of metals in ash and reduce slagging in the WtE plants. The characterization techniques utilized to examine the effect of additive to the ash are X-ray diffractometer, Energy dispersive X-ray, thermogravimetric and evolved gas analyzers.

This presentation will show comparative results for several conditions and discuss possible reaction mechanisms leading to the change in the oxidation state of the metals and the evolution of gases.

91g On-Site Electricity Generation Using Trashology's Pur-Gen Technology

William Wang, Matthew Young, Stephen Armstrong

Current waste management practices are one-dimensional and follow a singular approach regardless of waste stream composition, geographical location, and source-residential, commercial, or industrial. Traditional waste management for municipal solid waste disposal in a landfill is logistically inefficient, has limited capacity, and releases biogas and leachates. Long-term, this approach is not sustainable and detrimental to the environment.

Trashology is a waste-to-energy company currently focused on converting solid waste to electricity on-site at the point of generation providing electricity behind-the-meter in a distributed manner. With applications for the military operating in remote locations and commercial/industrial entities seeking to attain ESG goals, reduce waste disposal costs, and reduce utility costs, Trashology's goal is to eliminate the negative impacts associated with solid waste disposal. Trashology's technology consists of 3 main components – Solid Waste Preprocessing, Gasification, and Electricity Generation – and is controlled by a fully automated PLC that operates with minimal human intervention. The gasifier can process 2-4 tons/day of solid waste in a patented air-blown downdraft gasifier to produce syngas that is suitable for combustion in a genset for electricity generation. The major issue surrounding solid waste gasification has been tar production and management. The use of a bubble column to remove tar from the syngas has been demonstrated and shown to be highly effective.

The major factors influencing where on-site waste-to-energy makes logical and economic sense, the function of each component with a focus on its impact on syngas quality and clean-up, results from major demonstrations with real solid waste from Fort Jackson (Columbia, SC) and simulated waste per Army Corps of Engineers formulation requirements, emissions profile, fate of pollutants, process simulations and economics, and process development beyond electricity will be presented.

91h Recovery of Resources from Food Waste to Synthesise Nutrient-Loaded Slow-Release Hydrogel

Yen Wah Tong

Food waste contains a significant amount of recyclable nutrients that are not fully utilized . In addition, the excessive and rapid release of chemical fertilizers contributes to low utilization rates and serious environmental concerns. To address these problems, hydrogel materials with excellent adsorption, loading and modification capabilities can be employed to recover nutrients from wastewater and produce nutrient-loaded slow-release carriers. However, to improve the efficiency of nutrient utilization and minimize loss, it is necessary to establish a release model that allows for the prediction of controlled release of nutrients and clarifies the mechanism of nutrient binding and release from the hydrogel. In this work, we prepared a biodegradable chitosan/polyvinyl alcohol/montmorillonite hydrogel through a cyclic freeze-thaw process and utilized the response surface model to construct a model for nutrient release. Additionally, we used DFT calculations to probe the interaction sites between nutrient elements and hydrogel as well as the change of electrostatic potential before and after loading to illustrate the release mechanism.

Based on the characteristics of recycled water and the nutrients required for vegetable growth, an artificial wastewater was prepared. Chitosan and polyvinyl alcohol with varying mass ratios of montmorillonite were then added to the solution to construct hydrogels with different formulations and glutaraldehyde was used as the crosslinker.

We observed that the hydrogels prepared had similar gel contents, which accounted for about 80% of the solid content. As the chitosan content increased, the water absorption capacity of the hydrogels gradually increased to over 100 times. Additionally, as the chitosan content increased, the quality loss of the hydrogels caused by biodegradation also increased gradually. By the 11th week, the degradation had exceeded 50%. The FTIR and XRD analyses confirmed that each raw material was successfully cross-linked and that the nutrients were effectively loaded onto the hydrogels.
In conclusion, we have prepared a slow-releasing nutrient-loaded material using food waste that can be used as fertiliser, completing the circular economy of food waste to food.

94a Lessons Learned: Case Histories of Several Environmental Disasters and Their Implications for the Waste Plastics Crisis

Jeffrey Seay, Mary Ternes, Robert Peters

This presentation reviews and summarizes several environmental disaster case histories involving chemicals/materials such as friable asbestos, dichloro-diphenyl-trichloroethane (DDT), polychlorinated biphenyls (PCBs), bisphenol A, and phthalates. These materials were very successfully designed for a specific intended purpose. However, these materials/ chemicals are extremely stable, are very resistant to degradation, and persist in the environment for very long periods of time. This presentation describes and summarizes the environmental impacts of these materials, with particular attention devoted to the impact of these materials on human health, mammals, and ecosystems. Results of these investigations will be used to assess the potential impacts on the waste plastics crisis.

2023 AIChE ANNUAL MEETING
November 5 — November 10, 2023
Hyatt Regency Orlando
Orlando, FL

Methods For AI-Enabled Water Treatment

Da Yan [1], Robert W. Peters [1], Yang Zhou [2], Mohamed K. Mostafa [3]

[1]Department of Civil, Construction and Environmental Engineering, University of Alabama at Birmingham, Birmingham, AL, USA

[2] Auburn University, Auburn, AL

[3] Faculty of Engineering and Technology, Badr University in Cairo (BUC), Cairo, Egypt

Abstract

Water treatment aims to ensure the quality and sustainability of water resources, which is critical to drinking water safety, fishing, agriculture, and leisure activities. Methods for treatment of water can be classified as drinking water treatment and wastewater treatment. For drinking water treatment, water works commonly use coagulation, flocculation, sedimentation, and filtration to reduce turbidity, total organic carbon (TOC), disinfectant by-products (DBP), etc. It is also important to control the lead and copper levels during water distribution due to the corrosion of water distribution pipes. For wastewater treatment, wastewater treatment plants (WWTP) use anoxic reactions and aerobic reactions to improve the biochemical oxygen demand (BOD) and reduce total suspended solids (TSS), total dissolved solids (TDS), total nitrogen (TN) of the effluent, etc. The water treatment process is a complex dynamic process that can be regarded as an adaptive control problem. For example, the influent water properties may be impacted by anthropogenic pollutants and seasonal shifts, resulting in different alkalinities, temperatures, pH, and dissolved oxygen (DO) concentrations at different times and requiring different degrees of treatment. Overtreatment may cause corrosion of water pipes due to DBP, and additional material cost and power consumption; while sudden events such as massive rainfalls and flash floods can overload the system and cause large changes in influent water properties, requiring quick reaction in treatment. Two aspects in water treatment are important: sensing and control. This work surveys recent advancements in applying artificial intelligence (AI) technology (especially the deep learning models) to both aspects, and envisions new opportunities to apply AI models. Specifically, a system for water treatment often consists of cascaded subsystems with feedback control, where the various water quality variables in the influent and effluent of each subsystem can be monitored by sensors or sampled using laboratory analytical tests. For continuous monitoring and early reaction, soft-sensing is advocated by recent works to forecast future variable values based on historical readings. Models such as Long Short-Term Memory (LSTM), Gated Recurrent Units (GRU), and even Transformers have been applied and shown promising results in forecasting accuracy. As for feedback control, deep learning models have been proposed to approximate the ideal control law represented as non-linear dynamic equations. When simulation is available, deep reinforcement learning can also be applied to learn a good control policy. AI models can be trained either with simulated data or real historical data. Benchmark simulation model (BSM) No. 1 and No. 2 are examples where simulated data can be obtained for AI model training and testing, while real data can be collected at water works or WWTPs during their operation, or even through a mobile lab such as the one owned and operated by the Birmingham Water Works Board which allows for testing out various new treatment solutions (e.g., new materials) at various water sources (lakes and rivers) by pumping water. Finally, federated learning is a promising direction to collectively learn AI models at various water works, WWTPs, and sensors deployed at different places in lakes and rivers. Federated learning utilizes the data at various sites to better train an AI model without the need of direct data sharing among the sites.

Keywords: Artificial intelligence, deep learning, sensing and control, water treatment

1. Introduction

Artificial intelligence (AI) is currently used in many applications such as blockchain, cloud computing, intelligent decision-making, and the fourth industrial revolution [1]. Artificial intelligence is emerging in water treatment that can be used to integrate the whole process of water treatment. According to the market research, water industries are expected to invest in AI about $6.3 billion by 2030 [2]. The water industries are expecting to save between 20% to 30% of the operational expenses by optimizing the use of chemicals in water treatment, and thus decreasing the cost [2]. AI includes many techniques; the most popular technique is artificial neural network (ANN) [3], [4]. The AI techniques can be used to diagnose, assess, and simulate the treatment plant operation. They can also be used to manage wastewater treatment operations and enable water saving and reuse [5], [6]. The AI works in the water field through establishing a relationship between different variables [3], [4], [7]–[11]. For example, in the adsorption process, the AI is used to create a link between the input variables (adsorbent dosage, the initial concentration, stirring rate, temperature, pH, etc.) and the output variable (the adsorption capacity or the removal efficiency) [12]–[14]. In most cases, the AI results are in close agreement with the experimental results, which suggest the potential applications of AI in improving the efficiency of real water treatment systems [15]. To date, different models have been successfully applied to predict target variables (pH, biological oxygen demand (BOD), chemical oxygen demand (COD), total suspended solids (TSS), total nitrogen (TN), and total phosphorus (TP) including artificial neural network (ANN), long short-term memory (LSTM), adaptive neuro-fuzzy inference system (ANFIS), and genetic algorithm (GA) [16].

2. Predictive control of pollutants

ANN models have been used many published studies as a suitable method to describe the adsorption performance [17]. For instance, Mahmoud et al. [18] used ANN to simulate the phosphate removal efficiency at different operational conditions (adsorbent dosage, solution pH, and mixing speed). The results showed that the pH was the most influential attribute and the model predictive accuracy was R^2: 0.976 [18]. Hamdy et al. [15] investigate the adsorption of methylene blue (MB) onto nanoscale zero-valent iron (nZVI) and used ANN to predict the MB removal efficiency at different operating conditions. The model was accurate (R^2: 0.9313) and the results of the model showed that the optimum pH was 6, which achieved a MB removal efficiency of about 100% at initial MB concentration 10 mg/L, nZVI dosage 10 g/L, reaction time 30 min, temperature 30°C, and stirring rate 150 rpm [15]. Fan et al. [19] used ANN to forecast copper removal from aqueous solution using reduced graphene oxide-supported nZVI magnetic nanocomposites. Using the ANN-particle swarm optimization (PSO) based tool, the copper removal efficiency was improved by 8.54% as compared to the response surface methodology (RSM) model [19]. Granata et al. [20] proved that the machine learning algorithms could be successfully used to predict the water quality status of urban wastewater effluents. The model can be further considered for scaling up and sizing the treatment facility [20]. Taoufik et al., [21] stated that the following input data (solid retention time (SRT), pH, hydraulic retention time (HRT), inlet BOD, and temperature) are required to be able to accurately predict effluent BOD using mechanism of applied model. Historical record of input parameters, such as TSS, operational conditions of settling tank, biomass concentration, dose of coagulant, and turbidity are required for total TSS estimation [21]. Overall, ANN models efficiencies for BOD and COD estimation is >99 %.

3. Challenges and Prospects

AI requires a sufficient amount of data for experimental training and testing to be able to offer improvement in the system and to predict future outcomes. The prediction of AI tools may deviate from the actual results under certain conditions. For example, a sudden change in water quality or operating parameters may result in wrong prediction by AI models. Special attention must be given to data selection for the training steps, as the random data selection may resulted un-trustable prediction.

4. Conclusions

AI tools can be used to integrate the whole process of water treatment including fault detection, contamination control, water discharge; transportation; economics, environmental impacts, management of sludge, and policymaking. However, efforts must be made to improve the prediction performance of AI models so that they can accommodate sudden changes in the input variables and can be employed under various conditions. Hybrid AI techniques can be applied to improve the accuracy of prediction, which leads to a reduction in operational cost. Finally, special attention should be made while collecting the data to keep data integrity.

References

[1] Y. Lu, "Artificial intelligence: a survey on evolution, models, applications and future trends," *J. Manag. Anal.*, vol. 6, no. 1, pp. 1–29, Jan. 2019, doi: 10.1080/23270012.2019.1570365.

[2] G. Alam, I. Ihsanullah, M. Naushad, and M. Sillanpää, "Applications of artificial intelligence in water treatment for optimization and automation of adsorption processes: Recent advances and prospects," *Chem. Eng. J.*, vol. 427, p. 130011, Jan. 2022, doi: 10.1016/J.CEJ.2021.130011.

[3] A. S. Mahmoud, N. Y. Mohamed, M. K. Mostafa, and M. S. Mahmoud, "Effective Chromium Adsorption From Aqueous Solutions and Tannery Wastewater Using Bimetallic Fe/Cu Nanoparticles: Response Surface Methodology and Artificial Neural Network," *Air, Soil Water Res.*, vol. 14, Jun. 2021, doi: 10.1177/11786221211028162/ASSET/IMAGES/LARGE/10.1177_11786221211028162-FIG7.JPEG.

[4] A. S. Mahmoud, R. A. SaryEl-deen, M. K. Mostafa, and R. W. Peters, "(655d) Artificial Intelligence for Organochlorine Pesticides Removal from Aqueous Solutions Using Entrapped Nzvi in Alginate Biopolymer | AIChE," *AIChE Annual Meeting Proceedings*, 2017. https://www.aiche.org/conferences/aiche-annual-meeting/2017/proceeding/paper/655d-artificial-intelligence-organochlorine-pesticides-removal-aqueous-solutions-using-entrapped (accessed Nov. 22, 2023).

[5] A. S. Mahmoud, M. K. Mostafa, and R. W. Peters, "A prototype of textile wastewater treatment using coagulation and adsorption by Fe/Cu nanoparticles: Techno-economic and scaling-up studies," *Nanomater. Nanotechnol.*, vol. 11, Sep. 2021, doi: 10.1177/18479804211041181/ASSET/IMAGES/LARGE/10.1177_18479804211041181-FIG10.JPEG.

[6] M. T. Kamar, H. Elattar, A. S. Mahmoud, R. W. Peters, and M. K. Mostafa, "A critical review of state-of-the-art technologies for electroplating wastewater treatment," *Int. J. Environ. Anal. Chem.*, Jul. 2022, doi: 10.1080/03067319.2022.2098486.

[7] R. S. Farag, M. M. Elshfai, A. S. Mahmoud, M. K. Mostafa, and R. W. Peters, "(592d) Green Synthesis of Nano Iron Carbide: Preparation, Characterization and Application for Removal of Phosphate from Aqueous Solutions," *Annual AIChE Meeting Proceedings; Pittsburgh, PA, October 28 - November 2*, 2018. https://www.aiche.org/conferences/aiche-annual-meeting/2018/proceeding/paper/592d-green-synthesis-nano-iron-carbide-preparation-characterization-and-application-removal (accessed Nov. 22, 2023).

[8] J. Heberling, Y. Adewuyi, A. S. Mahmoud, M. Mostafa, and R. W. Peters, "(12e) AOP Performance at Wastewater Treatment Plants | AIChE," *AIChE Annual Meeting Proceedings, David L. Lawrence Convention Center, Pittsburgh, PA, October 28 - November 2*, 2018. https://www.aiche.org/conferences/aiche-annual-meeting/2018/proceeding/paper/12e-aop-performance-wastewater-treatment-plants (accessed Nov. 22, 2023).

[9] A. S. Mahmoud, M. S. Mahmoud, N. Yousri, and M. K. Mostafa, "Advanced Wastewater Treatment by Using Entrapped nZVI into Alginate (Ag) Biopolymer: Adsorption Isotherm, Kinetic Models, and Statistical Analysis," *Key Eng. Mater.*, vol. 921, pp. 147–160, 2022, doi: 10.4028/P-XYGKON.

[10] M. M. El-Shafei, A. S. Mahmoud, M. K. Mostafa, and R. W. Peters, "Effects of entrapped nZVI in alginate polymer on BTEX removal," *Environmental Division 2016 - Core Programming Area at the 2016 AIChE Annual Meeting*, 2016. https://www.aiche.org/conferences/aiche-annual-meeting/2016/proceeding/paper/60b-effects-entrapped-nzvi-alginate-polymer-on-btex-removal (accessed Nov. 22, 2023).

[11] R. S. Farag, M. M. Elshfai, A. S. Mahmoud, M. K. Mostafa, A. Karam, and R. W. Peters, "(670d) Study the degradation and adsorption processes of organic matters from domestic wastewater using chemically prepared and green synthesized nano zero-valent iron," *AIChE Annual Meeting, Conference Proceedings*, 2019. https://www.aiche.org/conferences/aiche-annual-meeting/2019/proceeding/paper/670d-study-degradation-and-adsorption-processes-organic-matters-domestic-wastewater-using-chemically (accessed Nov. 22, 2023).

[12] A. H. Sadek, O. M. Fahmy, M. Nasr, and M. K. Mostafa, "Predicting Cu(II) Adsorption from Aqueous Solutions onto Nano Zero-Valent Aluminum (nZVAl) by Machine Learning and Artificial Intelligence Techniques," *Sustain. 2023, Vol. 15, Page 2081*, vol. 15, no. 3, p. 2081, Jan. 2023, doi: 10.3390/SU15032081.

[13] A. H. Sadek and M. K. Mostafa, "Preparation of nano zero-valent aluminum for one-step removal of methylene blue from aqueous solutions: cost analysis for scaling-up and artificial intelligence," *Appl. Water Sci. 2022 132*, vol. 13, no. 2, pp. 1–23, Dec. 2022, doi: 10.1007/S13201-022-01837-7.

[14] M. S. Mahmoud, A. S. Mahmoud, M. A. El-Said, and M. K. Mostafa, "Comparison of aluminum and iron nanoparticles for chromium removal from aqueous solutions and tannery wastewater, empirical modeling and prediction," *Emergent Mater.*, vol. 5, no. 6, pp. 1729–1744, Dec. 2022, doi: 10.1007/S42247-021-00320-4/METRICS.

[15] A. Hamdy, M. K. Mostafa, and M. Nasr, "Regression analysis and artificial intelligence for removal of methylene blue from aqueous solutions using nanoscale zero-valent iron," *Int. J. Environ. Sci. Technol.*, vol. 16, no. 1, pp. 357–372, Jan. 2019, doi: 10.1007/S13762-018-1677-Z/TABLES/4.

[16] N. K. Singh *et al.*, "Artificial intelligence and machine learning-based monitoring and

design of biological wastewater treatment systems," *Bioresour. Technol.*, vol. 369, p. 128486, Feb. 2023, doi: 10.1016/J.BIORTECH.2022.128486.

[17] M. K. Mostafa, A. S. Mahmoud, M. S. Mahmoud, and M. Nasr, "Computational-Based Approaches for Predicting Biochemical Oxygen Demand (BOD) Removal in Adsorption Process," *Adsorpt. Sci. Technol.*, vol. 2022, 2022, doi: 10.1155/2022/9739915.

[18] A. S. Mahmoud, M. K. Mostafa, and M. Nasr, "Regression model, artificial intelligence, and cost estimation for phosphate adsorption using encapsulated nanoscale zero-valent iron," *Sep. Sci. Technol.*, vol. 54, no. 1, pp. 13–26, Jan. 2019, doi: 10.1080/01496395.2018.1504799.

[19] M. Fan, J. Hu, R. Cao, K. Xiong, and X. Wei, "Modeling and prediction of copper removal from aqueous solutions by nZVI/rGO magnetic nanocomposites using ANN-GA and ANN-PSO," *Sci. Reports 2017 71*, vol. 7, no. 1, pp. 1–14, Dec. 2017, doi: 10.1038/s41598-017-18223-y.

[20] F. Granata, S. Papirio, G. Esposito, R. Gargano, and G. de Marinis, "Machine Learning Algorithms for the Forecasting of Wastewater Quality Indicators," *Water 2017, Vol. 9, Page 105*, vol. 9, no. 2, p. 105, Feb. 2017, doi: 10.3390/W9020105.

[21] N. Taoufik, W. Boumya, M. Achak, H. Chennouk, R. Dewil, and N. Barka, "The state of art on the prediction of efficiency and modeling of the processes of pollutants removal based on machine learning," *Sci. Total Environ.*, vol. 807, p. 150554, Feb. 2022, doi: 10.1016/J.SCITOTENV.2021.150554.

94c Valorization of Water Treatment Plant Residuals to Biodegradable Deicers

Alexander Mathews

Water softening residuals disposal from the chemical treatment of water supplies is a national problem in the U.S. About 3.1 million tons/yr of lime softening sludge containing calcium carbonate, magnesium hydroxide and other organic and inorganic impurities are generated in the U.S from lime-soda softening to reduce the hardness of water supplies. The current disposal methods principally involve the application of sludges on to agricultural lands or landfilling. These options are expensive due to the high dewatering and transportation costs involved, and the potential deleterious effects on the environment. This research is aimed at the valorization of water plant residuals in the production of a biodegradable and non-corrosive deicing chemical that can be applied onto roads for snow and ice control in winter. Calcium and magnesium acetate deicers can be produced from water plant residuals by reacting the residuals with acetic acid. This reaction is controlled by the mass transfer rate of acetic acid to the calcium carbonate solid, surface reaction rate, and mass transfer rate of the product from the surface. This paper will report results obtained from a study of the reaction process to optimize operating conditions to maximize conversion rates. A freezing point depression model will be presented to predict freezing point depression of the deicer from basic colligative property relationships. This model is found or provide results in good agreement with experimental data.

94d Bio-Electrochemical Treatment of Dairy Manure for in-Situ Sulfide Remediation

Lingkan Ding, Bo Hu

As a highly toxic and corrosive gas with a characteristic rotten-egg odor, hydrogen sulfide (H_2S) emission spikes are often detected during dairy manure storage because of high levels of organic content (e.g., sulfur-containing proteins) and sulfate-containing minerals in feed and bedding materials. Most large sized farms collect liquid manure and slurry in a reception pit (or transition pit) before manure is pumped to the long-term storage or processed via anaerobic digestion, and these reception pits are often close in proximity to the barn animals and farm workers. Thus, this can be an emission point of H_2S and offensive odor, especially when manure agitation and pumping events occur. In addition to the toxicity to farm workers and animals, H_2S in the air of pits as well as in the subsequent biogas can be very corrosive for the equipment and infrastructure. All these clearly show that mitigating the H_2S generation during dairy manure storage and treatment is important and urgent. Conventional H_2S mitigation strategies typically include biogas cleaning technologies using biofilters or aqueous solutions (e.g., sodium hydroxide, ferrous chloride, or ferric hydroxide), which always require a separate unit housing the facilities and are chemical- and energy-intensive. Therefore, an in-situ H_2S mitigation method is critical to the sustainable manure management with less energy and chemical input as well as an easier operation. In the present study, a bio-electrochemical unit was integrated into the dairy manure storage and anaerobic digestion treatment for in-situ H_2S remediation. Through the lab-scale tests, a series of materials, including sacrificial metals (i.e., low carbon steel, stainless steel, aluminum, and copper) and non-sacrificial materials (i.e., graphite and titanium), were examined for the efficiency of H_2S removal, among which low carbon steel anodes showed the most effective H_2S removals (over 95%). This implied that the precipitation of insoluble sulfide minerals seems to be the major contributor to the sulfide remediation, rather than anodic sulfide oxidation. Subsequently, an experiment to determine the optimum applied voltage to the electrode pairs (anode: low carbon steel; cathode: stainless steel with a longer lifespan) for sulfide removal were conducted. It was recorded that an applied voltage of 0.6 V was high enough to release ferrous ions from the low carbon steel anode into dairy manure for ferrous sulfide precipitation as well as lead to higher methane production. In contrast, higher applied voltages over 0.8 V resulted in faster anode material consumption and triggered significant electrochemical hydrogen production with the inhibition of methane generation. Then, at the optimum voltage of 0.6 V applied to the low carbon steel-stainless steel electrode pair, strategies of intermittent electrochemical treatment were examined aiming to reduce the electricity and anode material consumption. It was observed that the electrochemical treatment of 6h/6h (on/off) and 12h/12h (on/off) were efficient in H_2S remediation, whereas that of 3h/3h (on/off) did not reduce the H_2S release effectively, indicating that a continuous electrochemical treatment period over 3 h was definitely required to properly function. With all the selections and optimizations above, a pilot-scale electrochemical unit was accordingly designed and then installed in the dairy manure pit in a local dairy farm in Minnesota, and its effects in in-situ H_2S remediation in a real application scenario were documented. In summary, this proposed bio-electrochemical system can successfully reduce the odor and improve the safety of a dairy farm during manure storage and treatment.

94e A Novel Three-Stage Process to Treat Sewage Sludge with High Phosphorus Recovery and Bioenergy Production

Lingkan Ding, Daniel Jurado Pineda, JAMES Postiglione, Bo Hu

Many municipal and industrial wastewater treatment plants (WWTPs) have a biological phosphorus (P) removal process to concentration P into the activated sludge and also employ anaerobic digestion (AD) to reduce sludge volume. However, these two processes negatively affect each other, causing issues that impact performance and increase operating costs. For instance, P accumulated in the sewage sludge through a bio-P process will be released during AD, resulting in high recycled P loading and P mineral scale in reactors and pipes, which further cause poor dewaterability and require additional equipment with costly maintenance. Meanwhile, P recovery is more difficult by its mixture with a high concentration of biological solids, which requires expensive separation equipment (e.g., fluidized bed reactors) and limits its application only to a few large metro WWTPs. In this study, a novel integrated process, including a first-stage thermophilic acid AD (TAAD), a second-stage P mineralization and precipitation, and a third-stage mesophilic high-rate AD (HRAD), was proposed to promote nutrient removal/recovery and renewable bioenergy production in WWTPs. In the first-stage TAAD at 55 °C, thickened waste activated sludge (TWAS) collected from a local WWTP in Minnesota was used as the feedstock loaded to a continuous stirred-tank reactor (CSTR). After over 400-day operation at a hydraulic retention time (HRT) of 2-5 days, the P solubilization in first stage reached up to 900-2000 mg P/L, corresponding to over 80% of the total P in the TWAS. Meanwhile, whey permeate concentrate was used as an organic waste additive to TAAD with the purposes of controlling the pH around 6 and facilitating the degradation of sludge with increased volatile fatty acids (VFAs) release. Subsequently, the second-stage P precipitation from the TAAD effluent was conducted through a standardized jar test, and the maximum P removal was achieved with a molar dosage of 1:1.9 (phosphate to magnesium ions) in the solution. With this dosage, up to 80% of the soluble phosphate was recovery in the form of P-bearing mineral precipitates (e.g., struvite). These precipitates with mainly P-bearing minerals could be sold as a biofertilizer with a high P purity and create a revenue stream, thus significantly improving the economic feasibility of this process. After that, the supernatant with a high level of soluble COD and VFAs was pumped to an up-flow anaerobic sludge blanket (UASB) reactor operated at 37 °C, and most of the residual organics were converted to biogas at an HRT of 1-3 days via this third-stage HRAD process. Therefore, at an overall HRT of 3–8 days, this integrated system achieved a significant sludge treatment performance with very efficient P recovery and biogas production that outperformed a conventional wastewater sludge AD process with an HRT of 20-30 days. In conclusion, this novel three-stage process treating sewage sludge has the potential to improve the efficiency and effectiveness of P recovery and enhance the generation of bioenergy in the form of biogas with a shorter retention time, thus benefiting current WWTP operations.

94f A Systematic Study of Co-Digestion of Fishery Sludge and Food Waste for Biogas Production

Reza Khorasani, Jin Wang, Q. Peter He

The global population is projected to reach 8.5 billion by 2030 and 9.7 billion by 2050 (UN DESA, 2022). This growth will be accompanied by a projected 40–75% increase in total protein demand. Compared to other animal production systems, aquaculture has the greatest overall efficiency (D'Abramo, 2021). As a result, intensification has emerged as a promising pathway for increasing aquaculture production. However, aquaculture intensification leads to increased demand for feed and increased waste production. In terms of waste, only about 30% of the nutrients supplied to the farmed aquaculture animals are converted to product, while the rest is excreted as soluble nutrients or settleable solids (e.g., ammonia and feces, respectively) (Chiquito-Contreras et al., 2022). The waste produced by aquaculture farms is often discharged into the environment to maintain water quality in the aquaculture systems. Such waste disposal practice can cause eutrophication of the surrounding areas of fish farms and threaten the sustainability of local ecosystems (Bannister et al., 2016; Pędziwiatr, 2017). Therefore, sustained intensification of aquaculture production must be coupled with a sufficient supply of sustainable nutrients (feed) and efficient waste management. If feed can be produced from waste, it solves two problems with a single solution.

Previously, we have proposed to convert aquaculture waste into single cell protein (SCP) as aquafeed supplements via a two-step process: 1) utilizing commercially proven anaerobic digestion (AD) to produce biogas from fishery sludge (FS); 2) cultivating microbial biomass as feedstock for SCP production. In this work, we systematically study co-digestion of fishery sludge (FS) and food waste (FW) for optimizing biogas production. Previous studies have found that carbon/nitrogen ratio (C/N ratio) and percentage solid content (PSC) of the feed are two of the most important factors that affect AD operation and performance. The optimal C/N ratio for biogas production is about 15 - 25, while the optimal PSC is 8% - 12% (Linville et al., 2015). FS usually has a low C/N ratio (~4-12) and low PSC (~0.5-5%), while FW in general has a high C/N ratio (~18 - 30) and high PSC (~ 15-25%)(Shi et al., 2018). It has been reported that co-digesting low C/N waste with high C/N waste often produces significantly improved biogas yield and productivity (Angelidaki and Ellegaard, 2003; Mshandete et al., 2004; El-Mashad and Zhang, 2010). Therefore, in this work, we perform a comprehensive experimental study to determine the optimal FS/FW feed ratio and PSC to optimize AD performance. The AD performance of different combinations of FS/FW ratios and PSC values were assessed through multiple metrics, including biogas yield, biogas productivity, BOD reduction rate, COD reduction rate, etc. A response surface model was constructed to capture the impacts of operation conditions on biogas production performance, which serves as the model to guide the biogas production step in the overall waste-to-feed (W2F) application.

References:

Angelidaki, I., and Ellegaard, L. (2003). Codigestion of manure and organic wastes in centralized biogas plants. *Appl Biochem Biotechnol* 109, 95–105.

Bannister, R. J., Johnsen, I. A., Hansen, P. K., Kutti, T., and Asplin, L. (2016). Near-and far-field dispersal modelling of organic waste from Atlantic salmon aquaculture in fjord systems. *ICES Journal of marine Science* 73, 2408–2419.

Chiquito-Contreras, R. G., Hernandez-Adame, L., Alvarado-Castillo, G., Martínez-Hernández, M. de J., Sánchez-Viveros, G., Chiquito-Contreras, C. J., et al. (2022). Aquaculture—Production System and Waste Management for Agriculture Fertilization—A Review. *Sustainability* 14, 7257.

D'Abramo, L. R. (2021). Sustainable aquafeed and aquaculture production systems as impacted by challenges of global food security and climate change. *J World Aquac Soc* 52, 1162–1167.

El-Mashad, H. M., and Zhang, R. (2010). Biogas production from co-digestion of dairy manure and food waste. *Bioresour Technol* 101, 4021–4028. Available at: http://www.sciencedirect.com/science/article/pii/S0960852410000842.

Linville, J. L., Shen, Y., Wu, M. M., and Urgun-Demirtas, M. (2015). Current state of anaerobic digestion of organic wastes in North America. *Current Sustainable/Renewable Energy Reports* 2, 136–144.

Mshandete, A., Kivaisi, A., Rubindamayugi, M., and Mattiasson, B. (2004). Anaerobic batch co-digestion of sisal pulp and fish wastes. *Bioresour Technol* 95, 19–24.

Pędziwiatr, P. (2017). Aquaculture waste management. *Acta Innovations*, 20–29.

Shi, S., Li, J., Guan, W., and Blersch, D. (2018). Nutrient value of fish manure waste on lactic acid fermentation by Lactobacillus pentosus. *RSC Adv* 8, 31267–31274.

UN DESA (2022). World Population Prospects 2022. *https://www.un.org/development/desa/pd/sites/www.un.org.development.desa.pd/files/wpp2022_summary_of_results.pdf.*

94g One-Pot Synthesis of a Novel Chitosan-Based Hydrogel By C-Mannich Reaction and Screening of Its Cu (II) Uptake

John Rey Romal, Dr. Say Kee Ong

The persisting industrialization continues to worsen the dilemma of metal-induced water pollution. Water treatment by adsorption is known to be one of the most efficient methods, but the harsh manufacturing conditions of activated charcoal and the non-sustainability of hydrocarbon-based commercialized adsorbents imposed extra costs to this treatment strategy. More recent efforts are driven towards the utilization of polysaccharide-based hydrogel adsorbents as sustainable alternatives. This work presents the fabrication, characterization, and screening of Cu^{2+} uptake of a novel chitosan-based hydrogel achieved in a single step by C-Mannich reaction. The Mannich base crosslinked-chitosan hydrogel (CS-MB) was successfully fabricated in one-pot with chitosan (CS), glutaraldehyde (Glu), and 4-hydroxycoumarin (4-HC) in 3% *v/v* acetic acid at 22 ^{0}C.[1] The novel CS-MB exhibits a temperature and pH-dependent swelling behavior unique from other hydrogel systems with an optimal swelling at pH = 4 and 22 ^{0}C. Using this optimal swelling condition, Cu^{2+} uptake was conducted in batch mode and achieved a maximum adsorption capacity of 12.0 mg/g with 1.2 g/L adsorbent dosage after 36 h with agitation. The uniform surface adsorption of Cu^{2+} was verified by EDS and the adsorption site was identified by FT-IR. The one-pot fabrication method of the novel CS-MB at ambient conditions using a sustainable polysaccharide and its activity in Cu^{2+} uptake provides a promising cheap and sustainable alternative to the commercialized adsorbents for the removal of Cu^{2+} in water by adsorption.

References

[1] Romal, J.R.A*.; Ong, S.K. Single-Step Fabrication of a Dual-Sensitive Chitosan Hydrogel by C-Mannich Reaction: Synthesis, Physicochemical Properties, and Screening of its Cu^{2+} Uptake. *Processes* **2023**, *11*, 354. https://doi.org/10.3390/pr11020354

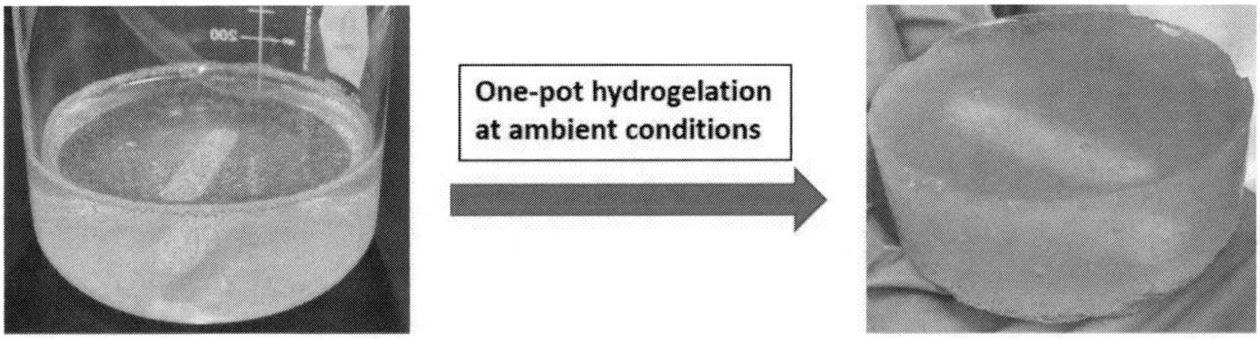

219a Engineering Aqueous Amine Solvents Based on Steric Effects for CO_2 Capture from First Principles Simulations

Bohak Yoon, David C. Calabro, Lisa S. Baugh, Sumathy Raman, Gyeong S. Hwang

Carbon dioxide (CO_2) emissions from fossil fuel consumption to the atmosphere remain a grand challenge and a major threat to humanity. Currently, the amine scrubbing method based on chemical absorption with aqueous amine solvents is the most widely implemented technology for the post-combustion CO_2 capture process, which is governed by two competing reactions leading to carbamate or bicarbonate formation. Sterically hindered alkanolamines with various degrees of hydrophobicity are experimentally found to influence the formation of these products, but the underlying mechanisms responsible for such preferential product formations are not well understood. In this work, we will present the CO_2 absorption behavior of several sterically hindered amines, focusing on the effects of different degrees of methylation and temperature. Our *ab initio* molecular dynamics with enhanced sampling simulations show that the degree of steric hindrance strongly influences the free-energy barriers for both carbamate and bicarbonate reaction pathways and, thus, impacts their kinetic favorability. Our analyses further demonstrate that various degrees of steric effects cause considerably different temperature-dependent solvation of the nitrogen atom of hindered amines, which may directly affect CO_2 accessibility to form carbamate. At low absorption temperatures, the enhanced solvation resulting from increased steric hindrance leads to kinetically facile amine protonation, promoting the formation of bicarbonate. Our study provides insights into the effects of various degrees of steric hindrance, with different temperature-dependent solvation natures governing reaction mechanisms and rates during CO_2 capture, thereby aiding in designing optimal solvents for the CO_2 capture process.

219b Theoretical Studies and Experimental Investigations on the Removal of Trace Metal Contaminants from Syngas Using Monometallic and Bimetallic Adsorbents

Dwijraj Mhatre, Divesh Bhatia

Introduction

To utilize the syngas derived from the gasification or pyrolysis of carbonaceous feedstocks for the production of valuable chemicals, the syngas has to be free from various trace contaminants such as Hg, As, Se, P, Cd, and Pb. These elements get released during gasification or pyrolysis in their elemental form, hydride form (AsH_3, H_2Se, PH_3), or sulfide form (PbS). Since they are volatile at temperatures above 400 °C and are potential poisons to the catalyst(s) employed downstream for syngas processing, they need to be eliminated. Various commercial carbon and sulfur-based adsorbents are used for syngas cleanup, but are inefficient for the removal of trace elemental contaminants at high temperatures. Hou et al. (2014) have reported that the strongest adsorption of one such contaminant viz. gas-phase elemental mercury (Hg^0) occurs on Pd-based adsorbents and on a few bimetallic adsorbents. In the present work, the adsorption of various gas-phase contaminants, viz., Hg^0, AsH_3, H_2Se, PH_3, Cd, Pb, and PbS is studied on monometallic and novel bimetallic surfaces. Insights are obtained into the dissociation of AsH_3, H_2Se, and PH_3 on metals and into the segregation behaviour of Hg, As, Se, P, Cd, and Pb in alloys. Further, the most promising monometallic adsorbents are synthesized and their efficacy for Hg^0 adsorption for different operating conditions is given.

Methodology

First-principles density functional theory (DFT) calculations are performed to calculate the adsorption energy and to evaluate the most stable geometric configurations for the adsorption of gas-phase Hg^0, AsH_3, H_2Se, PH_3, Cd, Pb, and PbS on various metal surfaces. The calculations are carried out using the Vienna ab initio simulation package (VASP) and by employing the Perdew-Burke-Ernzerhof (PBE), optPBE-vdW, optB88-vdW, and PBE-D3 functionals of the generalized gradient approximation along with an energy cut off of 400 eV. The reaction energy pathway for AsH_3 dissociation on various metals is studied using the CI-NEB method. AsH_3 dissociation on the (1 1 1) surface of various metals is considered to occur by the sequence of the elementary reactions described by Eqs. (R1)-(R4), where the subscript 'gas' represents the gas-phase species and the subscript 'ads' represents the adsorbed surface species.

$AsH_{3\,(gas)} \rightarrow AsH_{3\,(ads)}$ (R1)

$AsH_{3\,(ads)} \rightarrow AsH_{2\,(ads)} + H_{(ads)}$ (R2)

$AsH_{2\,(ads)} \rightarrow AsH_{(ads)} + H_{(ads)}$ (R3)

$AsH_{(ads)} \rightarrow As_{(ads)} + H_{(ads)}$ (R4)

Similarly, calculations are performed considering the step-wise abstraction of H atoms from H_2Se and PH_3 as well and the reaction energy barriers and the reaction energies for the elementary reactions are calculated.

The Pt-based monometallic adsorbents shortlisted from DFT are prepared using the incipient wetness method using doped aluminas (SIRAL-40, SASOL), metal oxides (Ceria, zirconia, titania) and silica-based molecular sieve (SBA-15) as support materials.

Results and Discussion

DFT calculations show that the adsorption of Hg^0, Cd, Pb, and PbS is the most favourable on the hollow sites of most metal surfaces, whereas the adsorption of hydrides AsH_3, H_2Se, and PH_3 is favourable on the top sites of most metals considered in this study. The binding energy calculations indicate that the adsorption of gas-phase Hg, AsH_3, H_2Se, PH_3, Cd, Pb, and PbS is exothermic on various metals. The calculations reveal that Rh, Ru, Pd, Pt, and Ir are the most promising metals for the removal of these contaminants from gas streams.

AsH_3 dissociation on various metals is studied and the reaction energy diagram for AsH_3 dissociation is shown in Figure 1. The analysis of the calculated reaction energies and activation barriers for the catalytic dissociation of AsH_3 on various metal surfaces shows that its dissociation is more facile on Rh, Ru, Pd, Ir, and Pt as compared to Cu, Ag, and Au. Further, the dissociation of AsH_3 on Ir, Pt, and Pd is found to occur without the formation of the intermediate adsorbed species AsH. The dissociation on these metal surfaces occurs in two steps with the first dissociation step given by Eq.(1) and the second dissociation step as $AsH_{2\,(ads)} \rightarrow As_{(ads)} + 2\,H_{(ads)}$. Similar results are observed for the dissociation of H_2Se and PH_3 on various metals. Further, segregation energy calculations predict the tendency of elements Hg, As, Se, P, Cd, and Pb to segregate towards the surface of Ru, Rh, Pt, Ir, Pd, Cu, Ag, and Au after alloy formation. Furthermore, the formation energy calculations indicate the possibility of the formation of alloys between metals Ag, Au, Cu, Ir, Pd, Pt, Rh, and Ru and elements Hg, As, Se, P, Cd, and Pb. The calculations of the partial density of states for different alloys containing Hg, As, Se, P, Cd, and Pb provide information on the alterations in the electronic properties of the alloy surface and the effect of these contaminants on specific orbitals.

DFT calculations are performed to evaluate the binding of the trace contaminants on 56 different compositions of bimetallic adsorbents. The binding energy of AsH_3 on the (1 1 1) plane of various bimetallic surfaces is given in Figure 2. It is found that bimetallic adsorbents with an overlayer of Ru, Rh, Ir, or Pt on Ag or Au exhibit a higher affinity for AsH_3 as compared to the monometallic adsorbents. Similar results are obtained for the adsorption of Hg, H_2Se, PH_3, Cd, and Pb. The adsorption of these contaminants on the overlayered bimetallics is stronger than that on monometallic surfaces due to the ligand and strain effects. Bimetallic adsorbents comprising of a single overlayer of Rh, Ru, Pd, Pt, and Ir on Ag, Au, or Cu are the novel and cost-effective adsorbents identified for the removal of gas-phase contaminants Hg, AsH_3, H_2Se, PH_3, Cd, and Pb from syngas.

The d-band center, d-band width, and upper edge of the d-band for various monometallic and bimetallic surfaces are calculated and correlated to the adsorption energies of various contaminants . The increase in adsorption energy of Cd on various monometallic and bimetallic surfaces is correlated to the upward shift in the d-band center (Figure 3). Similar results are obtained for Hg, AsH_3, H_2Se, PH_3, and Pb adsorption on various monometallic and bimetallic surfaces. Thus, the adsorption of these contaminants on various surfaces is found to be consistent with the d-band model.

The initial Hg^0 adsorption efficiency and the adsorption capacity of the lab-synthesized Pt adsorbents Pt/SBA-15, Pt/S-40 (SIRAL-40), Pt/ZrO_2, Pt/TiO_2, and Pt/CeO_2 are evaluated. The adsorption efficiency of these Pt-based adsorbents at 270 °C for various compositions of syngas is shown in Figure 4. The Hg^0 adsorption efficiency of Pt/SBA-15 is more than 95% in a N_2 atmosphere. Under these conditions, the adsorption efficiency of Pt/SBA-15 is the highest followed by Pt/S-40, Pt/ZrO_2, Pt/TiO_2, and Pt/CeO_2. The adsorption efficiency over Pt/S-40, Pt/ZrO_2, Pt/TiO_2, and Pt/CeO_2 reduces significantly in the presence of syngas components H_2 and H_2O thus showing their inhibiting effect towards Hg^0 adsorption. In contrast, the adsorption efficiency of Pt/SBA-15 is retained above 90% in the presence of H_2 and H_2O. Characterization results from XRD and H_2-TPR techniques indicate the presence of Pt nanoparticles in the elemental state on Pt/SBA-15, Pt/S-40, and Pt/TiO_2, and the presence of Pt nanoparticles in the oxidized form on Pt/ZrO_2 and Pt/CeO_2. Various other characterization techniques such as UV-vis spectroscopy,

BET surface area analysis, Field emission-SEM, and TEM are used to study the physical and chemical nature of the adsorbents, and their correlation with the adsorption activity, which would be described in the full manuscript. The bimetallic adsorbents identified by DFT will be prepared and the effect of operating conditions will also be discussed in the full manuscript.

References

[1] Hou, W., Zhou, J., Yu, C., You, S., Gao, X., Luo, Z. (2014). Pd/Al_2O_3 Sorbents for Elemental Mercury Capture at High Temperatures in Syngas. *Ind. Eng. Chem. Res. 53*, 9909-9914.

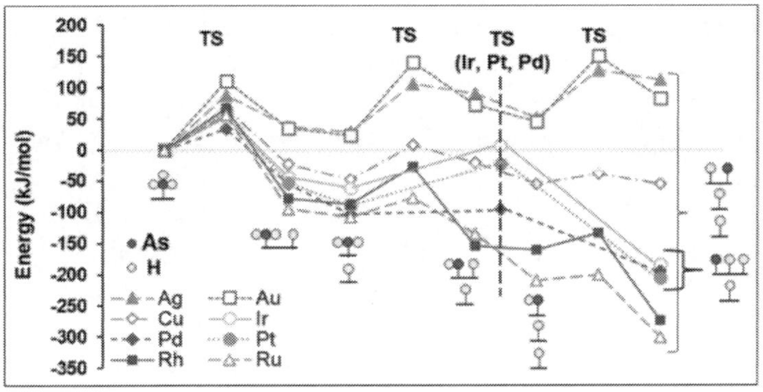

Figure 1. Reaction energy diagram for AsH_3 dissociation on the (1 1 1) surface of various metals

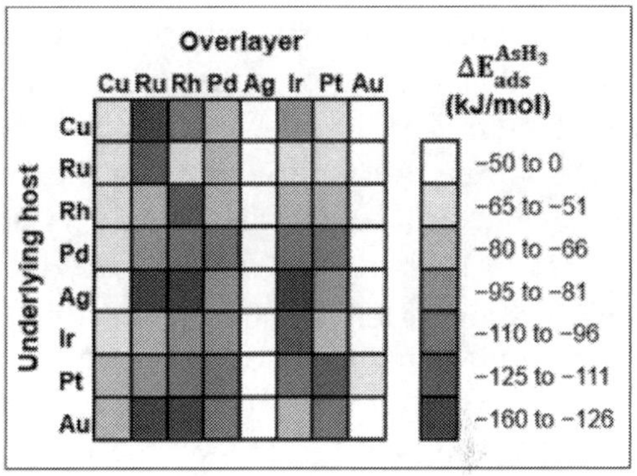

Figure 2. AsH_3 adsorption energy ($\Delta E_{ads}^{AsH_3}$) on various overlayered bimetallic surfaces

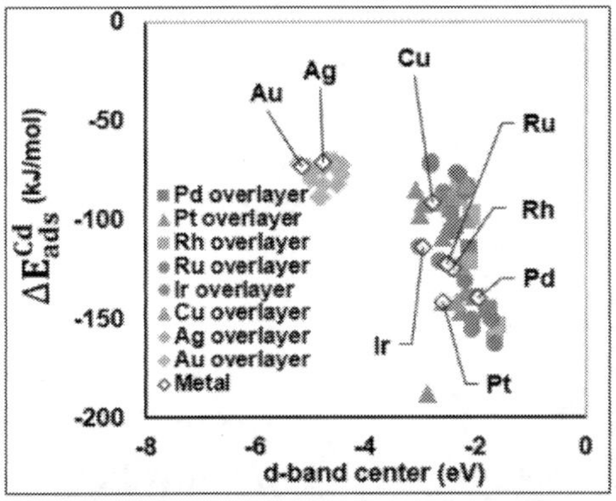

Figure 3. Correlation between Cd adsorption energy (ΔE_{ads}^{Cd}) and d-band center of monometallic and bimetallic surfaces

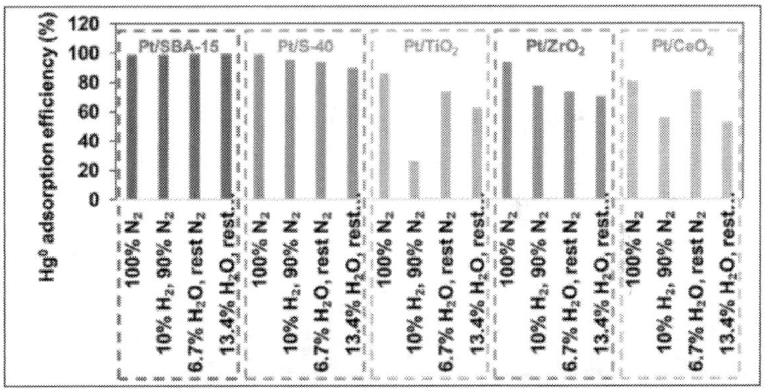

Figure 4. Hg^0 adsorption efficiency for various Pt-based adsorbents under different syngas conditions at 270 °C

219c Optimization of a Greenhouse Farming System Under Food-Energy-Water Nexus System Considerations

Marcello Di Martino, Patrick Linke, Efstratios Pistikopoulos

Food security remains a major challenge for many countries due to a globally increasing population. Additionally, urbanization and soil degradation reduce the amount of available agricultural land, which further stresses food supply systems. Moreover, for semi-arid and arid water scarce regions with high temperatures and harsh external conditions, outdoor agriculture is neither a viable nor a sustainable option [1]. One possibility of overcoming these challenges is the utilization of closed systems in the form of greenhouses, which provide a controlled environment to reduce the risks related to plant development [2]. While the control of the greenhouse farming systems is a well-researched area, literature regarding the design optimization of a greenhouse is scarce. In fact, the greenhouse design is almost exclusively assumed to be fixed.

This work focuses on modeling, optimization and design of a greenhouse farming system dependent on a geographic location, available crops, and utilized technologies. Climate data in the form of solar DNI, ambient air temperature and humidity are considered to incorporate geographic dependencies [3]. Accordingly, the influence of hot or cold external climate profiles on the greenhouse productivity, resulting in different heating or cooling greenhouse profiles, is investigated. Further, various crops with different temperature, humidity and irrigation requirements are taken into account, which dictate the operating point of the greenhouse [4]. An array of viable cooling and heating technologies are evaluated for the optimal design and operation of the greenhouse, such as natural ventilation, evaporative cooling, or conventional HVAC systems. A food-energy-water nexus approach is employed with the objectives to minimize cost, while maximizing water and energy resource utilization [5, 6]. This approach results in a multi-objective mixed-integer optimization problem enabling the trade-off analysis between completely isolated greenhouses and glass greenhouses.

References

[1] F. Mahmood, R. Govindan, A. Bermak, D. Yang, C. Khadra, T. Al-Ansari. Energy utilization assessment of a semi-closed greenhouse using data-driven model predictive control, Journal of Cleaner Production, vol. 324, p. 129172, 2021, https://doi.org/10.1016/j.jclepro.2021.129172

[2] I. Ghiat, F. Mahmood, R. Govindan, T. Al-Ansari. CO2 utilisation in agricultural greenhouses: A novel 'plant to plant' approach driven by bioenergy with carbon capture systems within the energy, water and food nexus, Energy Conversion and Management, vol. 228, p. 113668, 2021, https://doi.org/10.1016/j.enconman.2020.113668

[3] J. Chen, J. Yang, J. Zhao, F. Xu, Z. Shen, L. Zhang. Energy demand forecasting of the greenhouses using nonlinear models based on model optimized prediction method, Neurocomputing, vol. 174, pp. 1087–1100, 2016, https://doi.org/10.1016/j.neucom.2015.09.105

[4] P. Van Beveren, J. Bontsema, G. Van Straten, E. Van Henten. Minimal heating and cooling in a modern rose greenhouse, Applied Energy, vol. 137, pp. 97–109, 2015, https://doi.org/10.1016/j.apenergy.2014.09.083

[5] M. Di Martino, R. C. Allen, E. N. Pistikopoulos. The Food-Energy-Water Nexus in Sustainable Energy Systems Solutions, Handbook of Smart Energy Systems, Springer, 2022,

https://doi.org/10.1007/978-3-030-72322-4_168-1

[6] M. Di Martino, S. Avraamidou, J. Cook, E. N. Pistikopoulos. An Optimization Framework for the Design of Reverse Osmosis Desalination Plants under Food-Energy-Water Nexus Considerations, Desalination, 503, 2021. https://doi.org/10.1016/j.desal.2021.114937

219d A Data-Driven Numerical Framework for the Richards Equation for Sustainable Irrigation and Food Production

Zeyuan Song, Zheyu Jiang

Accurately modeling and predicting soil moisture is crucial for a variety of applications such as precision agriculture, smart irrigation, and drought prevention. Estimating root zone soil moisture from the surface or near-surface soil moisture data is typically achieved by solving a hydrological model that describes water movement through soils. Most existing agro-hydrological models are based on the Richards equation [1], which captures irrigation, precipitation, evapotranspiration, runoff, and drainage dynamics in soil. The Richards equation is a highly nonlinear, degenerate elliptic-parabolic partial differential equation (PDE) that is notoriously difficult to solve. State-of-the-art Richards equation solvers rely on spatial discretization based on finite difference, finite element, or finite volume method (FVM). Unlike finite difference and finite element methods, FVM adopts an integral form of the Richards equation, which offers some valuable physical insights (e.g., mass conservation) about water flow dynamics. However, these insights are often lost again in the process of transforming the discretized equations into a matrix equation. Furthermore, the resulting matrix is stiff and sparse and is computationally challenging to solve. To address these numerical difficulties, several linear iterative schemes have been proposed, among which the L-scheme exhibits the nice property of unconditional convergence which is irrespective of the choice of initial guess [2,3]. Nevertheless, in the standard L-scheme, the linearization parameter is an arbitrarily chosen static constant that is not adaptive to any change in pressure head. This greatly impacts the performance of the L-scheme. In the first part of this talk, we introduce a novel adaptive L-scheme whose linearization parameters dynamically adjust themselves for each discretized cell, time step, and iteration, thereby significantly improving the stability and robustness of the L-scheme.

To further incorporate the underlying physics of water flow dynamics and conservation in soil, in the second part of this talk, we introduce the first data-driven global random walk algorithm to solve the FVM-based adaptive L-scheme. A key assumption made in existing global random walk algorithms for solving the Richards equation is that the pressure head is proportional to the number of particles in a discretized cell [4]. Nevertheless, we have shown that this assumption is invalid, and the relationship between the pressure head and the number of particles may not be continuous, smooth, or explicit. Instead, we propose a novel data-driven approach and used two neural networks to accurately learn the mapping and inverse mapping between the pressure head and the number of particles. Coupling this with the adaptive L-scheme, we show that our data-driven framework not only is the first-of-its-kind that can solve 3-D Richards equation, but also outperformed commercial and state-of-the-art solvers in accurately capturing the underlying physics.

References

[1] L.A. Richards, Capillary conduction of liquids through porous mediums, Physics, 1931, 1(5): 318-333.

[2] K. Mitra, I. Pop, A modified l-scheme to solve nonlinear diffusion problems, Computers & Mathematics with Applications, 2019, 77(6): 1722-1738.

[3] F. List, F. Radu, A study on iterative methods for solving Richard's equation, Computational Geoscience, 2016, 20: 341-353.

[4] N. Suciu, D. Illiano, A. Prechtel, F. A. Radu, Global random walk solvers for fully coupled flow and transport in saturated/unsaturated porous media, Advances in Water Resources, 2021, 152: 103935.

219e Heroma VV Sensor, Water Toxicity Sensor Challenge

ANA Vega, Valentina Mayorca Gonzalez, Maria Paula Heredia Fuentes

Life revolves around water, so a strict control on its quality is vital to guarantee human health and protection of biodiversity. However, nowadays, industrialization accelerated by human population growth and people's lack of information about possible environmental problems end up polluting water bodies with biological and artificial pollutants. Including, the presence of heavy metals in drinking water can have serious repercussions on health such as intestinal and kidneys diseases, accompanied even by severe brain damage. Therefore, this work aims to develop a biosensor that is at the service of detecting two harmful heavy metals, mercury, and lead. Among biosensors, the amperometric enzymatic biosensor was selected due to the advantages they offer over others, for instance, simplicity, short time responses and miniaturization potential are some of these features. At same time, an innovative design will allow to detect heavy metals in an easy and cost-effective way without neglecting the ease of transport and handling with a series of simple instructions that allow access not only to scientists but also to standard or unskilled users. The acting principle includes the analysis of urease enzyme behavior in the presence of specific heavy metals since it is inhibited at -0.03 V to 0.0 V reaching the formulation of calibration curves that show the concentration of an analyte by considering the variation in the current flow. The sensor has maximum dimensions of 76.2 mm long, 39.54 mm high and 62 mm wide with an approximate weight of 218 grams. The initial cost per measured sample is estimated to be approximately 12 USD.

219f Sequestration of Pfas Surfactants from Water Using Functional Polymers

Paschalis Alexandridis, Samhitha Kancharla, Aditya Choudhary, Dengpan Dong, Dmitry Bedrov, Marina Tsianou

Surface active per- and polyfluoroalkyl substances (PFAS) find niche applications because of their high chemical and thermal stability, high interfacial activity, and incompatibility with both water and hydrocarbons. However, several widely used PFAS surfactants have been found extremely resistant to degradation, accumulate in the environment, and have long half-lives in humans, consequently generating great concern [https://doi.org/10.1016/j.jhazmat.2021.128137]. To sequester PFAS from aqueous media, we develop materials and processes that are based on PFAS binding to functional polymers. We present here examples on how the notorious PFAS surfactants perfluorooctanoate (PFOA) and 2,3,3,3-tetrafluoro-2-(heptafluoropropoxy) propanoate (GenX) bind to poly(ethylene oxide) (PEO) polymers [https://doi.org/10.1016/j.jcis.2021.10.176]) in aqueous solution, and to polymer networks [https://doi.org/10.1021/acsmaterialslett.2c00168] [https://doi.org/10.1021/acsmacrolett.2c00383]. We also discuss how PFAS-adsorbent materials can be regenerated by disrupting the PFAS-polymer binding [https://doi.org/10.1039/D1CP00049G]. From the analysis of complementary experimental (small-angle neutron scattering, SANS, with contrast variation) and modeling (molecular dynamics, MD) results, a detailed description emerges on how PFAS molecules organize around the different parts of the polymers. This information is used to rationalize the capacity of these polymers for PFAS sequestration, and to design polymers with optimal binding and selectivity.

219g Evaluating Green Novel Deep Eutectic Solvents for Pfas Removal from Water; COSMO Simulation with Experimental Validation

Laura Fronchetti Guidugli, Toufiq Reza

In recent years, perfluoroalkyl substances (PFAS), also known as forever chemicals have been identified as global pollutants. Due to their unique chemical properties, such as high solubility in water and extreme persistence in the environment, PFAS have raised concerns about their potential human toxicity. As a result, there is a growing need to understand the fate, transport, and remediation of these chemicals in the environment as conventional water treatment methods, such as coagulation/flocculation, filtration, and chlorination are ineffective in removing them. Consequently, it is crucial to develop an efficient method that can remove PFAS effectively. This study used a computational approach to systematically screen over one hundred novel and environmentally friendly deep eutectic solvents (DES) using the Conductor-like Screening Model for Real Solvents (COSMO-RS) to investigate the removal mechanisms of two specific PFAS compounds, perfluorooctane sulfonic acid (PFOS) and perfluorooctanoic acid (PFOA), from water. DES are a mixture of two compounds (a hydrogen bond donor or HBD and a hydrogen bond acceptor or HBA) that conjugate through hydrogen bonding, resulting in a deep depression of the melting point of the mixture and an increase in solvent properties. The DES-based separation was evaluated by predicting various thermodynamic properties, such as sigma profiles, sigma potentials, logarithmic activity coefficient ($\ln(\gamma)$), and excess enthalpy (H_E) of absorption. The predicted thermodynamic properties are further validated with experimental results. In addition, an environmental health and safety assessment was performed on the DES components to evaluate their toxicity and impact on the ecosystem. Overall, tetra-alkyl ammonium bromide-based DES showed higher separation efficiency due to their affinity for PFOS and PFOA.

219h Yeast-Laden Hydrogel Capsules for Scalable Trace Lead Removal

Patritsia M. Stathatou, Devashish Gokhale, Christos E. Athanasiou, Patrick Doyle

Traces of heavy metals in water resources, due to mining activities and e-waste discharge, pose a global threat, with lead being one of the most abundant and toxic trace contaminants. As a result of various incidents of lead-contaminated drinking water, relevant water regulations are being revised, while according to the US Environmental Protection Agency no level of lead in drinking water is considered safe. Conventional treatment processes fail to remove trace lead from drinking water in a resource-efficient manner. We have previously shown that by using the yeast *Saccharomyces cerevisiae* we can effectively remove trace lead from water *via* a rapid mass transfer process, called biosorption, achieving an uptake of up to 12 mg lead per gram of biomass in solutions with initial lead concentrations below 1 part per million. We have also found that equilibrium is achieved within the first 5 minutes of contact. The rapid and high lead uptake is advantageous for the large-scale application of this inexpensive and abundant biomaterial for the removal of trace heavy metals from water.

The main limitation for scaling up the developed biosorption approach is the requirement for additional treatment steps to remove the added yeast. In this presentation, we demonstrate the encapsulation of yeast cells in hydrogel microparticles to overcome this limitation. Yeast-containing poly(ethylene glycol) diacrylate (PEGDA) hydrogel capsules are synthesized using an off-the-shelf microfluidic device through a scalable approach. Scanning electron microscopy and confocal fluorescence imaging have been used to measure the capsules' porosity and assess the distribution of yeast cells inside the capsules. Through kinetic and equilibrium experiments we characterize the stoichiometry, equilibrium, and selectivity of the hydrogel capsules under different initial lead concentrations (in the range of 30 – 1000 ppb). Residual lead concentrations have been measured using inductively coupled plasma - mass spectrometry. Data shows that hydrogel capsules are effective vehicles for holding yeast cells, being sufficiently large to be easily separated from water under the effect of gravity and sufficiently porous to not limit the adsorption capacity of the yeast or the kinetics of lead removal. Finally, we perform biomechanical static and dynamic compression tests to assess the mechanical robustness of the capsules and guide the design of a cm-scale, packed-bed bioreactor operated as a flow-through filter to serve as a proof-of-concept of our approach. Our work overcomes separation and structural stability issues that limit biosorption scalability, opening a new generation of environmentally friendly, highly effective and sustainable biosorbents targeting emerging contaminants.

117a Reversing Eutrophication with a Low-cost Chemisorbent – from Lab Research to Industrial Scale and Field Studies

Latif Ullah, Eva Björkman, Johan Larsson, Mats Lindberg, Kenneth Winroth, Maria B. Gustavsson, Niklas Hedin

The Baltic Sea suffers from a high phosphorus (P) load, which has led to severe eutrophication and the loss of aquatic life. P-removal is installed in many sewage water treatment stations around the Baltics, but there are large amounts of historic P in the seabed that is reactivated cyclically. Hence, a dynamic treatment is sought, and activated limestone[1] is a promising adsorbent to sequester mobile P in the seabed. The thermally activated limestone is produced from a cheap side product (marl) of the lime quarries. Here, we show that activated limestone can speed up the recovery of the Baltic Sea by binding P in to the sediment. The hypotheses are that the activation enhances the specific surface area and that this area is defining the action of the sorbents. The phosphate ions chemisorb on the calcium carbonate fraction or co-precipitate with a dissolving fraction of calcium hydroxide. We demonstrate that when activated limestone are prepared and used at the 30+ ton scale, it lowered the phosphate level in the deep water in a local bay in the Baltics significantly immediately after spreading. It should be noted that the use of marl, to produce the activated limestone, also reduced the amount of dumped marl on land, as this is commonly regarded as waste. The use of activated limestone could potentially lead to the revival of the Baltic Sea, benefiting the fishing, tourism, and limestone mining industries and creating new business opportunities. There are numerous other lakes, rivers, and arms of oceans that suffer from P-related eutrophication, where this approach could be relevant for remediation.

This project was initiated in 2012 by Assoc. Prof. Sven Blomqvist Stockholm University, (who unfortunately is no longer part of the project).

[1] SE1951309 (A1) "Binding of phosphorus in the bottom sediment" Björkman, E., Person, H., Ekeroth, N., and Blomqvist, S.

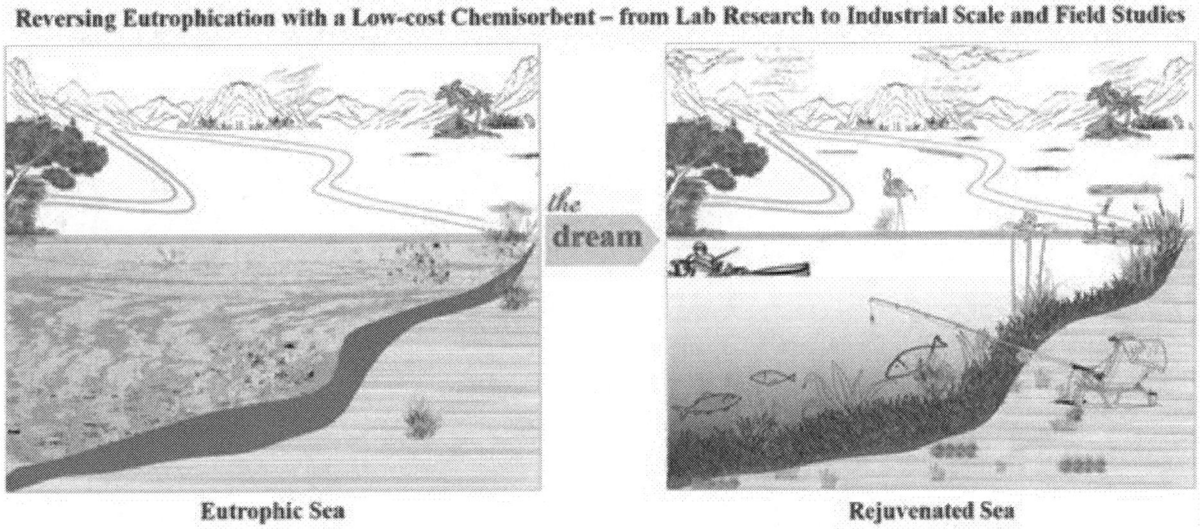

117b Occurrence of Antibiotics and Antibiotics-Resistant Bacteria and Its Mitigation in Drainage Discharge Channel

Gaoshen Zhao, Malvin Subroto Pamudji, Ka Kit CHENG, Yifan LI, Sin Hang Wong, Long Ting Luk, Wei Han, King Lun Yeung

Antibiotic was discovered dated back in 1928, has helped to save millions of people from the infectious disease that was once considered fatal in the 19th century. Nevertheless, the overuse of antibiotics such as clinical misuse and inside livestock could induce the potential emergence of antibiotics-resistance gene (ARG) inside the human body or even in the environment. In 2018, World Health Organization has reported ARG as one of the enormous threats to global health, food security and development. This study was examining the concentration of two antibiotics of interest, Clarithromycin and Ciprofloxacin in one of the drainage discharge channels that serves the treated wastewater. It was reported that wastewater treatment plant is the major source of the antibiotic present in the water stream, whereas the emergence of the antibiotic-resistant bacteria is observed at the middle part of the discharge channel, coming from the wet market influent, located 260m away from the wastewater inlet. Continuous 18 months of Ciprofloxacin and Clarithromycin measurement throughout the discharge channel indicated that there is no external pharmaceutical inflow into the channel, and the average concentration are 1.947 mg/L and 0.5051 mg/L respectively. Nevertheless, discharge affluent from the wet market was found out to contain a significant amount of ARG bacteria, with the range of 40-50 CFU/mL and 50-2000 CFU/mL for Ciprofloxacin and Clarithromycin resistant respectively.

Resistance gene measurement indicates that approximately 0.27%, 2.97% and 0.34% of the total microorganism inside the discharge channel are resistant to Ciprofloxacin, Clarithromycin and both Ciprofloxacin and Clarithromycin respectively. Multifunctional hydrogel, developed by HKUST team in 2015, has been utilized to eradicate the ARG bacteria developed throughout the discharge channel. The hydrogel encapsulates metabolic inhibitor that would be continuously released once the hydrogel is in contact with water. It is reported to reduce the total aerobic bacteria by 60-80% and total anaerobic bacteria by 60-65% during the two trial studies. On the same test, the hydrogel is also capable to curtail 70-75% and 65-70% of Ciprofloxacin-resistant and Clarithromycin-resistant bacteria respectively.

117c Liquidgel Technology for Disinfection and Odor Control in Waste Handling Facilities

Long Ting Luk, Ka Kit CHENG, Sin Hang Wong, Malvin Subroto Pamudji, Wei Han, King Lun Yeung

Malodor, in particular, hydrogen sulfide (H2S) has been reported as one of the major air pollutants produced by anaerobic digestion of organic matter in waste collecting processes such as sewer systems and wastewater treatment plants. In view of this, an innovative product, Liquid Gel (LG), has been developed, it is an active liquid colloidal sol that contains disinfecting and deodorizing compounds. The sol phase of liquid gel transforms into an active micro-gel triggered by temperature, pH, and salt concentration, which traps disinfecting and deodorizing compounds within a gel network. It provides immediate odor control by liquid phase bio-reaction and a sustained release of disinfecting and deodorizing compounds after micro-gel formed that can last for over 48 hours. Additionally, it offers an 80-90% H2S remediation performance over the existing method used in Stonecutter Island Sewage Treatment Work (SCISTW). Liquid gel dosing also requires 80% less chemical amount compared to the existing method, resulting in significant cost savings for the treatment plant. Besides, liquid gel dosing is certified by HOKLAS lab, a non-corrosive to the treatment plant equipment, which can extend the lifespan of the plant's infrastructure and reduce the need for equipment repairs and replacements. With its liquid form, liquid gel dosing is flexible and compatible with the existing equipment in SCISTW, making it an easy and efficient solution for wastewater treatment plant operators. Overall, liquid gel dosing technology provides a reliable, efficient, and sustainable solution for odor control in wastewater treatment plants.

117d New Odor and Corrosion Control Technologies for Concrete Stormwater Drainage and Sewer System

Malvin Subroto Pamudji, Sin Hang Wong, Ka Kit CHENG, Long Ting Luk, Wei Han, King Lun Yeung

Hydrogen Sulfide gas and Sulfuric Acid are the two materials that coexist together inside the sewer or drainage system because of the metabolic digestion of sulfur-containing materials inside the foul water or contaminated stormwater. The former would cause the odor nuisance to the surrounding neighborhood whereas the latter would corrode the water pipe that was constructed using the basic concrete materials through acid-base reactions. This study is aimed to develop a sustainable and durable solution to tackle the long-lasting issues in both drainage and sewer systems. The proposed approach is with the application of malodor-control (MOC) hydrogel that encapsulates metabolic inhibitors to curtail both Sulfur-Oxidizing bacteria (SOB) and Sulfur-Reducing bacteria (SRB) and capsulize the concrete façade from the biogenic sulfuric acid.

Laboratory research study demonstrated that MOC hydrogel is capable of reducing both SRB and SOB by approximately 90% throughout the 25 days of the measurement along with the release profile test of the metabolic inhibitors. In the practical experimentation in one of the drainage channels, the result indicated a 90% curtailment of the hydrogen sulfide gas with 82% of the reduction in SRB population. For the corrosion prevention, two types of epoxy coating were employed in the study, which are the commercial epoxy coating and the epoxy coating that is modified with anti-microbial ingredients that equipped the epoxy coating with self-sanitizing capabilities. Utilization of the commercial epoxy coating inside the severely corroded sewer manhole was capable to curtail 50% (90 days) and 71% (180 days) of the corrosion rate compared with the uncoated concrete. On the other hand, the modified epoxy coating exhibits a higher curtailment rate, with 67% and 78% respectively.

117e Techno-Economic Optimization Scenario of Three-Stage Membrane Separation and CO_2 Liquefaction Process

Jung Moo Lee, Junwoo Park, Ki Bong Lee

Achieving carbon neutrality requires efficient treatment of greenhouse gases, particularly CO_2. Several technologies, including absorption, adsorption, and membrane separation, are available for CO_2 treatment. Among these, membrane separation has advantages such as low energy consumption, ease of operation, and scalability. This study utilizes advanced optimization techniques to perform techno-economic optimization on a three-stage membrane separation and liquefaction process. The performance of the membrane, including CO_2/N_2 selectivity and permeance, was assumed and analyzed, along with future carbon credit scenarios, to determine the economically viable process case.

Choosing The Best Decentralized Wastewater Treatment Systems to Be Applied In The Developing Countries

Mohamed Hazem[1], Ahmed S. Mahmoud[2], Robert W. Peters[3], and Mohamed K. Mostafa[4]

[1]Civil Engineering Department, Valley Higher Institute for Engineering & Technology, Al Qalyubia, Egypt
[2] Scientific Research Development Unit, Egyptian Russian University (ERU), Badr, Egypt
[3] Department of Civil, Construction, and Environmental Engineering, University of Alabama at Birmingham (UAB), Birmingham, AL
[4]Faculty of Engineering and Technology, Badr University in Cairo (BUC), Cairo, Egypt.

Abstract
As a result of population growth and urbanization, the traditional centralized wastewater treatment (CWWT) is unable to meet the requirements of wastewater flows, in addition to the high capital and operating costs that burden governments, so we discuss the various technologies of decentralized surface wastewater treatment (DWWT). The economic value of decentralized systems lies in their lower capital costs compared to centralized systems, because they require less construction, smaller and less complex networks, and they also require lower operating costs, because they use less energy, lower labor requirements, and are usually cheaper and fewer chemicals and materials to process. Some of them have already been implemented and are in operation in many countries. Studies showed that different methods achieved better removal efficiencies for some pollutants depending on the wastewater source and system design. The highest economic value was recorded for (the aeration tank system + sedimentation tank) and the lowest value was recorded for (the septic tank + anaerobic filter) system. In this paper, we discuss a novel wastewater treatment system using 100% ferric chloride as the coagulant was successfully developed and tested. The study also explores replacing traditional chlorine disinfection methods with a low-cost synthesis of Nano-Zero Valente Iron (nZVI) for tertiary wastewater treatment. The system was found to be more efficient and cost-effective than traditional wastewater treatment systems. The treated wastewater met the requirements of the Egyptian code for reuse in different agricultural applications.

1. Introduction
The increasing global population and rising water consumption are leading to a significant environmental challenge in urban areas, resulting in significant amounts of wastewater daily. This wastewater, if not treated properly, can pollute rivers, lakes, and oceans, posing a threat to aquatic life and human health, (Islam and Tanaka,2004, Liu and Diamond,2005, M. Abdeldayem et al.,2020). Therefore, it is essential to implement effective wastewater management systems that can treat and reuse this water resource efficiently. By adopting sustainable practices, such as water recycling and efficient water use, we can ensure a more sustainable future for our cities and protect our precious water resources (DEWATS) for resource recovery in rural and peri-urban areas. MDPI, Water, 2020). With 2.4 billion people living in cities and towns and an estimated 2 billion more people expected to gain access to sanitation facilities in developing countries (Daily and Ehrlich, Gleick, Lisburman, Mustafa, et al.), the volume of wastewater generated is expected to increase dramatically in the coming years. This poses a threat to both human health and the environment, as untreated wastewater can contaminate water sources, leading to the spread of diseases and the degradation of ecosystems (DEWATS): A review. Journal of Environmental Management, 2010). Therefore, it is crucial for governments and organizations to invest in effective wastewater management systems and technologies to ensure the safe and sustainable disposal or reuse of this valuable resource (Langergraeber and Müllegger,2005, Mostafa and Peters et al.,2015). This pollution not only affects aquatic ecosystems but also poses serious health risks to

humans who rely on these water sources for drinking and recreation. In addition, untreated wastewater can contain harmful pathogens and chemicals that can contaminate food crops and lead to the spread of diseases. By investing in wastewater management systems, governments, and organizations can mitigate these risks and ensure the long-term sustainability of water resources for both human and environmental health. To address this issue, effective wastewater treatment and management strategies are needed. Traditional plumbing systems are not suitable for industrial and developing countries, leading to the focus on ecological sanitation (EcoSan) methods, which offer sustainable closed-loop systems connecting health and agriculture (Flores et al, 2009, Morsy et al, 2020.). EcoSan methods involve the use of innovative technologies and practices that promote the safe and efficient reuse of wastewater. These methods include the use of constructed wetlands, bio filtration systems, and composting toilets, which treat and recycle wastewater for various purposes such as irrigation and fertilization. By implementing EcoSan methods, not only can we reduce the pollution and depletion of water resources, but we can also improve agricultural productivity and enhance food security in regions with limited access to clean water. Overall, adopting sustainable wastewater treatment and management strategies like EcoSan is crucial in ensuring the long-term availability and quality of water resources for the well-being of both humans and the environment. The EcoSan system aims to minimize labor-intensive processes involved in nutrient and water cycles, promoting a more sustainable approach to wastewater management (winblad et al.2004, Simha and Ganesapillai, 2017, Mustafa and Peters, 2016). Furthermore, EcoSan methods prioritize the reuse of nutrients and water, reducing the dependence on external inputs and minimizing the environmental impact caused by conventional wastewater treatment systems. By integrating health and agriculture, EcoSan systems can provide a source of safe and nutrient-rich fertilizers for crop production, promoting food security and sustainable agricultural practices. Additionally, these closed-loop systems can also contribute to the reduction of water scarcity and pollution, making them an attractive option for industrial and developing countries striving for sustainable development, (DEWATS) Water Science and Technology, 2008). The researcher developed a new wastewater treatment process that combines coagulation, nano adsorption, and disinfection. The coagulation process uses a ferric chloride to remove suspended solids, turbidity, and organic matter from wastewater. The nano adsorption process uses low-cost synthesized nZVI to remove bacteria and viruses from wastewater. The disinfection process uses UV radiation to kill any remaining microorganisms.

2. Materials and Methods

2.1. Chemicals and Reagents

The following Chemical substance were used in this study: Ferric chloride ($FeCl_3.6H2O$, Loba Chemie, Extra pure), Sodium hydroxide (NaOH, 98% pure, Oxford Co.), Extra soft black tea , Sulfuric acid (H_2SO_4, 97%, Fisher chemical), Ethanol (C_2H_6O, 98% pure, Chem lab), Zinc sulfate ($ZnSO_4$, 98.9, Winlab) and standard 4% sodium hypochlorite solution.

2.2. Samples collection

In May 2023, 40 composite water and wastewater samples were collected from a certain location at 30.12706, 31.70292 at Badr University in Cairo, Egypt. The samples were collected on two consecutive days and then transported to the chemistry lab at the Housing and Building National Research Centre (HBRC) in Egypt for analysis and treatment.

2.3. Preparation of nZVI

The process for extracting reducing agents from soft black tea being rich in polyphenols, and can serve as a source of reducing agents for the reduction of metal cation from soluble salt Ferric chloride ($FeCl_3.6H2O$) for synthesis of nZVI (nano Zero Valent Iron), (Mahmoud et al. 2019, Mahmoud et al. 2020).

2.4. Batch studies

The first experiment involved treating a wastewater sample by mixing it with five coagulants called $AlCl_3$,Al2$(SO_4)_3$, $FeCl_3$, $FeCl_3$/ H_2O_2 and $ZnSO_4$. The mixture underwent a coagulation process for 4 minutes with stirring at 160 rpm, followed by settling for an hour and 15 mins. After precipitation, the adsorption process was carried out using nZVI. The operating conditions for the adsorption process were determined based on previous studies. The batch experiments were conducted under different pH levels, doses, contact times, and stirring rates as showed in table (1). nZVI was used as the adsorbent, and its efficiency in removing contaminants was evaluated by measuring the remaining concentrations of COD, BOD, TSS, TN, and TP. The removal efficiency and the amount of sorbed contaminants were calculated using specific equations. The experiments were performed in triplicate, and the results were consistent with laboratory measurements (Mahmoud et al., 2020, Mahmoud and Mahmoud 2021, Rabie Farag and Peters, 2019).

$$\text{sorption } (\%) = \frac{C_0 - C_E}{C_0} \times 100$$

The equilibrium concentration in solution (Ce) is an important parameter in various chemical and biological processes. It represents the concentration at which the forward and reverse reactions occur at the same rate, resulting in a stable system. Ce is influenced by several factors, including the initial concentration (Co) of the reactant. By understanding the relationship between Co and Ce, we can gain insights into the behavior of the system and make informed decisions in various applications (mg/L).

$$Q_e(mg/g) = \frac{(\mathrm{C0} - \mathrm{Ce}) \times \mathrm{V}}{m}$$

In the formula, qe is the equilibrium adsorption capacity (mg/g), V is the volume of aqueous solution (L), m Dry weight of adsorbent (mg) (Mahmoud et al., 2020b, Mahmoud et al., 2019b).

Table 1. Summary of batch experiments at different materials and different operating parameters.

Effect	Dose (g)	Contact time (min)	PH	Stirring rate (rpm)	Temperature (^{0}C)
$AlCl_3$,$Al_2(SO_4)_3$, $FeCl_3$, $FeCl_3$/ H_2O_2 and $ZnSO_4$					
Material selection	Dose (1g/L)	5min mixing/25min flocs formation/50min settling	pH (7)	---	24
$FeCl_3$/ H_2O_2					
Effect of pH	0.4	5min mixing/25min flocs formation/50min settling	2-9	---	24
Effect of adsorbent dose	0.2-0.7	5min mixing/25min flocs formation/50min settling	6	-----	24
nZVI					
Effect of pH	0.6	50	7-10	260	24
Effect of adsorbent dose	0.2-0.9	50	9	260	24
Effect of contact time	0.6	15-90	9	260	24
Effect of stirring rate	0.6	50	9	150-450	24

3.Result and dissections

The research study focuses on developing a cost-effective and efficient wastewater treatment system using a combination of coagulation/precipitation and nano-disinfection techniques. The study explores the effectiveness of different coagulants and optimal operating conditions for the treatment process.

3.1. Characterization

The passage in figure (1) describes the characterization of black tea synthesized Nano zero-valent iron (nZVI) particles using X-ray diffraction (XRD), scanning electron microscopy (SEM), and energy-dispersive X-ray spectroscopy (EDAX). XRD analysis revealed that the nZVI particles have a predominantly alpha iron structure, with two peaks at $2\theta = 44.5°$ and $64.9°$ corresponding to the 110 and 200 planes. As shown in figure (1 A) is a low-magnification image that shows the overall morphology of the nZVI particles. The particles are irregularly shaped and have a size range of 30-80 nanometers. Figure (1B) is a higher-magnification image that shows the surface structure of the nZVI particles. The particles have a rough and porous surface. Figure (1C) is a higher-magnification image that shows the agglomeration of the nZVI particles. The particles tend to clump together, which can reduce their reactivity. Figure (1D) is a high-magnification image that shows the individual nZVI particles. The particles have a variety of shapes, including spherical, rod-shaped, and irregular shapes. This suggests that the nZVI particles are not pure iron, but rather a mixture of different iron compounds (Mahmoud et al., 2017).

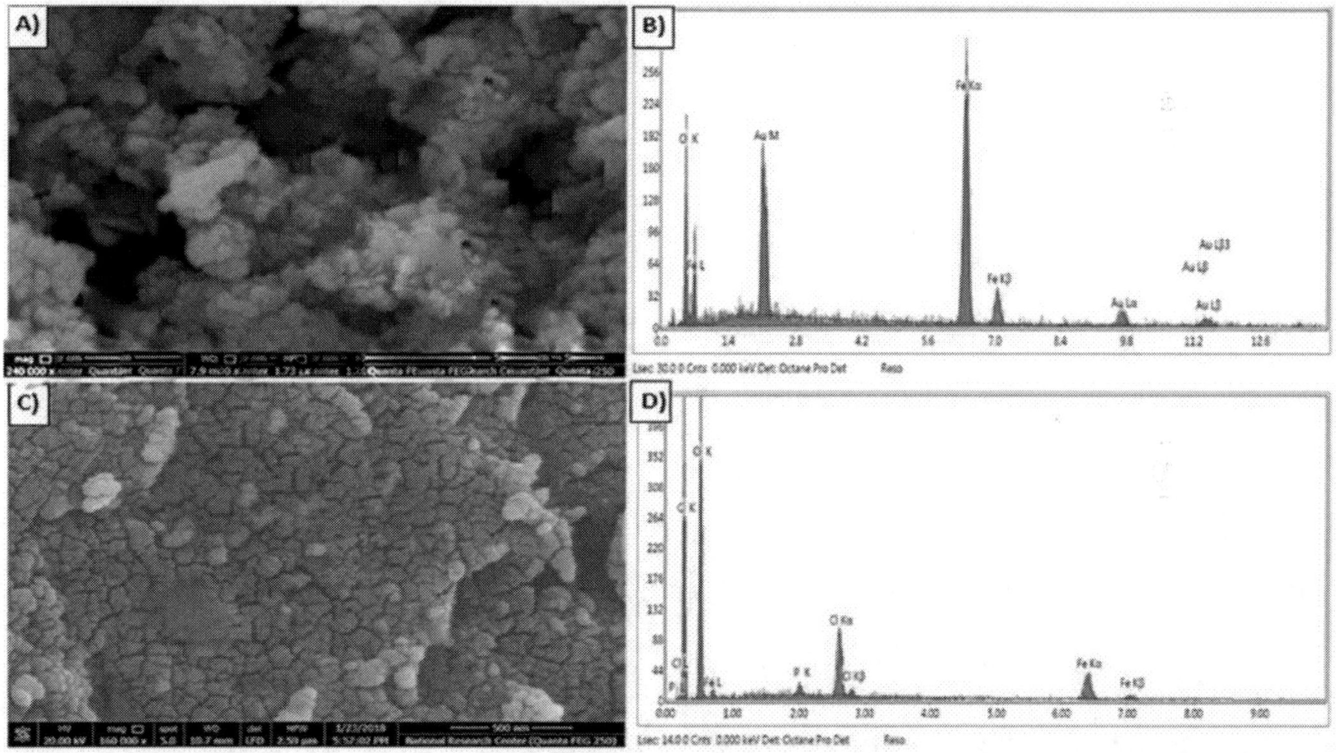

Figure 1: **(A)** SEM results of nZVI with sizes between 31 and 60 nm, **(B)** EDAX analysis of as-prepared nZVI, **(C)** SEM analysis of nZVI with sizes between 60 and 80 nm, and **(D)** EDAX analysis of prepared nZVI. EDX stands for energy dispersive X-ray; nZVI, green synthetic nano zero-valent iron; nZVI, nano-zero-valent iron; SEM, scanning electron microscope.

3.2. Selecting the coagulant

The use of FeCl3/H2O2 Fenton coagulant for wastewater treatment can significantly reduce the levels of COD, BOD, TSS, TN, TP, oils, and sulfides. As shown in Figure 2, the COD removal rate can be as high as

79.1% for the initial COD concentration of 200 mg/L. The BOD removal rate can be as high as 85.3% for the initial BOD concentration of 100 mg/L. TSS removal can also be significant, with removal rates of up to 97%. In addition, the FeCl3/H2O2 Fenton coagulant can also remove nitrogen and phosphorus compounds, with removal rates of up to 60% and 70%, respectively. The FeCl3/H2O2 Fenton coagulant is also effective at removing oils and sulfides from wastewater. Studies have shown that oil removal can be as high as 95%, while sulfide removal can be as high as 99%. Overall, the $FeCl_3/H_2O_2$ Fenton coagulant is a very effective wastewater treatment technology. It can remove a wide range of contaminants at a relatively low cost. It is also easy to implement and operate.

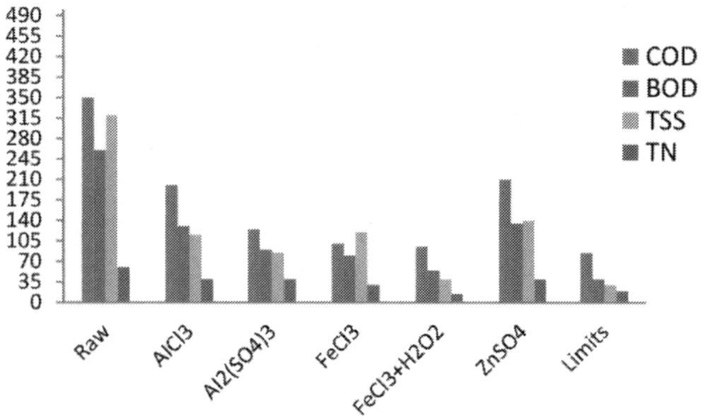

Figure 2: Effect of different coagulants on wastewater treatment Reduction of COD, BOD, TSS, and TN concentrationsconcentrations.

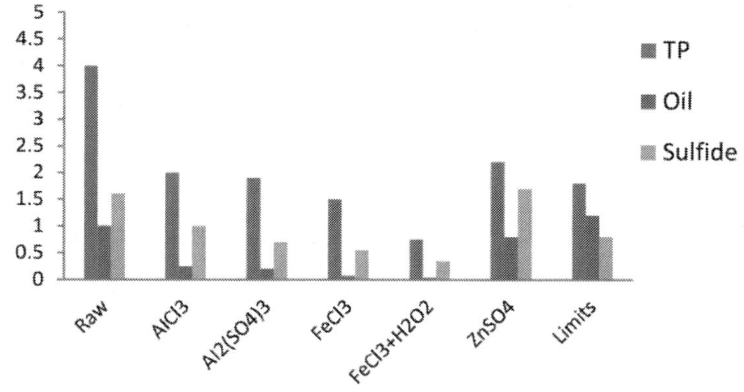

Figure 3: Effect of different coagulants on wastewater treatment Reduction Reduction of TP, oils, and total sulfide concentrations.

3.3. Selecting optimum conditions for the coagulation process

3.3.1. Effect of pH

The effect of pH on the coagulation process using $FeCl_3/H_2O_2$ Fenton coagulation for wastewater treatment was studied. The results showed that neutral pH (7-8) was best for degrading and precipitating a wide range of contaminants. However, the process also worked well in acidic and alkaline conditions for organic contaminants. The choice of pH depended on floc formation and removal efficiency. Other studies supported these findings, showing that pH 8 was optimal for orthophosphate removal, and slightly acidic pH (5) resulted in faster coagulation, (Syu and Chen, 1998). and alkaline pH was effective for TOC and turbidity removal. The optimal pH for coagulation depends on the specific contaminants and desired removal efficiency. Table 2 and Figure 4 show the wastewater treatment removal %, (Qasim et al., 1992).

Table 2: Effect of pH on wastewater treatment by using FeCl3/H2O2

pH	3	4	5	6	7	8	9	10
COD removal %	71	72	74	75	77	76	75	74
BOD removal %	75	76	75	77	78.0	78.0	79	76
TSS removal %	72	75	76	78	79	79	77	77
Fe removal %	88	89	89	90	91	91	91	91
Al removal %	100.0	100.0	100.0	100.0	100.0	100.0	100.0	100.0
TP removal %	60	62	64	68	71	70	68	66
TN removal %	50	53	52	52	57	56	56	55
Zn removal %	69	73	73	79.0	80	80	81	77
Phenol removal %	100.0	100.0	100.0	100.0	100.0	100.0	100.0	100.0

CN removal %	53	58	68	68	79	85	85	90.
Oil removal %	95	95	95	96	97	97.0	98	95.0
Sulfide removal %	39	38	41	44	53	53	50	45

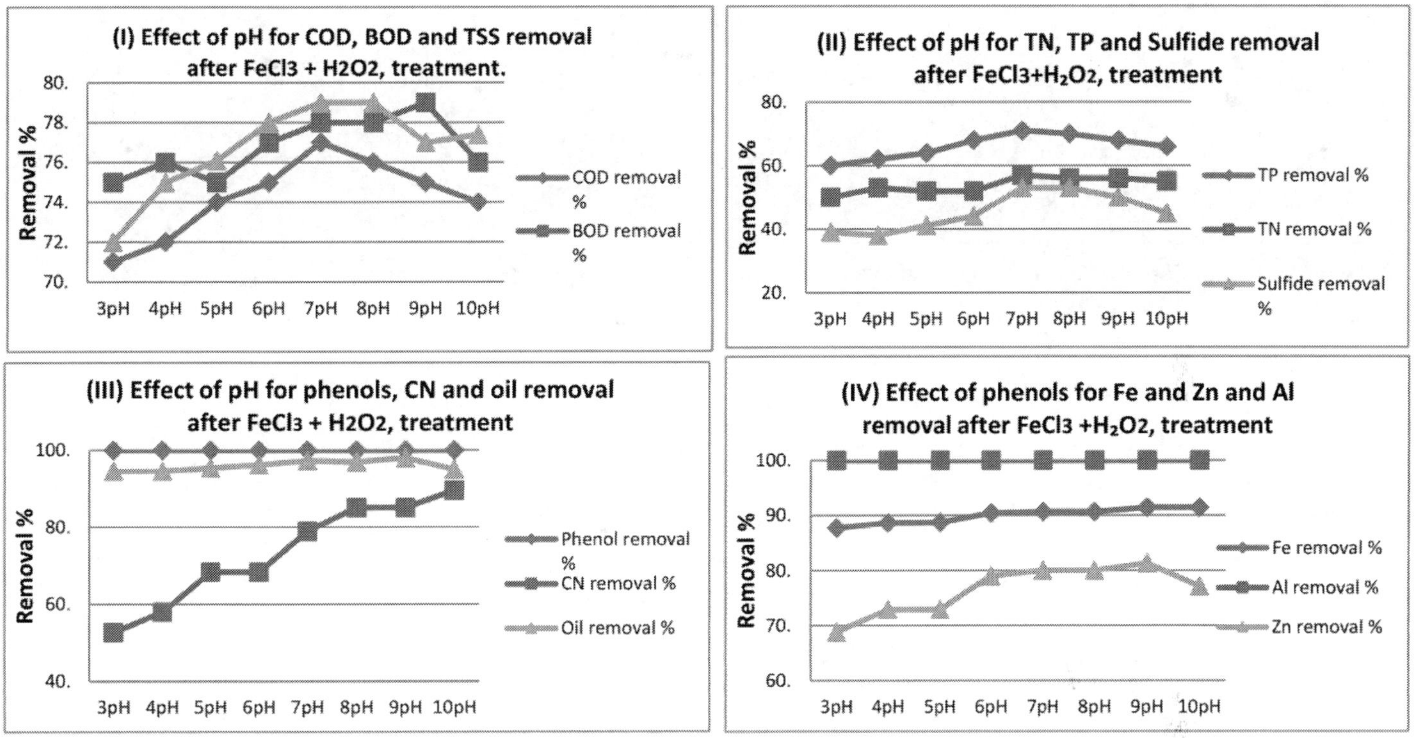

Figure 4. Effect of pH on wastewater treatment using FeCl3/FeCl2/H2O2 coagulation technology I) Reduce COD, BOD, and TSS concentrations, II) Reduce TP, TN, and total sulfide concentrations, III) Reduce phenols, cyanide, and oil concentrations, IV) Removal of iron and zinc concentrations.

3.3.2. Effect of coagulation dose

The study examined the impact of different doses of a mixed coagulant (FeCl3/H2O2) on the removal of various contaminants in university wastewater. The minimum effective dose was found to be 0.6g/L. The results showed that the coagulant was successful in removing carbonaceous organic contaminants (COD, BOD, TSS) with removal percentages of 90.60%, 91.60%, and 92.10% respectively, as shown in Figure 5. It also demonstrated acceptable removal rates for nitrogen (55%), phosphorous (80%), and sulfur (50%). Small concentrations of oil, CN, phenols, iron, and zinc were present in the wastewater but could be eliminated using the minimum coagulant dose. Overall, the study suggests that the mixed coagulant at a dose of 0.6g/L effectively removes a range of contaminants commonly found in university wastewater.

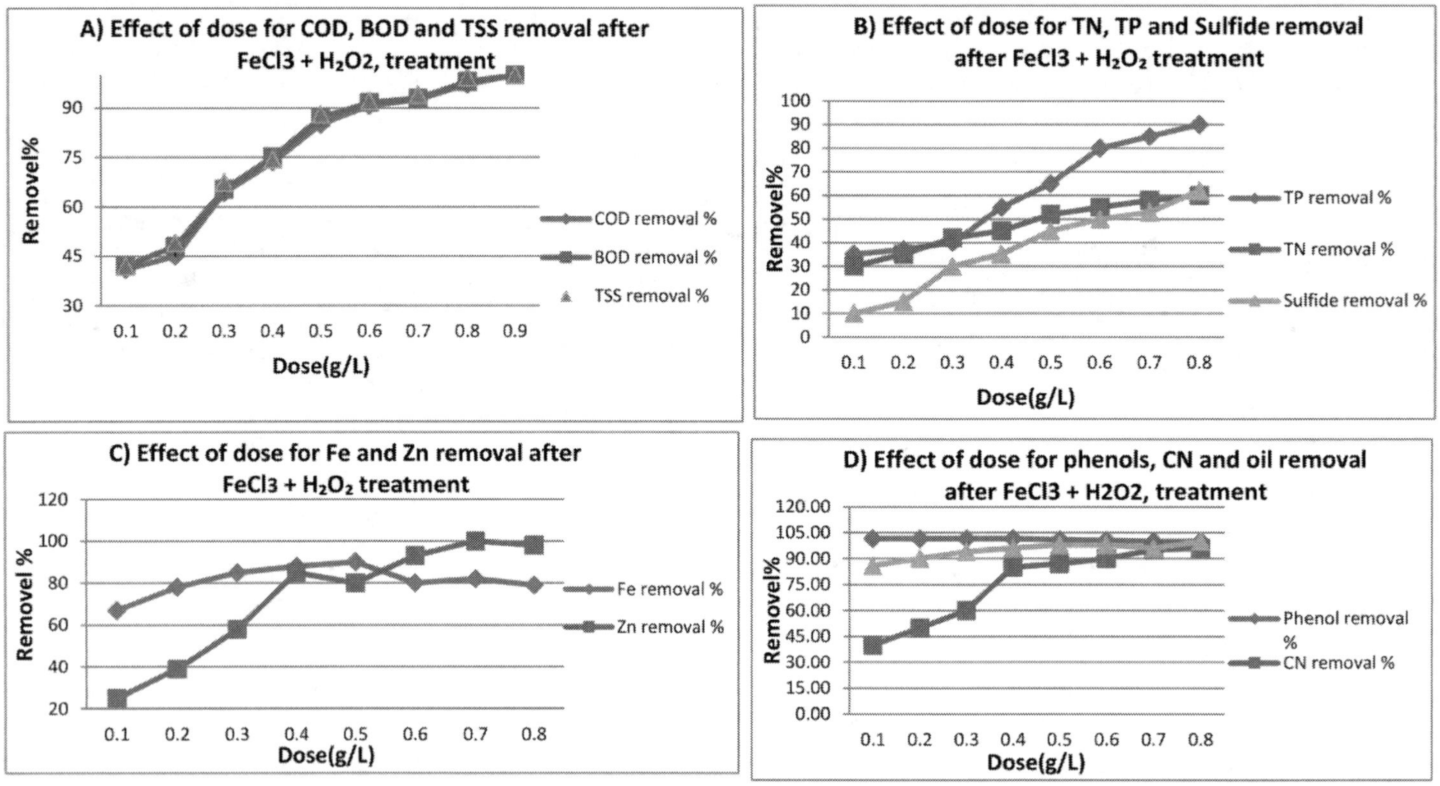

Figure 5. Effect of coagulant dosage on wastewater treatment A) reduction of COD, BOD and TSS concentrations, B) Reduce TP, TN and total sulfide concentrations, C) Effect of contact time, and D) Effect of stirring rate.

3.4. Selecting optimum operating conditions for the nZVI treatment process

nZVI is a promising technology for wastewater treatment. It is effective in removing a wide range of contaminants, including heavy metals, dyes, and organic contaminants (Mahmoud and Mahmoud, 2021, Mahmoud et al., 2021a, Rabie S. Farag and Peters, 2019. The optimal operating conditions for the nZVI treatment process are pH 10, dosage 4 g/L, contact time 40 minutes, and stirring rate 400 rpm. These conditions can be modified depending on the specific contaminants being treated and the characteristics of the wastewater. Because this is one of the most important factors that must be taken into consideration when determining the optimal conditions, it is the pH, dose, time, and stirring rate, (Maha M. Elshfai and Elsaid, 2021, Mahmoud et al., 2019a)., as shown in Figure 6.

3.4.1. Effect of pH

pH is important for removing organic contaminants from wastewater. The most effective pH is 8, and the removal percentage of COD, BOD, TSS, TN, and TP is 63%, 66%,66%,30%, and 59% at pH 8 in the same order, like figure 6 A.

3.4.2. Effect of dose

This paper showed that nZVI is an effective material for phosphate removal. The effective dose of nZVI is around 0.7 g/L. Farag et al., 2018 studied aqueous phosphate chemisorption onto nZVI and the obtained results indicated that the effective dose to decrease 65% from orthophosphate 9mg/L solution was 0.5g/L (Rabie S. Farag and Peters, 2018), like figure 6 B.

3.4.3. Effect of contact time

The minimum contact time required for the nZVI to reach equilibrium is 60 minutes. This is consistent with the findings of a previous study by (Mahmoud et al.2019) at different times namely at 10, 20, 30, 40, 50, 60, 70, and 80min as shown in Figure 6C.

3.4.4. Effect of stirring rate

GT-nZVI wastewater treatment requires at least 250 rpm agitation to break up nanoparticle clumps and increase the surface area for pollutant adsorption and degradation as shown in Figure 6D.

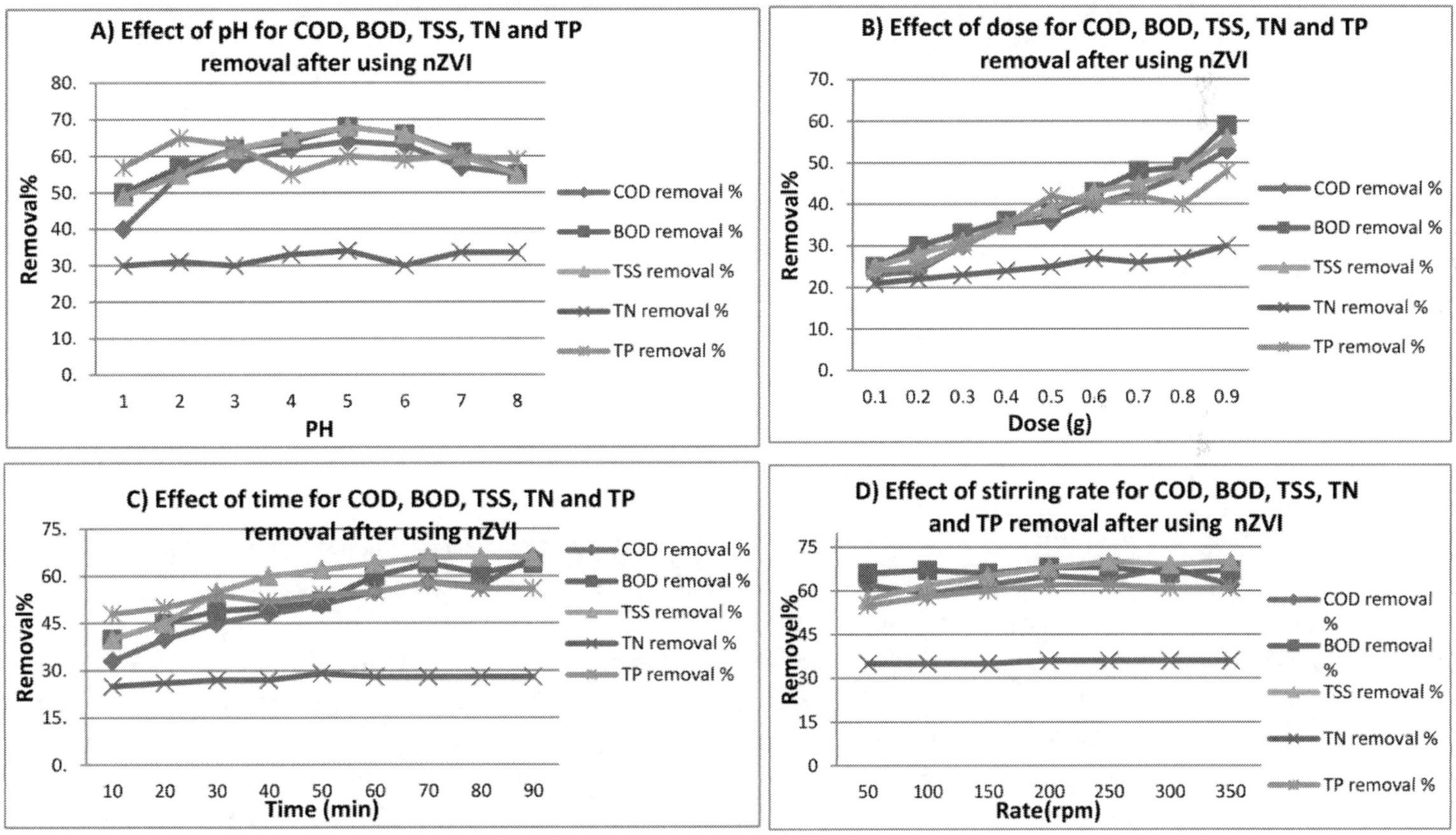

Figure 6. Effect of operating conditions on advanced wastewater treatment A) Effect of pH value, B) Effect of dosage, C) The effect of contact time and D) the effect of stirring speed.

4. Conclusion

In conclusion, the study focuses on efficient municipal wastewater treatment at a low cost through the utilization of coagulation/precipitation followed by nano-disinfection. The researchers conducted a comparison of various coagulants $AlCl_3$,$Al_2(SO_4)_3$, $FeCl_3$, $FeCl_3$/ H_2O_2 and ZnSO4 and determined that a combination of ferric chloride exhibited superior efficacy in treating domestic wastewater compared to other coagulants. The effect of coagulation pH and dose were studied and the obtained results indicated that 0.4 g of coagulant at pH 6 can degrade and precipitate more than 75% for carbonaceous organics, more than 65 and 75% from phosphorous and nitrogen organic loads, respectively, and more than 55% from the sulfur organic load. The advanced level of the treatment process was proceeded by adding 0.6 g/L of freshly prepared nZVI at pH 7, contact time 55 min, and stirring rate 260 rpm. Additionally, they successfully substituted conventional chlorine disinfection with a cost-effective synthesis of nano-Zero Valent Iron (nZVI) for tertiary wastewater treatment. The study meticulously optimized the operational parameters for both the coagulation process and the nZVI treatment process, encompassing pH, dosage, duration, and agitation rate. The findings demonstrated that the proposed system effectively eliminates diverse contaminants from wastewater, including COD, BOD, TSS, TP, TN, and total sulfide concentrations. Consequently, the study proposes that this system presents a viable and sustainable alternative to conventional wastewater treatment technologies, particularly in decentralized wastewater treatment units.

Acknowledgements

The authors gratefully acknowledge the financial support given by the Joint Academy of Scientific Research and Technology (ASRT)/the Bibliotheca Alexandrina (BA) Research Grants (Grant Number: 1469).

References

DAILY, G. C., & EHRLICH, P. R. (2000). On the relationship between economic development and human well-being. Proceedings of the National Academy of Sciences, 97(25), 12548-12553.

DEWATS , Water Science and Technology, 2008.

DEWATS, for resource recovery in rural and peri-urban areas. MDPI, Water, 2020.

FLORES, A. E., MARA, D. D., & SPELLMAN, F. R. (2009). Sustainable sanitation in developing countries. IWA Publishing.

GLEICK, P. H. (2009). Water in crisis: Science, policy, and management. Oxford University Press.

ISLAM ,M.A, &TANAKA,Y. (2004). Wastewater treatment and reuse for sustainable irrigation in developing countries: case studies. Journal of Environmental Engineering, 130(3), 354-361.

LANGERGRAEBER, G., & MULLEGGER, E. (2005). Ecological sanitation—a way to solve global sanitation problems? Environment, Development and Sustainability, 7(1), 273-285.

LISBURMAN, D., MUSTAFA, M. A., & PETERS, S. (2010). (DEWATS): A review. Journal of Environmental Management, 91(7), 2141-2149.

LIU, L., & DIAMOND, J. (2005). China's water crisis. Nature, 436(7047), 328-329.

M ABDELDAYEM, O., ELDAGHAR, O., K MOSTAFA, M., M HABASHY, M., HASSAN, A. A.,

MAHMOUD, H., M MORSY, K., ABDELRADY, A. & PETERS, R. W. 2020. Mitigation plan
and water harvesting of flashflood in arid rural communities using modelling approach: A case
study in Afouna village, Egypt. Water, 12, 2565.

M. ABDELDAYEM, AM El-SHAFAI, MS El-GENDY, MI ABDEL-AZIZ, MA Al-HAZMI, & SR MAHMOUD (2019). Wastewater treatment and reuse for irrigation in arid and semi-arid regions: Challenges and opportunities. Journal of the Saudi Society of Agricultural Sciences, 18(4), 487-499.

MAHMOUD, A. S., FARAG, R. S. & ELSHFAI, M. M. 2020a. Reduction of organic matter from municipal wastewater at low cost using green synthesis nano iron extracted from black tea: Artificial intelligence with regression analysis. Egyptian Journal of Petroleum, 29, 9-20.

MAHMOUD, A. S., SARYEL-DEEN, R. A., MOSTAFA, M. K. & PETERS, R. W. Artificial Intelligence for Organochlorine Pesticides Removal from Aqueous Solutions using Entrapped nZVI in Alginate Biopolymer. 2017 Annual AIChE Meeting; Minneapolis, MN, October 29 - November 3, 2017,2017.

MAHMOUD, M. & MAHMOUD, A. S. 2021. Wastewater treatment using nano bimetallic iron/copper, adsorption isotherm, kinetic studies, and artificial intelligence neural networks. Emergent Materials, 1-9.

MORSY, E.M., Eid, H.M., & El-TAWEEL, G.H. (2020). Ecological sanitation: A sustainable solution for wastewater management in rural areas. Environmental Technology, 41(17), 2202-2211.

MOSTAFA, M.A., & PETERS, S. (2016). Ecological sanitation: A sustainable solution for wastewater management in developing countries. In Sustainable Wastewater Management in Developing Countries (pp. 39-54). Springer.

PETERS, S. (2013). Wastewater treatment and reuse in developing countries: A review of literature and implementation issues. Frontiers in Environmental Science & Engineering, 1(1), 1-15.

QASIM, S., HASHSHAM, S. & ANSARI, N. 1992. TOC removal by coagulation and softening. *Journal of Environmental Engineering,* 118, 432-437.

RABIE S. FARAG, M. M. E., AHMED S. MAHMOUD, MOHAMED K. MOSTAFA, AHMED KARAM, & PETERS, R. W. Study the Degradation and Adsorption Processes of Organic Matters From Domestic Wastewater using Chemically Prepared and Green Synthesized Nano Zero-Valent Iron. 2019 Annual AIChE Meeting; Orlando, FL, November 10 - 15, 2019.

RABIE S. FARAG, M. M. E.-S., AHMED S. MAHMOUD, MOHAMED K. MOSTAFA, & PETERS, R. W. Green Synthesis of Nano Iron Carbide: Preparation, Characterization and Application for Removal of Phosphate from Aqueous Solutions. Proc. 2018 Annual AIChE Meeting, Pittsburgh, PA, (October 28 – November 2). 2018.

RIJSBERMAN, F. R. 2006. Water scarcity: fact or fiction? *Agricultural water management,* 80, 5-22.

SARYEL-DEEN, R. A., MAHMOUD, A. S., MAHMOUD, M., MOSTAFA, M. K. & PETERS, R. W. Adsorption and Kinetic Studies of using Entrapped Sewage Sludge Ash in the Removal of Chemical Oxygen Demand from Domestic Wastewater, with Artificial Intelligence Approach. 2017 Annual AIChE Meeting, 2017.

SIMHA, P. & GANESAPILLAI, M. 2017. Ecological Sanitation and nutrient recovery from human urine: How far have we come? A review. *Sustainable Environment Research,* 27, 107-116.

SIMHA, P., & GANESAPILLAI, M. (2017). Ecological sanitation: A sustainable approach to wastewater management in developing countries. In Sustainable Wastewater Management (pp. 39-54). Springer.

SYU, M.-J. & CHEN, B.-C. 1998. Back-propagation neural network adaptive control of a continuous wastewater treatment process. *Industrial & engineering chemistry research,* 37, 3625-3630.

WENG, C. N. A critical review of Malaysia's accomplishment on water resources management under agenda 21. Proceedings of International Conference on Environmental Management, Centre for Graduate Studies, National University of Malaysia (UKM), Bangi, 2002. 315-331.

WINBLAD, U., KENGNE-LPUGO, A., SIMHA, P., & CALVERT, P. (2004). Ecological sanitation (EcoSan): A new paradigm for water recycling and safe reuse. Water Science and Technology, 50(3), 263-271.

WINBLAD, U., SIMPSON-HÉBERT, M., CALVERT, P., MORGAN, P., ROSEMARIN, A., SAWYER, R., XIAO, J., RIDDERSTOLPE, P., ESREY, S. A. & GOUGH, J. 2004. *Ecological sanitation.*

Effect of Using Nanomaterial in Concrete Mixture Production

Mohamed Mostafa[1], Mahmoud A. Hekal[1], Abdelrahman M. Dyab[1], Mohamed I. Atwa[1], Ahmed Mahmoud[2], Ahmed Abodonya[1]

[1] *Faculty of Engineering and Technology, Badr University in Cairo (BUC), Cairo, Egypt*

[2] *Institute of Environmental studies, Arish University, North Sinai, Egypt*

Abstract

This study aims to investigate the effect of using green nanomaterial in the mechanical properties of concrete mixtures. Nano materials are used to improve the concrete mechanical properties, because it fills in the voids generated in concrete production. Green nanomaterials, such as nano iron oxide, are used in different proportions in the concrete mix. Both fresh and hardened concrete tests were performed, and results are compared to the control concrete mixture that contains zero nano material. The experimental plan tests control concrete mixture with grade 30 MPa. Slump cone test is performed for fresh concrete. Also, the concrete mechanical properties: compressive, and flexural strength at ages of 7 and 28 days are evaluated. The results showed an improvement in the compressive strength of cubes containing green nanomaterials. Results also showed that the efficiency of using green nanomaterials in concrete mixtures is similar to using other types of nanomaterials reported in literature such as, nano-clay. The main challenge facing using nanomaterials in the concrete industry is the expected increase in the cost. By using green nanomaterials, the cost is expected to reduce comparing with using other types of nanomaterials.

Keywords: Compressive strength, concrete mixture, flexural strength, nanomaterial

1. Introduction

Concrete has been used as a building material for decades. Ancient civilizations like the Mesopotamians and Ancient Egyptians are well known for using concrete in its earliest forms [1]. To build primitive structures, they combined straw, clay, and mud. However, the Roman Empire saw one of the most important developments in the usage of concrete. Roman concrete or "pozzolanic concrete" is a superior variety of concrete that the Romans invented [2]. They created a strong, long- lasting substance by combining volcanic ash, lime, and water. Aqueducts, bridges, and iconic buildings like the Pantheon and the Colosseum were all built with Roman concrete. Throughout history, concrete has been used extensively, with different civilizations and cultures perfecting its composition and methods. Concrete has become one of the most extensively utilized building materials in the world because of modern technological breakthroughs and increased awareness of the qualities of concrete.

Cement is the core binder in the concrete matrix. Cement is very harmful to the environment. According to Borenstein [3], cement pollution rates went from 1.4 tons in 2002 to 2.9 tons in 2022. Cement industry is responsible for 7% of global CO_2 emissions [3].

Improving concrete properties, such as both compressive and flexural strength, durability and workability, have both economic and environmental effects. In general, there are several types of additives used in the industry to improve concrete quality, such as Chemical additives, Mineral admixtures, Fiber reinforcement, Nanotechnology addition & Hybrid approaches. The nano-material is added to concrete as a replacement of cement. Figure 1 presents the different scales of materials used in concrete. It can be concluded from the literature review that the use of a single nano additive is better than using binary and tertiary additives, in terms of compressive strength and capillary permeability. It also stated that the combination of more than one additive is still to be studied more and needs more testing. It also showed that the best single additive is aluminum followed by iron then silica, however, the cost factor was not taken into consideration. Nano materials have many applications in the field of engineering. It currently used for wastewater treatment applications [4-14]. In this study, the mechanical properties of concrete mixtures will be tested after integrating the nanomaterial in the mixture.

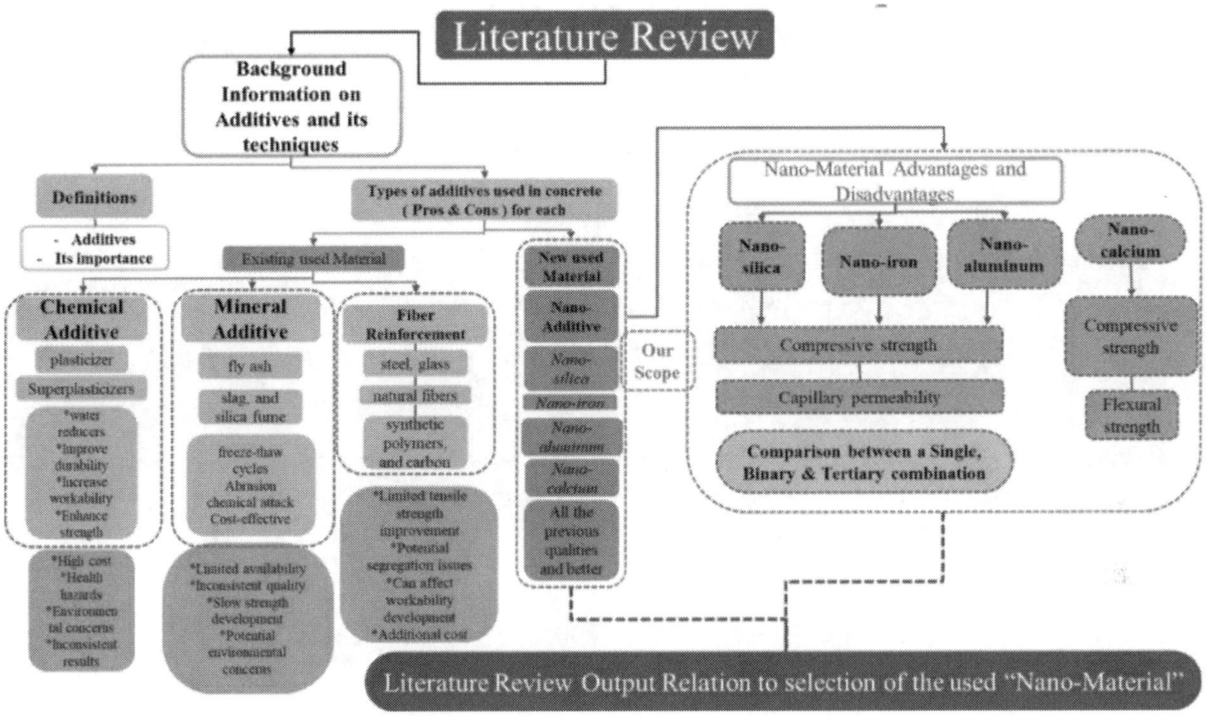

Figure 1 Summary on literature findings and paper scope

2. Material and Methods

Green nano-material is used in this study in different percentages as a replacement of cement to investigate the effect of adding nano-material to concrete. Figure 2 summarizes the testing methodology, samples preparation, and used nano-materials. The nano-materials are used in three percentages (0.5, 1, 2%) from the cement content used in the control mix. The nano-material used is namely nano-silica where slump cone test was used to determine the concrete workability. Moreover, hardened concrete tests such as compressive strength, and flexural strength are performed. All results are compared to a control mix that contains zero percent nano-material.

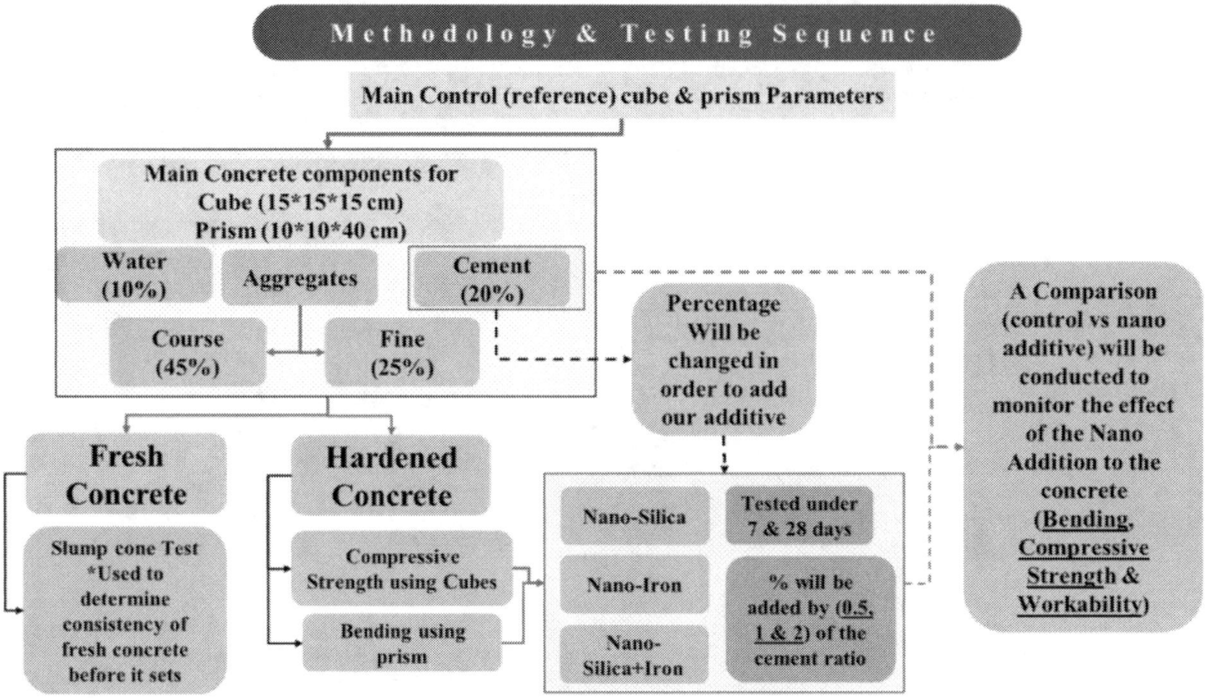

Figure 2 methodology and testing

3. Result and discussions

The tests were performed according to the American Society for Testing and Materials guidelines [15]. Figure 3 represents the results obtained for the concrete slump. It can be clearly seen that as the percentage of the nano-silica increases the concrete slump decreases. The reason for such a performance is that as the percentage of nano-silica increases, the cement hydration process is improved and nano-silica's surface area increases which absorbs mixing water in higher rates.

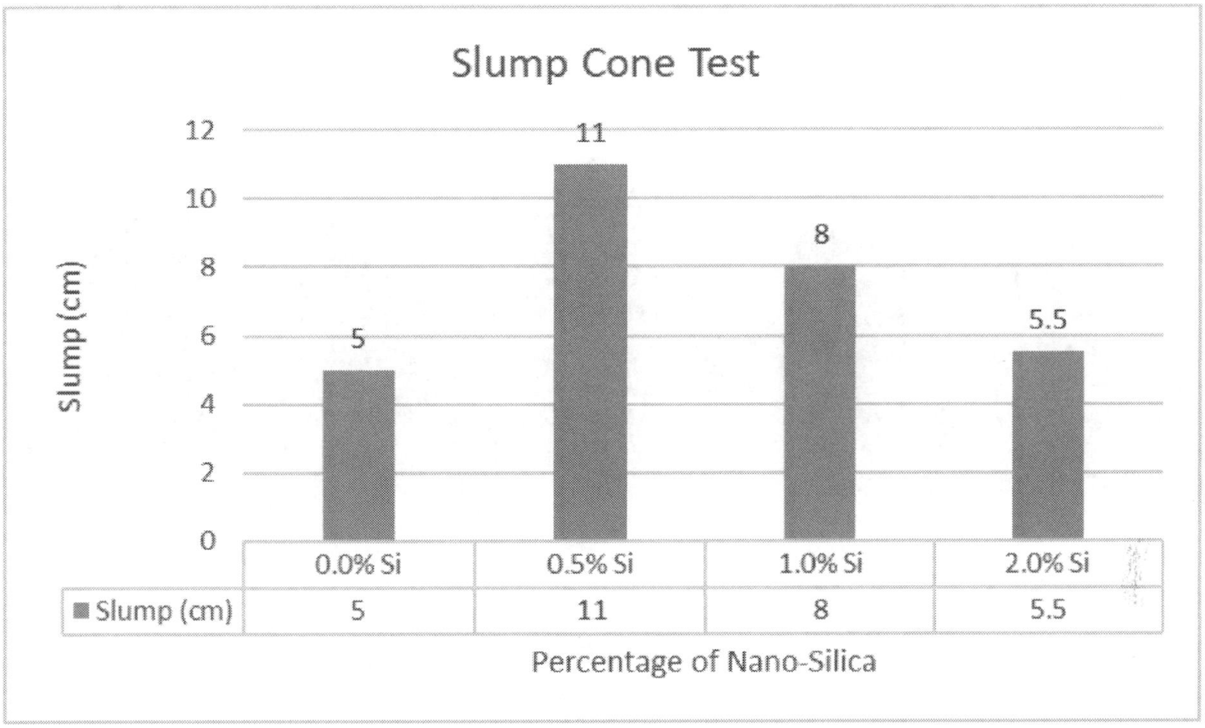

Figure 3 slump cone test

The hardened concrete is also tested. Both 7- and 28-days compressive strength test was performed and results are reported in Figure 4. The results show an increase in the compressive strength as the percentage of nano-silica increases where as the change in the 7-days compressive strength between the control specimen and specimens with 0.5 and 1.0% is negligible. An increase of 15%, 16%, 47% can be observed in the 28-days compressive strength for 0.5%, 1%, 2% replacement in nano-silica.

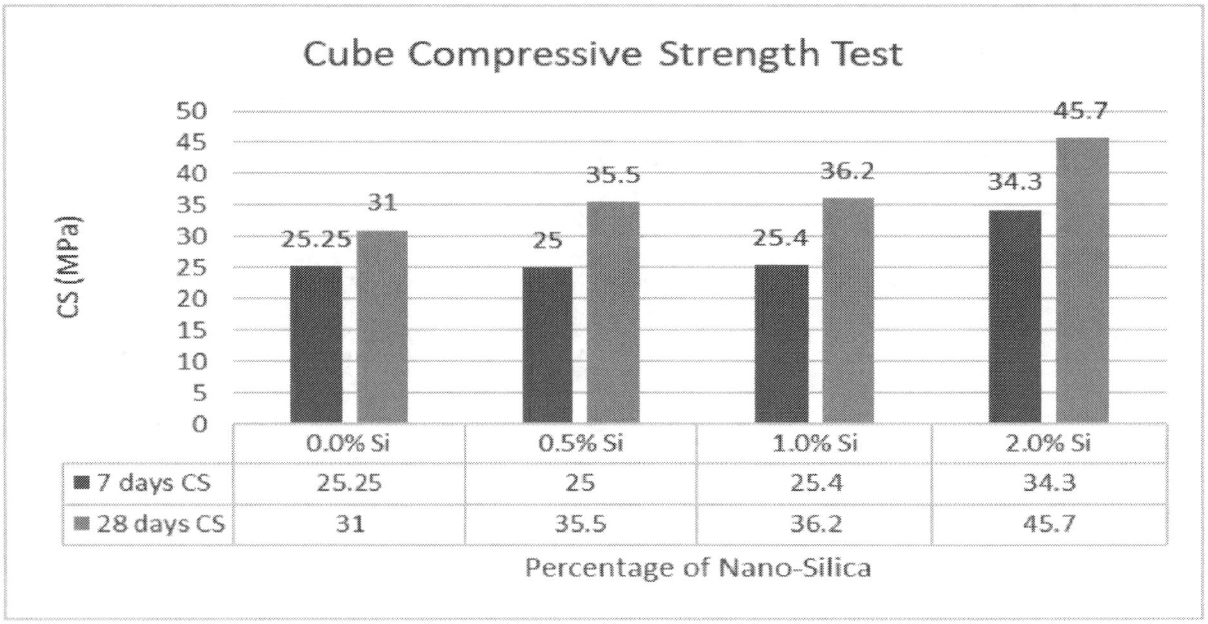

Figure 4 cube compressive strength for 7 and 28 days

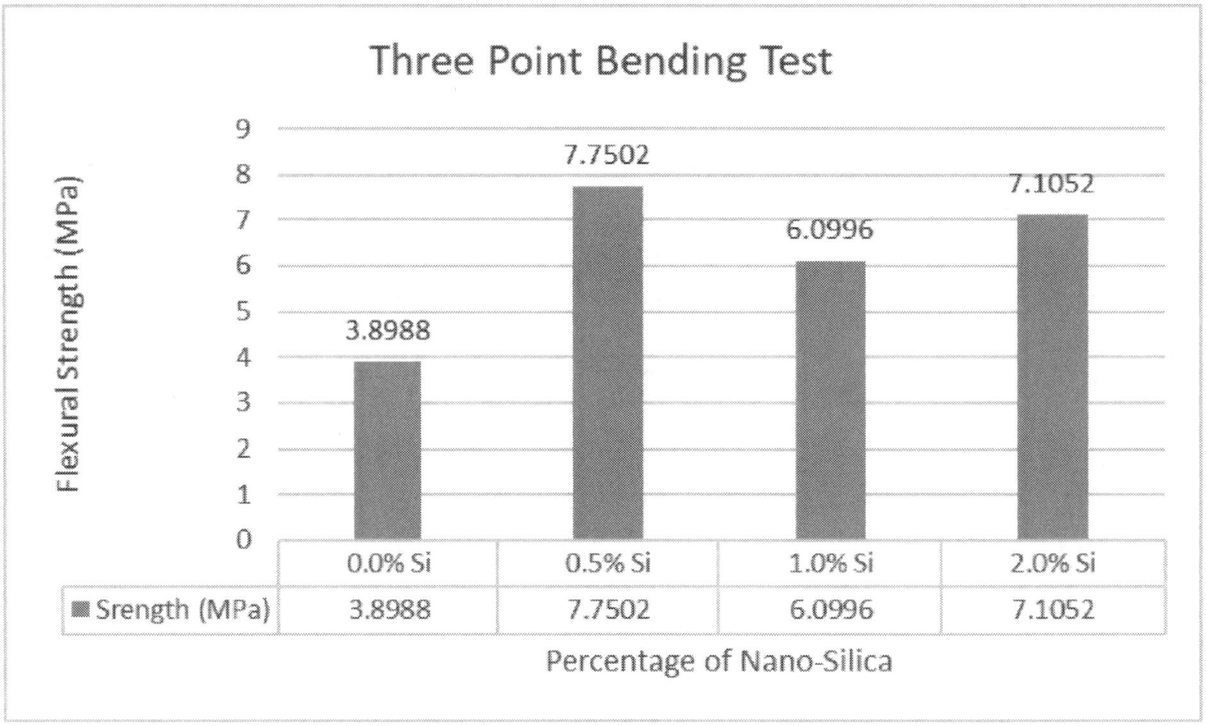

Figure 5 three-points bending test - flexural strength

The flexural strength was also measured for a prism using the three-point bending test. The results shown in Figure 5 show an increase in flexural strength between the control specimen and the specimens containing nano-silica. Meanwhile, there is no major change in flexural strength for different percentages of nano-silica.

4. Conclusions

This research studies the effect of using green nano-material in concrete. A control specimen was produced in order to compare its results with other mixes having nano-silica of (0.5, 1, 2%) replacement of the cement content. The slump cone test shows an increase in slump for 0.5% nano-silica compared to the control mixture and the slump is decreasing with increase percentage of nano-silica. The 7-days cube compressive strength test results did not show a major improvement in the results, while the 28-days compressive strength shows a drastic increase in strength around 33% increase for 2% nano-silica. Moreover, the flexural strength was evaluated using a three-point bending test and results show an increase of 80% of the control flexural strength. Meanwhile, there is not a noticeable improvement in flexural strength for different nano-silica percentages. Finally, the use of green nano-material shows promising improvement in concrete mechanical properties, and further tests are required to build more knowledge on the chemical properties of the concrete such as durability testing.

References

[1] Twelvetrees, W. (2007). Concrete and Reinforced Concrete. Nature. *Nature*.
[2] Sētiņa, J., Gabrene, A., & Juhņeviča, I. (2013). Effect of Pozzolanic Additives on Structure and Chemical Durability of Concrete. Procedia Engineering.
[3] Borenstein, S. (2022). Cement carbon dioxide emissions quietly double in 20 years. US: Ap News.
[4] El-Shafei, M.M. et al. (2016) Effects of entrapped nZVI in alginate polymer on BTEX removal, Environmental Division 2016 - Core Programming Area at the 2016 AIChE Annual Meeting. Available at: https://www.aiche.org/conferences/aiche-annual-meeting/2016/proceeding/paper/60b-effects-entrapped-nzvi-alginate-polymer-on-btex-removal (Accessed: 22 November 2023).
[5] Farag, R.S. et al. (2018) (592d) Green Synthesis of Nano Iron Carbide: Preparation, Characterization and Application for Removal of Phosphate from Aqueous Solutions, Annual AIChE Meeting Proceedings; Pittsburgh, PA, October 28 - November 2. Available at: https://www.aiche.org/conferences/aiche-annual-

meeting/2018/proceeding/paper/592d-green-synthesis-nano-iron-carbide-preparation-characterization-and-application-removal (Accessed: 22 November 2023).
[6] Farag, R.S. et al. (2019) (670d) Study the degradation and adsorption processes of organic matters from domestic wastewater using chemically prepared and green synthesized nano zero-valent iron, AIChE Annual Meeting, Conference Proceedings. Available at: https://www.aiche.org/conferences/aiche-annual-meeting/2019/proceeding/paper/670d-study-degradation-and-adsorption-processes-organic-matters-domestic-wastewater-using-chemically (Accessed: 22 November 2023).
[7] Heberling, J. et al. (2018) (12e) AOP Performance at Wastewater Treatment Plants | AIChE, AIChE Annual Meeting Proceedings, David L. Lawrence Convention Center, Pittsburgh, PA, October 28 - November 2. Available at: https://www.aiche.org/conferences/aiche-annual-meeting/2018/proceeding/paper/12e-aop-performance-wastewater-treatment-plants (Accessed: 22 November 2023).
[8] Kamar, M.T. et al. (2022) 'A critical review of state-of-the-art technologies for electroplating wastewater treatment', International Journal of Environmental Analytical Chemistry [Preprint]. Available at: https://doi.org/10.1080/03067319.2022.2098486.
[9] Mahmoud, A.S. et al. (2017) (655d) Artificial Intelligence for Organochlorine Pesticides Removal from Aqueous Solutions Using Entrapped Nzvi in Alginate Biopolymer | AIChE, AIChE Annual Meeting Proceedings. Available at: https://www.aiche.org/conferences/aiche-annual-meeting/2017/proceeding/paper/655d-artificial-intelligence-organochlorine-pesticides-removal-aqueous-solutions-using-entrapped (Accessed: 22 November 2023).
[10] Mahmoud, A.S. et al. (2021) 'Effective Chromium Adsorption From Aqueous Solutions and Tannery Wastewater Using Bimetallic Fe/Cu Nanoparticles: Response Surface Methodology and Artificial Neural Network', Air, Soil and Water Research, 14. Available at: https://doi.org/10.1177/11786221211028162/ASSET/IMAGES/LARGE/10.1177_11786221211028162-FIG7.JPEG.
[11] Mahmoud, A.S. et al. (2022) 'Advanced Wastewater Treatment by Using Entrapped nZVI into Alginate (Ag) Biopolymer: Adsorption Isotherm, Kinetic Models, and Statistical Analysis', Key Engineering Materials, 921, pp. 147–160. Available at: https://doi.org/10.4028/P-XYGKON.
[12] Mahmoud, A.S., Mostafa, M.K. and Peters, R.W. (2021) 'A prototype of textile wastewater treatment using coagulation and adsorption by Fe/Cu nanoparticles: Techno-economic and scaling-up studies', Nanomaterials and Nanotechnology, 11. Available at: https://doi.org/10.1177/18479804211041181/ASSET/IMAGES/LARGE/10.1177_18479804211041181-FIG10.JPEG.
[13] Mahmoud, M.S. et al. (2022) 'Comparison of aluminum and iron nanoparticles for chromium removal from aqueous solutions and tannery wastewater, empirical modeling and prediction', Emergent Materials, 5(6), pp. 1729–1744. Available at: https://doi.org/10.1007/S42247-021-00320-4/METRICS.
[14] Mostafa, M.K. et al. (2022) 'Computational-Based Approaches for Predicting Biochemical Oxygen Demand (BOD) Removal in Adsorption Process', Adsorption

Science and Technology, 2022. Available at: https://doi.org/10.1155/2022/9739915.
[15] ASTM. (2018, Januarary 1). Standard Specification for Concrete Made by Volumetric Batching and Continuous Mixing. Retrieved from ASTM: https://www.astm.org/c0685_c0685m-17.html

117h A Systematic Study of Hexavalent Chromium Adsorption and Removal from Aqueous Environments Using Chemically Functionalized Amorphous and Mesoporous Silica Nanoparticles

Eun-Hye Jang, Seung-Pil Pack, Il Kim, Sungwook Chung

Silica-based materials have been an essential class of adsorbents widely employed due to their large surface areas and excellent chemical, thermal, and mechanical stabilities. Amorphous silica nanoparticles (ASNs) are one of the common nanomaterials being used in industrial, manufacturing, and biomedical applications due to their favorable physic-chemical properties, facile synthesis, and relatively easy surface modification.

Herein, we present the results of work aimed at preparing two types of bare and two types of 15 – 60 nm diameter NH_2-functionalized silica nanoparticles of similar sizes, that is, ASNs with an amorphous solid core and mesoporous silica nanoparticles (MSNs) with a nanoporous core. We demonstrate the adsorption of hexavalent chromium (Cr(VI)) ions on amino-functionalized ASNs (NH_2–ASNs) and MSNs (NH_2–MSNs) and their removal from aqueous environments. We show that the specific surface area (SSA) of NH_2–MSNs is four times as larger as that of NH_2–ASNs and that more than 70% of the total SSA of NH_2–MSNs is due to the presence of nanopores. Analyses of Cr(VI) adsorption kinetics on NH_2–ASNs and NH_2–MSNs exhibited relatively rapid adsorption behavior following pseudo-second order kinetics as determined by nonlinear fitting. NH_2–ASNs and NH_2–MSNs exhibited significantly higher Cr(VI) adsorption capacities of 34.0 and 42.2 $mg \cdot g^{-1}$ and removal efficiencies of 61.9 and 76.8% than those of unfunctionalized ASNs and MSNs, respectively. Their kinetics, temperature-dependent isotherms, and thermodynamics of Cr(VI) adsorption were systematically analyzed.

Our findings are significant because they suggest a structural and chemical rationale for enhanced Cr(VI) adsorption, removal, and conversion to relatively non-toxic Cr(III). Understanding the detailed mechanism of Cr(VI) to Cr(III) conversion represents a meaningful step toward the goal of removing toxic Cr(VI) from wastewater environments.

[1] Jang et al. *Sci. Rep.* **10**, 5558 (2020).

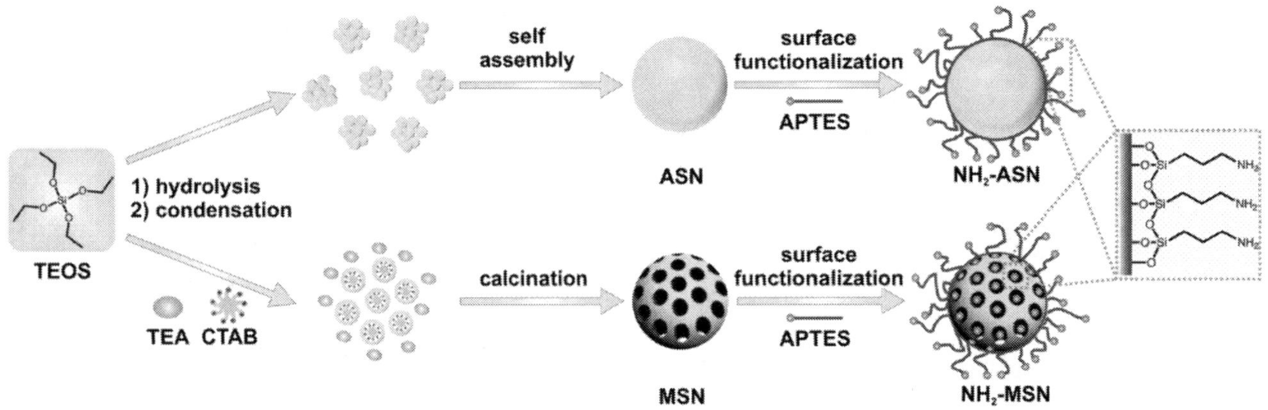

117i Sustainable Carbon Capture and Utilization Process Evaluation from Economic, and Environmental Perspectives

Ha-Jun Yoon, Tesfalem Atsbha, Taeksang Yoon, Dong-Kyoung Shin, Jihong An, Mohammadamin Zarei, Ali Cherif, Chul-Jin Lee

There is much research on carbon capture and utilization (CCU). Promising CCU products for many CCU reactions are being analyzed and evaluated. However, in most cases, CCU products are ranked by only market size or reaction stoichiometry. For example, A. Otto et al (2015) established special selection criteria to identify and rank CO_2 conversion reactions in the literature based on economic and environmental relevance. Based on their investigation, 23 bulk chemicals were ranked quantitatively. However, their selection criteria didn't consider process design.

In this study, process design for promising CCU processes was conducted. Process modeling is performed for eleven bulk chemical products that have more than 0.5 million tonnes per annum (MTPA) market size. Those are calcium formate, formic acid, formaldehyde, dimethyl carbonate, dimethyl ether, urea, ethylene oxide, acetic acid, methyl acrylic acid, and oxalic acid. Techno-economic analysis (TEA) and Life cycle assessment (LCA) for each process was computed. Here, this study proposes a novel indicator, called Sustainable Feasibility Index (SFI). SFI is an indicator which integrates market size, levelized cost of product (LCP), and global warming impact (GWI). Based on SFI, ranking and quantification of emerging CCU processes have been done. Finally, the case study was conducted to discuss the reduction of CO_2 through the application of CCU from three different perspectives: CO_2, H_2, and electricity source point of view.

From the investigated CCU processes, methanol, acetic acid, DME, formic acid, and calcium formate were identified as the most feasible CCU processes based on SFI indicator. Also, the case study shows that switching CO_2 sources has the critical impact on both LCP and GWI perspectives.

117j Development of 0.1 Ton/Day Plastic Pyrolysis-Combustion System of Circulated Fluidized Bed Reactor

Hyungseok Nam, Byung Wook Hwang, Daewook Kim, Yujin Choi, Doyeon Lee, Ho-Jung Ryu

With the modernization of industry and continuous global population growth, the demand for plastic consumption has been steadily increasing due to its light weight, ease of use, low cost, easy manufacturing, corrosive resistance, and structural characteristics. The huge amount of plastic usage is a serious issue, polluting environment and negatively accelerating GHG emissions. A chemical recycling of plastic waste is one good way to resolve the issue by converting wastes into reusable chemicals such as liquid fuel and naphtha chemicals over pyrolysis reaction. The major reactor of most commercial scale pyrolysis is kiln or screw type process that requires external energy from pyrolysis gas and oil or electricity by transferring heat from gas to solid waste (low heat transfer efficiency). In this study, a combined pyrolysis-combustion conceptual circulated fluidized bed system is proposed. Studies with four different experiment setups are designed and planned for the proof-of-concept. Once the concept is proof, over 0.1 ton/day scale system is scheduled to be developed in the following year (2022-2023). The goals for this project are, (1) auto-thermal process of pyrolysis (no solid char are being discharged and no external energy is supported for pyrolysis), (2) compact circulated fluidized bed system by applying inner loopseal system instead of outer one, (3) over 40% naphtha range chemicals, produced in the produced pyrolysis oil, and (4) CO_2 capture ready system. To achieve four goals, three experimental setups are developed and one experimental is planned to be constructed based on the results from the other setups. First, a hydrodynamic study is being conducted to investigate the controllable ranges of solid circulation rate, which is important factor to maintain the desired temperature of pyrolysis in bubbling fluidization chamber and combustion in turbulent chamber. Another important experiment with the hydrodynamic study is the gas mixing ratios between two reaction chambers. Second, a bubbling fluidized bed reactor was ready to investigate the plastic pyrolysis performance including the effect of waste feeding rate, residence time, temperature, fluidizing gas velocity, fluidizing bed material and plastic type. Third, the development of the heterogeneous catalyst for pyrolysis fuel reforming or upgrading is going on to find a proper catalyst to increase the light chemicals by breaking pyrolysis fuel wax. Last, a circulated fluidized bed system with inner-side loopseal is going to be designed based on the experiment results obtained with the three experimental setups. Then the system is built to evaluate the temperature balance between combustion of solid char and pyrolysis to verify the energy transfer rate simulation. This study suggested that plastic pyrolysis can be conducted using compact circulated fluidizing reactor to increase process efficiency, to maximize the liquid fuel product yield and not to discharge any solid black carbon.

117k Plastic Pyrolysis Oil in Fluidized Bed and Fixed Bed Reactors

Yujin Choi, Shuang Wang, Byung Wook Hwang, Hyungseok Nam, Ho-Jung Ryu, Doyeon Lee

In response to the increasingly prominent double crisis of the environment (especially the white pollution of plastic waste) and energy, in this study a sustainable development strategy for converting plastic waste into energy through pyrolysis technology has been proposed. Furthermore, in order to improve the working efficiency as well as the yield and quality of liquid oil, in the present work, a fluidized bed was designed for the pyrolysis of three plastic wastes (PP, LDPE, and ABS) for the first time. A traditional fixed bed was used as comparison. Although reactor type does not significantly affect the yield of liquid oil (PP: 61~62%, LDPE: 82~87%, ABS: 82~85%), it can greatly affect its quality. Compared with fixed reactor, fluidized reactor can transfer more middle oil (C_{11}-C_{22}) (and heavy oil (> C_{23})) to light oil (C_5-C_{10}) for ABS (LO: ~61%) and PP (LO: ~26%), due to its better heat transfer efficiency. However, for LDPE, due to its unsuitable pyrolysis temperature (520 °C used in this work), heavy oil was the main fraction (47~67%) regardless of the reactor type. Plastic oils over fluidized reactor showed a high similarity with diesel, especially ABS (ABS: 92% and PP: 69%), with several important properties such as HHV: 45~46 MJ/kg, viscosity: 1.12~2.65 cSt@40 °C and TAN: 0.0~1.5 $mg_{KOH}/g_{oi.}$ In addition, according to the naphtha standard (BP < 200°C), the contents of ABS and PP oil are 62% and 33% respectively. Furthermore, GC-MS analysis showed that ABS fluidized bed oil mainly contained aromatics (~71%), aromatic-N (~13%) and olefins (~8%), and LDPE fluidized bed oil mainly contained olefins (~66%) and paraffins (~30%), PP fluidized bed oil mainly contains olefins (~41%), paraffins (~21%) and a large amount of oxygenated compounds (e.g. esters, alcohols and ketones). Another thing worth noting is that although oxygen content of 0~2.5% exists in plastic waste, it converted into other oxygenated compounds instead of acids (eg. esters, alcohols and ketones) with a help of fluidized reactor, resulting in very low TAN value. Finally, the current study suggested that the fluidized bed has great potential for the pyrolysis application of plastic waste.

1171 *Chemical Sector Risk Management Agency Resources for Enhanced Resilience*

Ashley Pennington, Cheryl Louck, Janine Mason, Jacob Mehl

The US Department of Homeland Security's Cybersecurity and Infrastructure Security Agency (CISA) functions as the Sector Risk Management Agency (SRMA) for the chemical sector. The Chemical sector is one of the 16 critical infrastructure sectors(1) whose assets, systems, and networks, whether physical or virtual, are considered so vital to the United States that their incapacitation or destruction would have a debilitating effect on security, national economic security, national public health or safety, or any combination thereof.

To maintain environmental resilience safely and securely, it is imperative for the critical infrastructure sectors to function soundly and securely at full capability. Across CISA, teams work to support functions of the Chemical SRMA to ensure safe, secure, resilient, and sustainable functioning of the chemical sector to protect the homeland and associated environment. The impacts of the chemical sector influences the function of all critical infrastructure sectors. The team will present an overview of the Chemical SRMA and its role in capacity building, training, publications, supply chain, public-private partnerships, voluntary programs, and no-cost resources available to the chemical sector and within the four industry segments (basic chemicals, specialty chemicals, agricultural chemicals, and consumer products). The Chemical SRMA focuses on coordination and collaboration across the sector and with cross-sector critical infrastructure partners, both at the headquarters level and at the individual site level through the CISA regional offices. Among the resources related to supply chain discussed will be an overview of the 2022 Effective All-Hazards Chemical Sector Response Playbook(2) and webinar series which provides a Standard Operating Procedure (SOP) to assist the Chemical Sector in preparing for, responding to, and recovering from an all-hazards emergency; as well as information pertinent to the supply chain of water treatment chemicals. The team will also discuss how to contact your regional office for interaction and resources, an overview of the Homeland Security Information Network – Critical Infrastructure (HSIN-CI) Chemical Portal information, attendance at classified and unclassified briefings, and how to receive pertinent Chemical SRMA updates will be covered.

- *(1) Critical Infrastructure Sectors | CISA ; (2) Chemical Sector Playbook | CISA*

117m Reuse of Zeolites for Sorption of Pfas

Charles Ponge, Nathaniel Sheehan, Clarice Sabolay, David R. Corbin, Mark B. Shiflett

Per- and Polyfluoroalkyl substances(PFAS) widely produced since the 1940s have a broad range of uses in consumer and industrial products. The persistence of PFAS in the environment, and the associated health risks, have called for monitoring and remediation of PFAS in drinking water in the US and around the world. While traditional sorbent materials such as activated carbon and ion exchange resins are being used to remove PFAS from municipal drinking water, the materials are not easily regenerated in situ. Zeolites have been identified as sorbent materials with the potential to be regenerated and demonstrate selectivity for sorbing PFAS. Zeolites have a wide range of pore openings and tunable properties, making them ideal for removing both short- and long-chain PFAS from the environment. Currently, there is minimal research investigating zeolites for sorption of PFAS and in-situ regeneration. In this poster, several zeolites are reviewed for their ability to adsorb PFAS from contaminated water samples, after multiple uses. An understanding of the adsorption mechanisms and potential interferences must be elucidated for the widespread application of adsorbent materials in a broad range of environmental matrices. Due to the widespread contamination of PFAS at low concentrations, specialized analytical techniques, with Triple Quad-Mass Spec and material considerations, had to be used to get consistent and accurate measurements at the part-per-billion level.

117n Poster: Thin-Film Nanocomposite Membranes Functionalized with Fe-Based Metal-Organic Frameworks for Enhanced Removal of Small and Uncharged Urea from Water

Tin Le, Milad Esfahani

Urea is widely used in fertilizer production for agricultural purposes, which risks runoff into soil and water sources. Leaching of urea into water resources poses a concern due to algal bloom formation which can kill local wildlife and cause environmental eutrophication. Thin-film composite (TFC) membranes are among the best-performing membranes for water filtration; however, they do not show acceptable rejection for small and uncharged molecules such as urea. To combat this issue, adsorptive membranes combine the power of membrane filtration with adsorbent materials to provide enhanced selective removal, specifically for small and uncharged molecules. Metal-organic frameworks (MOF) are a new generation of effective adsorbents with excellent tunability in structure and compatibility with polymeric membranes. In this study, we synthesized MIL-100 (Fe), an iron-based MOF, using a mild room temperature method and tested for urea adsorption. The urea adsorption capacity of MIL-100 (Fe) was examined under varying experimental conditions such as pH, temperature, MOF concentration, and urea concentration. The results revealed the superfast adsorption (reached almost 90% removal within 2 min) of neutrally-charged urea molecules with a high adsorption capacity of 3321 mg/g and a BET surface area of 1174 m^2/g. To fabricate the adsorptive membranes, the MIL-100 (Fe) was grafted on the surface of the polyamide layer of thin-film composite membranes via covalent binding for enhanced removal of urea from water. The MIL-100 (Fe)-TFC membrane showed enhanced permeability (14.10 vs 0.87 LMH/bar) and no significant decrease (47% vs 41%) in Mg salt rejection, indicating successful grafting without damaging the PA layer. In addition, although both bare TFC and MIL-100(Fe)@TFC showed similar rejection of 63% of urea, the MIL-100(Fe)@TFC adsorbed a portion of the urea compared to the bare TFC, which rejected it. The quantification of adsorption versus rejection is currently under investigation.

117o TEMPO-Oxidized Cellulose Nanofibril Aerogels for Carbon Capture

Xakin Ramirez Isunza, Yufei Nan, Maria Soledad Peresin, Stephen Martin

The IPCC estimates that global surface temperature in the first two decades of the 21st century was approximately 1 °C higher than the 1850-1900 average, with well-mixed greenhouse gases as the main driver of atmospheric warming. Carbon dioxide (CO_2) is the most abundant greenhouse gas with an annual average atmospheric concentration of 410 ppm, and it is estimated to be responsible for approximately two thirds of the total warming effect of greenhouse gases. CO_2 capture from point sources, such as power or manufacturing plants, could reduce CO_2 emissions when combined with direct air capture (DAC) and storage technologies.

There has been a growing interest in the use of solid CO_2 sorbents for carbon capture due to their potential advantages over liquid adsorbents, such as a reduced energy for regeneration and greater sorption capacity, selectivity, and ease of handling. Commonly investigated sorbents include zeolites, metal-organic frameworks, and amine-functionalized sorbents. Cellulose nanofibril (CNF) aerogels present a sustainable option for CO_2 capture systems.

We have synthesized a series of TEMPO-oxidized cellulose nanofibril (TOCNF) aerogels, which were then functionalized with graphene oxide and polyethyleneimine as sustainable sorbents for carbon capture. Each of the aerogels was characterized by Brunauer-Emmett-Teller (BET) analysis, Scanning Electron Microscopy (SEM), and thermogravimetric analysis (TGA). The pure CO_2 sorption capacity was determined by gravimetric analysis and the working CO_2 capacity was obtained from breakthrough measurements in a packed-bed column.

ELEVATED TROPOSPHERIC OZONE IMPACTS ON SOYBEAN PRODUCTION IN THE UNITED STATES FROM 1985 TO 2015

Sharmin Akter[1], Caitlyn Cushman[2], Anne Keary[2], Thomas Pauly[2], Kristina Wagstrom[2]
[1] University of Connecticut, Storrs, CT 06269-3037
[2] University of Connecticut, Storrs, CT, 06269-3222

Introduction

Tropospheric or ground-level ozone (O_3) is one of the harmful air pollutants formed during photochemical reactions from precursor compounds nitrogen oxides (NOx) and volatile organic compounds [1][2]. While tropospheric ozone is most often associated with its impact on human health, it can also damage agricultural ecosystems. Specifically, elevated ozone levels can decrease crop yields and cause economic losses of important crops such as wheat, soybeans, and rice [3]. It is crucial to quantify the potential impacts of ozone on crop yield to ensure current and future food security for the increasing global population [3][2]. However, direct measurements of ozone impacts are challenging to obtain because variable factors such as precipitation, temperature, solar radiation, and differences in crop management practices also strongly impact crop productivity. In addition, the sparsity of ozone monitoring stations in rural areas also makes ozone impacts on crop production difficult to quantify [4]. The estimated lost revenue due to ozone-related crop loss in the United States was $11-18 billion in 2000 [5]. Due to increased emissions of ozone precursors, tropospheric ozone is projected to increase 40-60% by the year 2100 [6][7]. This is a serious threat to global crop production in the upcoming decades [6][8].

Ozone can cause visible injury, reduction of crop yield quantity and quality, alterations to carbon allocations, and decrease in photosynthesis of the crops [8][7]. Increased ozone can decrease plant growth and stomatal conductance [9][7]. Elevated ozone also increases the respiration rates of plants [10][9].

Soybeans are the 4th most important seed crop globally [11] and are very sensitive to elevated ozone levels [12][13]. Soybean seedlings are particularly vulnerable to elevated ozone levels early in the ozone season [3]. Soybeans are a staple crop and a potential protein source for animals and humans [1]; they are of high economic importance in the United States, Brazil, China, Argentina, and India [2]. Various influential abiotic factors such as air pollution, drought, and global warming all pose challenges for future soybean production [1].

Based on data from the year 2000, elevated ozone over agricultural regions is estimated to reduce crop yield 6-16% and result in an economic loss of $2.0-5.8 billion annually [2]. According to the estimates of the Intergovernmental Panel on Climate Change Special Report on Ozone Emissions Scenarios (IPCC SRES) for 2030, the upper and lower boundary of future global soybean yield loss will be 15-19% and 9.5-15%, respectively [12].

The objective of this study is to estimate the soybean yield and economic loss from 1985 to 2015 in the United States due to elevated tropospheric ozone concentrations throughout each of the soybean-producing states.

Methods

We used estimates of national, county-level soybean production for the years 1985- 2015 from the United States Department of Agriculture's (USDA) National Agricultural Statistics Service (NASS) [14]. We used ozone concentration estimates developed by the Center for Air, Climate and Energy Solutions (CACES) using v1 empirical models as described in Kim et al. [15]. Specifically, we used the annual averaged, ground-level ozone estimates covering the contiguous United States from 1985-2015 (the average during May through September of the daily maximum 8-hour moving average) at the county level. These estimates are developed using satellite measurements of ozone in conjunction with land use regression (LUR) modeling. The land use regression models consider information about land use, such as locations of major and minor roads, elevation, and whether an area is urban or rural, and satellite-based estimates of air pollution to predict concentrations at locations without any measurements [15].

We omitted all counties that yielded no soybeans according to the USDA NASS. We used an estimated pre-industrial ozone level of 15 ppb based on a review of prior estimates [16]. We defined the ideal soybean yield (SY_{ideal}) as the soybean yield expected at pre-industrial ozone levels using the equation

$$SY_{ideal} = -0.8535\left([O_3]_{pre-industrial} - [O_3]_{actual}\right) + SY_{actual}$$

where SY_{ideal} is the estimated ideal soybean yield (bushels/acre) based on what would be expected under pre-industrial ozone conditions, SY_{actual} is the actual county-level soybean yield (bushels/acre) reported by USDA for each year [14], $[O_3]_{pre-industrial}$ is the pre-industrial ozone concentration (15 ppb) [16], and $[O_3]_{actual}$ is the actual ozone concentration for each county and year [15]. This approach is a modified version of the equation developed by Fishman et al. [4]. Fishman et al. [4] collected soybean crop yield data for five years over the Midwestern United States in combination with ground-level and satellite ozone measurements. Fishman et al. [4] provided estimates of soybean yield losses on the order of 10% by using a multiple linear regression model. We calculated the ideal soybean yield for each soybean-producing county in the United States and each year from 1985-2015.

We estimated the national soybean loss by using the following equation:

$$NSL = \sum_{County} (SY_{actual} - SY_{ideal})$$

where *NSL* is National Soybean Loss (bushels/acre). We estimated the annual lost soybean bushels for each state from 1985 to 2015 by similarly summing over just the counties in each state.

To assess and model the economic impacts of ozone on soybean yields, we evaluated crop loss and economic loss using the following equations:

$$Bushels\ Lost\ per\ Year = \sum_{County} Bushels\ Lost \times \frac{Acres\ Planted}{County}$$

$$Dollars\ Lost\ per\ Year = \frac{Bushels\ Lost}{Year} \times (Market\ Year\ Cost\ of\ Soybeans)$$

Results and Discussion

Ozone exposures happen in the Northern Hemisphere due to higher ozone precursors emissions [17]. Figure 1 shows the ozone concentrations in the soybean-producing counties in the United States during 2015. All counties experienced ozone levels exceeding the pre-industrial estimates of 15ppb. Many significantly exceeded the levels at which crop yield reductions are likely. Specifically, Nebraska, Kansas, Ohio, Pennsylvania, and Illinois experienced the highest ozone concentrations in 2015.

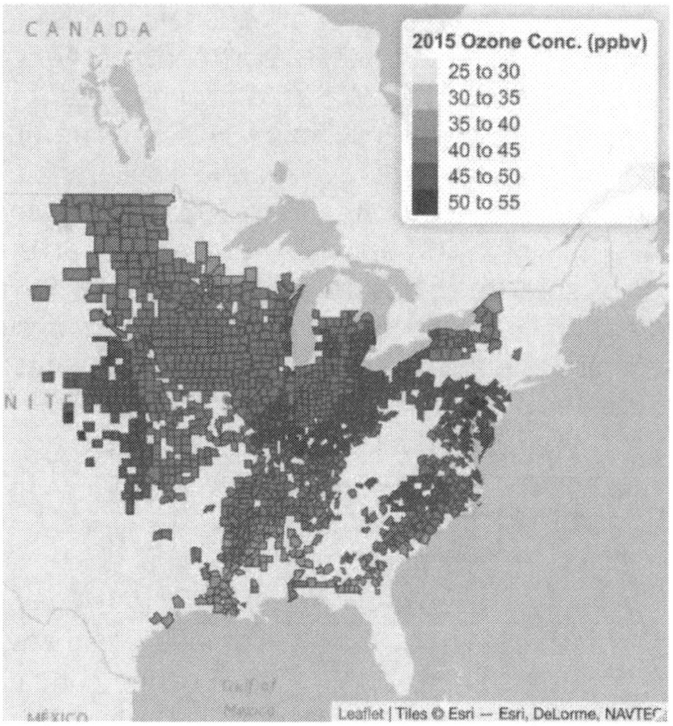

Figure 1. Ozone concentrations in ppbv on the county-level in all counties that grow soybeans within the United States in 2015.

Figure 2 shows the total millions of bushels of soybeans lost from 1985 to 2015 for each soybean-producing county. The biggest impacts are in the counties nearest to the urban areas, such as Chicago. This is the result of the transport of ozone and precursors from these urban areas to more rural areas. The larger losses are also associated with counties with large base ozone production rates. Some counties in the largest producing states (Illinois, Minnesota, Arkansas, Tennessee, and Mississippi) experienced estimated lost

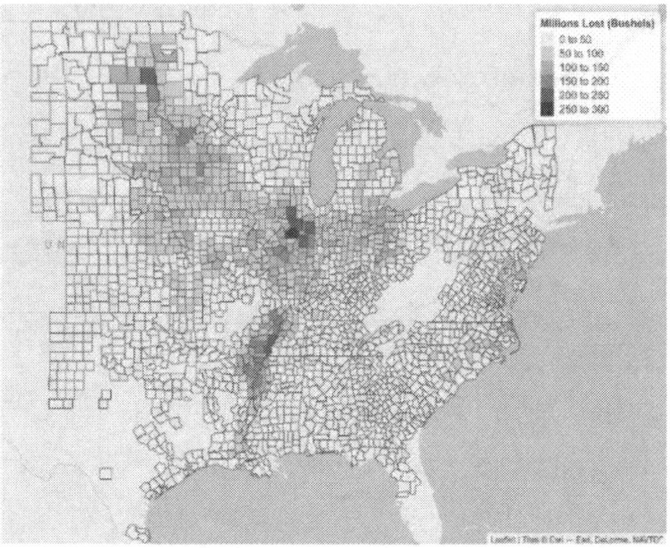

Figure 2. Total millions of bushels of soybeans lost from 1985 through 2015

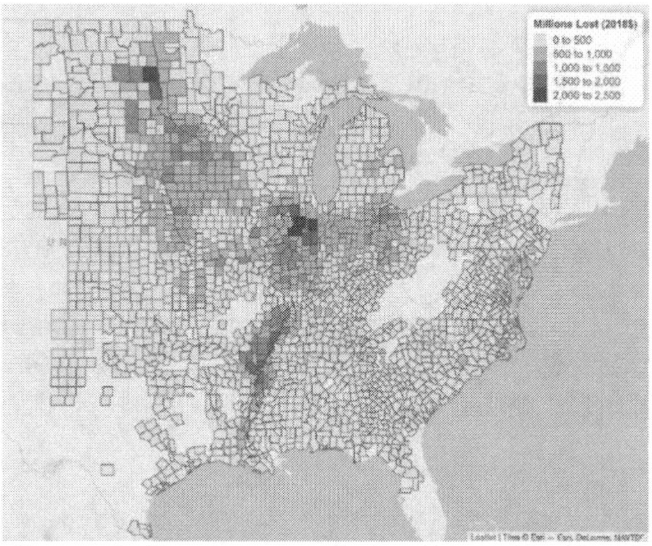

Figure 3. Total economic loss associated with soybeans lost from 1985 through 2015 (based on price of soybeans during 2018)

production exceeding 200 million bushels of soybeans. Several counties in Ohio lost between 50 and 200 bushels during those years.

Figure 3 shows total economic loss due to soybean loss from 1985 to 2015 for each soybean-producing county. We calculated the losses based on the market price of soybeans in 2018. Individual counties in Illinois, Minnesota, Arkansas, and Mississippi all experienced over $1.5 billion in economic losses over the time period.

Figure 4 represents billions of bushels of lost soybeans from 1985 to 2015 for each state in the US. We estimated the largest losses in South Dakota, Ohio, Nebraska, Missouri, Iowa, Indiana, and Illinois. We see the largest change in impacts in North Dakota between 1985 and 2015. Many of these changes may also result from an increase in actual soybean production because more areas in those states were growing soybeans by the 2010s when compared to the 1980s. We estimated the largest losses, more than 2 billion bushels, in 1988, 1998-2003, 2005, 2006, and 2012.

Iowa, Indiana, and Illinois are the states with the highest soybean production in the United States [4]. In our analysis, the loss of soybean bushels is also higher in these states just due to the higher production in those states. Ozone concentration has more than doubled compared to a century ago in the mid-latitudes of the Northern Hemisphere [11].

The time of exposure is also important to determine the damage to soybeans due to ozone, especially during the reproductive stages of growth. Pollutants enter the leaf via stomatal pores and interact with cell membranes, which yield reactive oxygen species. These can lead to tissue damage to soybeans and cell death [18].

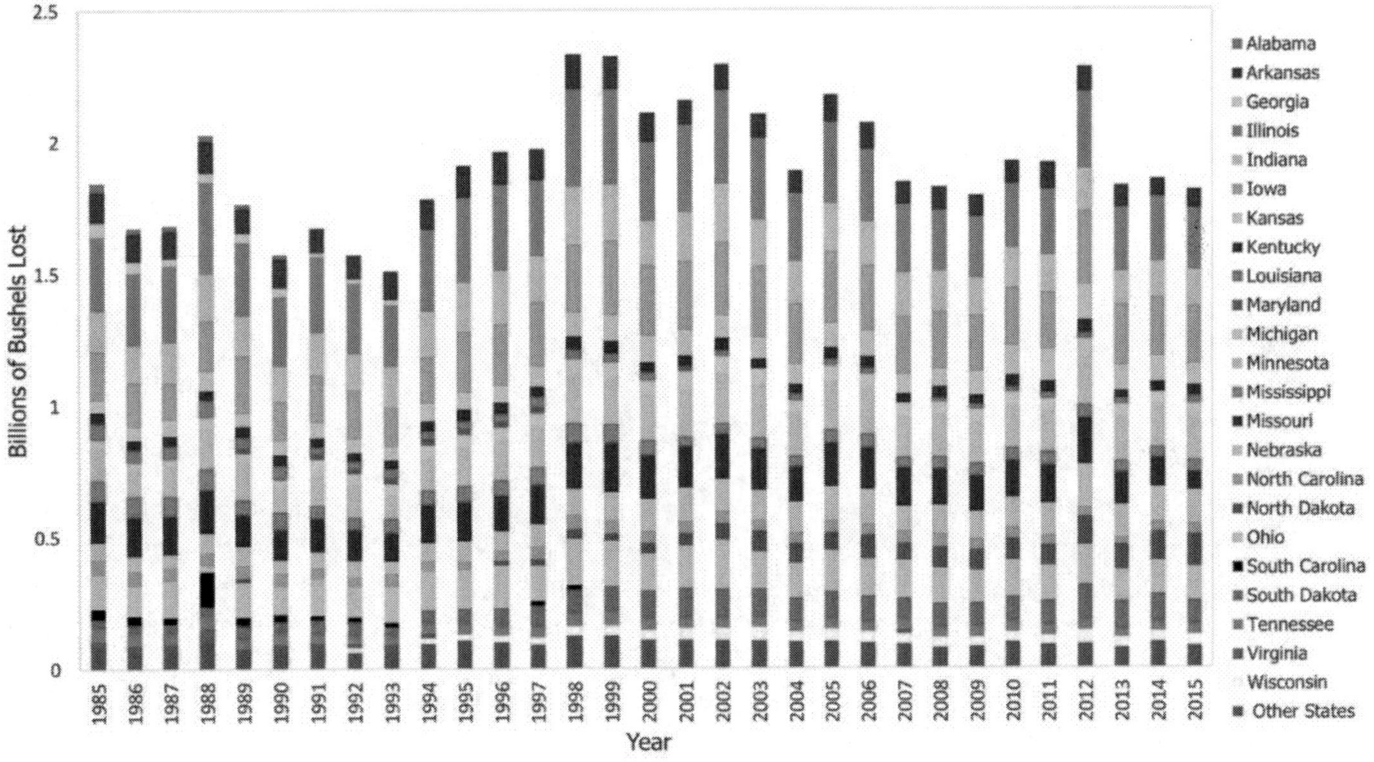

Figure 4. Annual estimates of billions of bushels lost in each state from 1985 through 2015.

Figure 5 represents the total billions of dollars loss of soybean during 1985 to 2015 years due to ozone impacts based on a 2018 market price of soybeans. In both 1988 and 2012, the estimated economic loss exceeded $30 billion. Hot, dry and stagnant air conditions are favorable for ozone formation. During 1988, meteorological conditions were favorable for ozone production [19]. Warm, sunny days with stagnant air are also conducive to ozone formation, in 2012 ozone concentrations were particularly elevated due to above average temperature and below average humidity conditions [20].

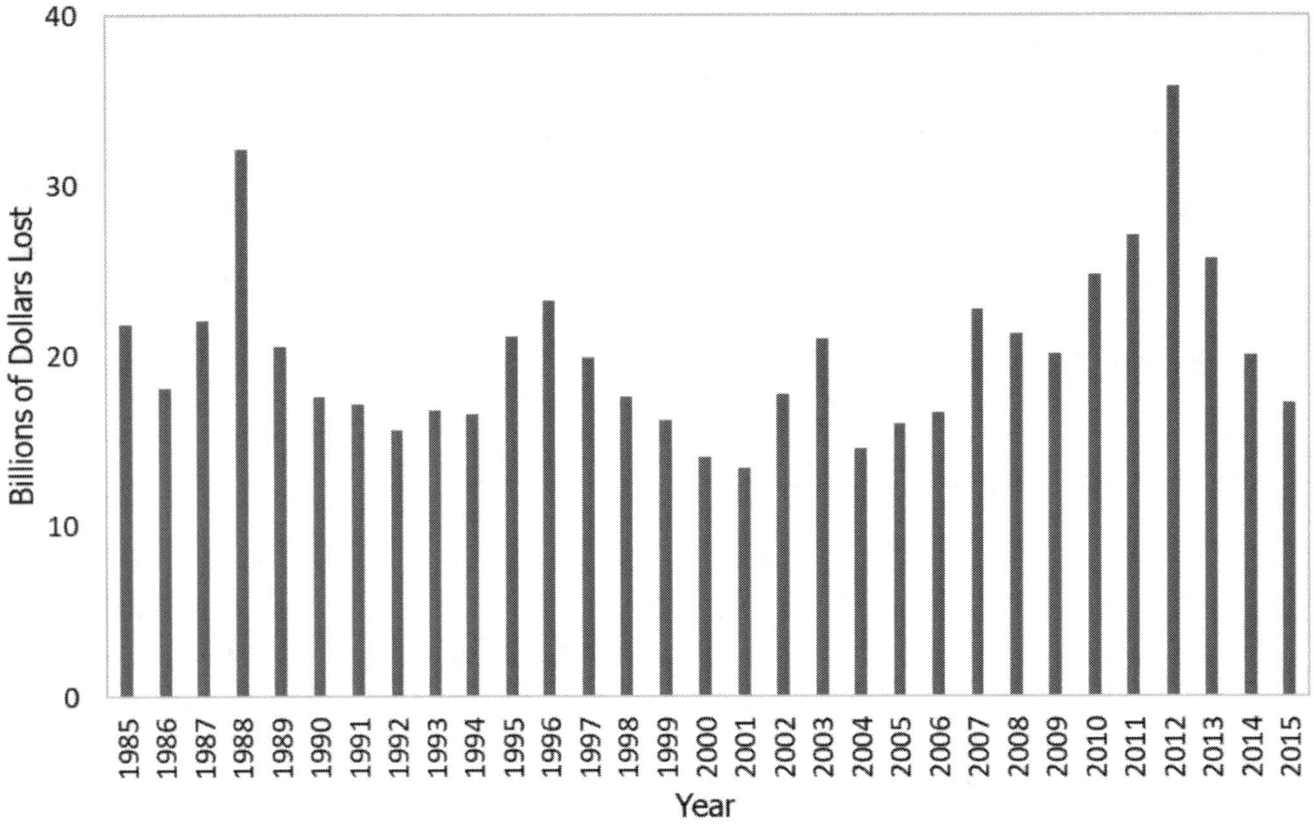

Figure 5. The total billions of dollars lost in the US for the years 1985 through 2015.

Conclusions

This study quantitatively estimated the impact of elevated ozone concentrations on soybean crop and economic loss in the United States from 1985 to 2015. The midwestern states (e.g. Illinoi, Iowa, Indiana, South Dakota, Nebraska, and Minnesota) experienced the highest soybean and economic loss due to elevated ozone concentrations during that studied period because that is where the majority of the soybean-production occurs in the United States. We estimate that over $35 billion were lost in 2012 alone due to soybean crop yield reductions from elevated ozone exposure.

Implementing mitigation measures can reduce agricultural ozone concentration exposures. However, in cases with limited volatile organic compounds, NOx reduction can actually increase ozone concentrations. Additional understanding and estimates of ozone impacts on soybeans is still needed including current and future research from global perspectives [18]. Soybeans grown in other global regions may demonstrate higher sensitivity to elevated ozone concentrations which would impact concentration response function used here [5]. In addition, the estimates here used maximum daily 8-hour averages, rather than daily averages. This could lead to a slight overestimate of the soybean losses. Changing crop calendars or crop varieties can be an alternative for some specific regions, but will not address all global concerns [18].

References

1. Li, C., Gu, X., Wu, Z., Qin, T., Guo, L., Wang, T., ... Jiang, G. (2021). Assessing the effects of elevated ozone on physiology, growth, yield and quality of soybean in the past 40 years: A meta-analysis. Ecotoxicology and Environmental Safety, 208(November 2020), 111644. https://doi.org/10.1016/j.ecoenv.2020.111644
2. Osborne, S. A., Mills, G., Hayes, F., Ainsworth, E. A., Büker, P., & Emberson, L. (2016). Has the sensitivity of soybean cultivars to ozone pollution increased with time? An analysis of published dose-response data. Global Change Biology, 22(9), 3097–3111. https://doi.org/10.1111/gcb.13318
3. Zhang, W., Feng, Z., Wang, X., Liu, X., & Hu, E. (2017). Quantification of ozone exposure- and stomatal uptake-yield response relationships for soybean in Northeast China. Science of the Total Environment, 599–600, 710–720. https://doi.org/10.1016/j.scitotenv.2017.04.231
4. Fishman, J., Creilson, J. K., Parker, P. A., Ainsworth, E. A., Vining, G. G., Szarka, J., ... Xu, X. (2010). An investigation of widespread ozone damage to the soybean crop in the upper Midwest determined from ground-based and satellite measurements. Atmospheric Environment, 44(18), 2248–2256. https://doi.org/10.1016/j.atmosenv.2010.01.015
5. Tai, A. P. K., & Val Martin, M. (2017). Impacts of ozone air pollution and temperature extremes on crop yields: Spatial variability, adaptation and implications for future food security. Atmospheric Environment, 169, 11–21. https://doi.org/10.1016/j.atmosenv.2017.09.002
6. Zhang, W., Wang, G., Liu, X., & Feng, Z. (2014). Effects of elevated O3 exposure on seed yield, N concentration and photosynthesis of nine soybean cultivars (Glycine max (L.) Merr.) in Northeast China. Plant Science, 226, 172–181. https://doi.org/10.1016/j.plantsci.2014.04.020
7. Singh, E., Tiwari, S., & Agrawal, M. (2009). Effects of elevated ozone on photosynthesis and stomatal conductance of two soybean varieties: A case study to assess impacts of one component of predicted global climate change. Plant Biology, 11(SUPPL.1), 101–108. https://doi.org/10.1111/j.1438-8677.2009.00263.x
8. Emberson, L. D., Pleijel, H., Ainsworth, E. A., van den Berg, M., Ren, W., Osborne, S., ... Van Dingenen, R. (2018). Ozone effects on crops and consideration in crop models. European Journal of Agronomy, 100(June), 19–34. https://doi.org/10.1016/j.eja.2018.06.002

9. Feng, Z., & Kobayashi, K. (2009). Assessing the impacts of current and future concentrations of surface ozone on crop yield with meta-analysis. Atmospheric Environment, 43(8), 1510–1519. https://doi.org/10.1016/j.atmosenv.2008.11.033
10. Dumont, J., Cohen, D., Gérard, J., Jolivet, Y., Dizengremel, P., & Le Thiec, D. (2014). Distinct responses to ozone of abaxial and adaxial stomata in three Euramerican poplar genotypes. Plant, Cell and Environment, 37(9), 2064–2076. https://doi.org/10.1111/pce.12293
11. Sun, J., Feng, Z., & Ort, D. R. (2014). Impacts of rising tropospheric ozone on photosynthesis and metabolite levels on field grown soybean. Plant Science, 226, 147–161. https://doi.org/10.1016/j.plantsci.2014.06.012
12. Tisdale, R. H., Zobel, R. W., & Burkey, K. O. (2021). Tropospheric ozone rapidly decreases root growth by altering carbon metabolism and detoxification capability in growing soybean roots. Science of the Total Environment, 766, 144292. https://doi.org/10.1016/j.scitotenv.2020.144292
13. Mills, G., Sharps, K., Simpson, D., Pleijel, H., Broberg, M., Uddling, J., ... Van Dingenen, R. (2018). Ozone pollution will compromise efforts to increase global wheat production. Global Change Biology, 24(8), 3560–3574. https://doi.org/10.1111/gcb.14157
14. (dataset) USDA National Agricultural Statistics Service (2017). NASS - Quick Stats. USDA National Agricultural Statistics Service. https://data.nal.usda.gov/dataset/nass-quick-stats. Accessed 2021-03-02.
15. Kim, S. Y., Bechle, M., Hankey, S., Sheppard, L., Szpiro, A. A., & Marshall, J. D. (2020). Concentrations of criteria pollutants in the contiguous US, 1979–2015: Role of prediction model parsimony in integrated empirical geographic regression. *PloS one*, *15*(2), e0228535.
16. Vingarzan, R. (2004). A review of surface ozone background levels and trends. *Atmospheric environment*, *38*(21), 3431-3442.
17. McGrath, J. M., Betzelberger, A. M., Wang, S., Shook, E., Zhu, X. G., Long, S. P., & Ainsworth, E. A. (2015). An analysis of ozone damage to historical maize and soybean yields in the United States. *Proceedings of the National Academy of Sciences*, *112*(46), 14390-14395.
18. Sampedro, J., Waldhoff, S. T., Van de Ven, D. J., Pardo, G., Van Dingenen, R., Arto, I., ... Sanz, M. J. (2020). Future impacts of ozone driven damages on agricultural systems. Atmospheric Environment, 231(October 2019). https://doi.org/10.1016/j.atmosenv.2020.117538
19. U.S. Environmental Protection Agency. (1991). National Air Quality and Emissions Trend Report, 1989. https://www.epa.gov/sites/default/files/2017-11/documents/trends_report_1989.pdf Accessed 2021-11-30
20. U.S. Environmental Protection Agency. (2021). Trends in Ozone Adjusted for Weather Conditions. https://www.epa.gov/air-trends/trends-ozone-adjusted-weather-conditions Accessed 2021-11-30

117q Deployable Pfas Effluent Treatment Systems

Justin Puhnaty, Scott Waisner, Jose Mattei-Sosa, Connor Clark, Jacob Lalley

Per- and Polyfluoroalkyl Substances (PFAS) are a group of chemicals that are found in many different consumer, commercial, and industrial products. PFAS compounds are known as "forever chemicals" because they do not naturally degrade over time. Perfluorooctanoic Acid (PFOA) and Perfluorooctane Sulfonate (PFOS) are the two most widespread and known toxic PFAS chemical compounds of concern in Aqueous Film Forming Foam (AFFF). AFFF has been used in military practices since the 1960s in combat or in fire training exercises. AFFF is made up of concentrated PFAS's that do a good job at suppressing fuel fires but can lead to negative environmental and health effects. The Engineering Research and Development Center (ERDC) has pioneered new innovative mobile and deployable systems to clean PFAS contaminated water. ERDC has developed two unique PFAS treatment systems, PFAS Effluent Treatment System (PETS) and the PFAS Water Treatment Research Reactor (PWTRR).

Following successful demonstrations at USAF Hurlburt Field, the original PETS was demonstrated at MCAS Futenma (Okinawa, Japan) for treatment of USTs for aircraft hangars and subsequently transferred to the USMC for continued use in theater. The PWTRR was developed to continuously treat and maintain water levels in a holding pond for a firefighting training area at USAB Misawa, Japan. The PWTRR was transferred to the USAF and continues to operate at Misawa.

117r Predictions of Hydrochar Properties Using Molecular Structure Determined By Solid State NMR

Karen Agro, Max Moran, Avery Brown, Matthew Rando, Klaus Schmidt-Rohr, Michael T. Timko

Hydrothermal carbonization (HTC) is a process where wet biomass is converted to a solid hydrothermal char by a chemical process under moderate temperature and autogenous pressure. Hydrochar can be used for a variety of environmental applications including water treatment, soil amendments, combusted as a solid bio-fuel, etc. A major challenge in the field is development of structure-property relationships that can be used to select an optimal application for a hydrochar based on its structure. A major step forward in the development of structure-property relationships is the finding of a molecularly detailed structure by collaborators at Brandeis University, based on solid state Nuclear Magnetic Resonance Spectroscopy (NMR) data. Solid state NMR provides molecularly detailed insight into hydrochar structure; however, NMR analysis is a time consuming, expensive, and complicated process that is not always practical. To determine similar information via simpler methods, predictions from the NMR were compared with other characterization methods such as Fourier Transform Infrared Spectroscopy (FT-IR), Raman Spectroscopy, Boehm Titration, and zeta potential measurements. A peak fitting process was used to quantify functional groups based on FT-IR bands which was compared to functional groups quantified in the NMR model structure. Density Functional Theory (DFT) analysis based on the NMR model structure was used to simulate its hypothetical Raman spectrum, which was then compared to an experimental Raman spectrum for glucose hydrochar. Acid surface sites measured experimentally through Boehm Titration were compared to those quantified in the NMR model structure. Finally, a predictive mathematical model was developed based on the NMR model structure data to determine if macroscopic hydrochar properties, like zeta potential, could be predicted based on surface characteristics. Experimental data and other characterization methods displayed good agreement with the NMR model structure, providing both confidence in the model structure and the use of simpler techniques to determine hydrochar properties. The results of this study are an important step in the design of hydrochars.

117s Effect of Metal Oxide Support on K/Mn-Based Catalysts for VOC Oxidation

Md Sarwar Kamal, Catherine B. Almquist

Cyclohexane is a common chemical used in consumer products, fuels, and as a solvent. It is used in the synthesis of adipic acid, which is a precursor to Nylon 6,6 and it is used in the production of caprolactam. Based upon its industrial relevance, cyclohexane was selected as a representative volatile organic carbon (VOC) with which to investigate the role of metal oxide supports in potassium/manganese (K/Mn) - based catalysts. Mn-based catalysts are interesting catalysts due to their several oxidation states, mobility of oxygen vacancies, and redox properties. Cryptomelane, a K/Mn-based catalyst material, has been shown by others to be an effective VOC oxidation catalyst. In this study, we are investigating K/Mn-based catalysts supported on various metal oxide materials for VOC oxidation. The performance of the catalysts will be investigated and related to their chemical and physical properties.

117u Meeting Increased Methane Emissions with Methanotrophs: Addressing Mass Transfer Problems in Reactor Design

Robert Bertrand, Emmanuel D. Revellame, Mark E. Zappi, Rafael Hernandez, William E. Holmes

Methane is the simplest hydrocarbon and the second most-released greenhouse gas, next to carbon dioxide. Anthropogenic methane emissions have been increasing in recent decades, and the remediation or capture of such emissions has become a global concern which rises with the increase in emissions volume. Burning of this methane at spot sources of high concentration has been an attractive choice for a long time, but this does not address the emission of carbon dioxide as a result. Carbon dioxide is chiefly an undesirable byproduct of the burning of methane-rich gas streams for heat and power. This is compounded by the fact that as these methane emissions increase, so will the amount burned, and thus the amount of carbon dioxide released. A different method of utilization of methane is biological upcycling. Biological upcycling utilizes methanotrophs, which are capable of deriving energy and carbon directly from methane. This process fixes a portion of the carbon fed into cellular biomass, preventing its release as greenhouse gas. In addition to this, biological upcycling provides carbon-rich byproducts such as single-cell protein, thermoplastics, lipids, and methanol. In addition to this, the foundational research regarding methanotrophs, their products, and their properties spans the better part of the last fifty years, and they have been noted as having the potential for industrial significance. Methanotrophs are not utilized in industry despite their benefits, and this is largely due to the inefficiency with which methanotrophic bioreactors operate. Methane has very low solubility in water, and this leads to low mass transfer rates, as the concentration driving force is low. Research into methanotrophic bioreactors universally notes them as being mass transfer limited for the methane substrate. Other bioreactor systems utilize carbon sources with much greater solubilities (sugars, etc.), but methanotrophs grow on a gaseous substrate with low solubility, which leads to low productivity. Much research in the last twenty years has been directed at using various techniques to improve the mass transfer properties of methanotrophic bioreactors. Improvements in the volumetric mass transfer coefficient (kLa) of these reactors has nearly always resulted in higher cell density, better product formation, and ultimately greater remediation of the methane pollutant. Several methods for improving the mass transfer properties of methanotrophic bioreactors have been investigated, including increasing area for mass transfer, increasing solubility of the substrate, utilization of a non-aqueous phase (NAP) to improve mass transfer, and even metabolic engineering of methanotroph strains for increased utilization of the substrate. The method investigated by this study for improvements to the kLa of these bioreactors is variation in the composition of gas fed to a methanotrophic bioreactor, and the utilization of an oil phase to improve mass transfer. Bioreactors utilizing NAP to improve mass transfer are called two-phase partitioning bioreactors (TPPBs). TPPB experimentation has generally revolved around utilizing an existing bioreactor or bioreactor design, and simply 'modifying' it by adding a small amount of an oil phase (10% by volume, usually). These studies are generally performed in stirred tank reactors, but some studies on bubble columns exist. Bubble columns are an interesting avenue of research for TPPBs, as they offer increased gas contact time, and avoid harmful shear rates inherent in stirred reactors. Additionally, bubble columns are very simple to construct and operate. Bubble columns will also benefit the most from the TPPB modification, as mass transfer rates are generally low. This research investigated the effect of oil selection and recirculation on the kLa of a bubble column TPPB, and additionally investigated the effect of gas composition, which was performed to determine the feed gas compositions which hindered the kLa of same. This study aims to determine the effectiveness of various oil additives, as no study has yet compared the primary oil-phase additions directly in the same experimental setup, nor has much investigation of novel modifications to this design been researched. This study hypothesized that the recirculation of the oil phase would have a beneficial effect on the kLa that goes beyond simple addition, and that this improvement would follow the overall solubility of methane in the

oil phase (that is, the improvement would be directly related to the increasing solubility). It was hypothesized that for the gas composition testing that the kLa for both methane and oxygen would be improved by increasing the balance of gas until it reached the saturation concentration, and that there would be limitations for methane kLa at high air composition, and vice versa. Experimentally, these studies were performed by determining the kLa of the bioreactor without any oil addition, and then testing with simple addition of various oil phases (silicone oil, paraffin oil, and mineral oil). Subsequently, testing of the mass transfer coefficient was performed with a recirculation of the oil phase, which consisted of a pumped loop which collected the oil at the top of the reactor and fed it into a modified sparger at the bottom, which was hypothesized to improve the contact between oil and water, and thus improve the mass transfer rate. For the gas composition experiments, 1 LPM total gas was introduced to the reactor by an air stone sparger at various ratios of air and methane, and the kLa measured for both methane and oxygen. The methane/oxygen kLa was measured at the extrema (>10% v/v for each gas) to determine limiting concentration for kLa. In all cases, the utilization of an oil phase improved the mass transfer coefficient significantly, which agrees with two-phase partitioning bioreactor research. Naturally, for the gas composition testing, there was no kLa observed for oxygen at 100% methane flowrate, and vice versa; however a dependence on composition was observed in the kLa for these experiments. In addition to this, it was observed that the recirculation of the oil phase offered a minor improvement in the mass transfer coefficient when compared to simple oil addition. One element of this study not tested was the effect of density on the mass transfer coefficient, as the differences between them were judged to be small. Improvements to the methane gas transfer coefficient and innovations in process design will drive methanotrophic bioreactors into an industrially-significant place, and allow them to produce useful products while simultaneously remediating methane emissions, all with less carbon dioxide release than simple burning.

117v Nanofiber Based Colorimetric Sensors for the Detection of Organic Analytes in Complex Environmental Media

Colton Duprey, Hadi Rouhi, Evan Wujcik

Drinking water is often exposed to contamination through industrialization. Trihalomethanes (THMs), are among the most common contaminants, being by-products of water chlorination/treatment. The carcinogenic and health effects of these compounds have motivated scientists to work on the accurate detection of THMs down to 80 ppb in treated water. Here, a superhydrophobic syndiotactic polypropylene (sPP) nanofiber mat produced via electrospinning is used to preconcentrate THMs in environmental water samples, and subsequently, detect them using a well-known colorimetric reaction chemistry. The reaction chemistry yields a visible red/pink chromophore under visible light absorption. This reaction occurs when the preconcentrated THM becomes trapped in the liquid phase of the reaction chemistry, on the surface of the sPP fibers. The resulting color change can be analyzed via a simple quantitative color intensity analysis utilizing widely-available software, measuring the THM content in water as low as 0.8 ppb. Similarly, polycyclic aromatic hydrocarbons (PAHs) are carcinogenic compounds resulting from oil spills and industrial pollution, and are capable of causing long-term health damage. Herein we demonstrate the first nanofibrous polymer-based on-site colorimetric sensor for the detection of PAHs at environmentally-relevant concentrations. Electrospinning is used to create silica nanofibers, formaldehyde, under acidic conditions, is used to colorimetrically detect the PAHs—reacting to form brown quinoid compounds on top of the white silica fiber mat. This robust and portable sensor is an uncomplicated alternative to the traditional, expensive analytical equipment (HPLC, GC-MS), and allows for on-site and rapid water monitoring.

117x Environmental Impacts of Direct Carbonation and Indirect Carbonation in Calcium Carbonation Manufacturing Using Slag

Kyeong Ho Kim, Min A Bae, Jae Ho Baek

The mineral carbonation is a technology that converts mineral resources into carbonate salts by directly or indirectly reacting with carbon dioxide [1]. However, the mineral carbonation process has problems that are difficult to widely apply due to crushing of raw materials, slow reaction speed, and distance from carbon dioxide emission sources [2]. Nevertheless, the mineral carbonation is a promising technology in the field of Carbon Capture, Utilization, and Sequestration (CCUS) because it can treat a large amount of carbon dioxide and can be a permanent carbon dioxide reduction technology [3]. Various industrial by-products such as steelmaking slag, cement kiln dust, waste ash and paper mill waste can be used as raw materials for the mineral carbonation [4]. Among them, steel slag generates more than tens of millions of tons every year and is used as an aggregate for mounding or concrete, but slag containing a large amount of alkali can be used as a raw material for mineral carbonation reactions [1]. In this study, we manufactured the calcium carbonate through direct and indirect carbonation reactions using slag, and compared the environmental impacts of two processes by Life Cycle Assessment (LCA).

The direct carbonation reaction produced calcium carbonate with a purity of 40% by reacting exhaust gas generated during the incineration process of municipal waste. In the indirect carbonation process, exhaust gas generated during the incineration of sewage sludge was used as a carbon dioxide raw material, and only calcium ions contained in slag were selectively separated to react with exhaust gas to produce calcium carbonate with a purity of 95%. Exhaust gas generated during the incineration of municipal waste and sewage sludge contains 9-10% carbon dioxide concentration. The direct carbonation facilities can produce 36 tons of calcium carbonate per day and the indirect carbonation facilities can produce 4 tons of calcium carbonate per day because of the addition of a process of extracting calcium ions in slag.

The scope of research to evaluate the environmental impact of direct carbonation and indirect carbonation reactions included from the process of slag entering the carbonation facility to the production of calcium carbonate. The functional unit is the manufacture of 1 kg of calcium carbonate, and input and output data collected the production of calcium carbonate for two months from June to July 2022. The environmental impact on the disposal of unreacted slag was excluded. LCA is a widely used method when evaluating the environmental impact of a product, and the program used in this study was SimaPro 9.2 version, and the methodology applied is IPCC 2013 GWP 100a.

The direct carbonation process not only has a small environmental impact on plant operation but also produces a large amount of calcium carbonate and sells it in bulk as a concrete material. However, indirect carbonation generates a large amount of carbon dioxide during plant operation because only calcium ions in slag must be selectively used in the reaction. However, it has the advantage of high economic feasibility because of the high price of calcium carbonate.

Acknowledgements

This work was supported by the Regulation-Free Special Zone (P0017177) and the Technology Development Program (S3025882) funded by the Ministry of SMEs and Startups (MSS, Korea) and Industrial Strategic Technology Development Program (00154970) funded By the Ministry of Trade, Industry & Energy (MOTIE, Korea) and the Korea Institute of Industrial Technology (KITECH, EH-

23-0020)

References

1. Minah Son, Gookhee Kim, Kunwoo Han, Min Woo Lee, Jun Taek Lim, Development status and research direction in the mineral carbonation technology using steel slag, Korean Chem. Eng. Res., 55(2), 141-155, 2017
2. Sanna, A., Uibu, M., Caramanna, G., Kuusik, R., Maroto Valer, M., A review of mineral carbonation technologies to sequester CO_2, Chem. Soc. Rev, 43, 8049-8080, 2014
3. Lackner, K. S., A guide to CO_2 sequestration, Science, 300, 2003
4. Li-Shan Xiao, Run Wang, Pen-Chi Chiang, Shu-Yuan Pan, Qing-Hai Guo, E. E. Chang, Comparative life cycle assessment (LCA) of accelerated carbonation processes using steelmaking slag for CO_2 fixation, Aerosol and Air Quality Research, 14, 892-904, 2014

117z Copper and Gallate MOF for Electrosorption of Cadmium Ions

Yonghwan Kim, Hyunjung Kim, Kwiyong Kim, Ho Hyeon Eom, Xiao Su, Jae Lee

There is a growing need for environmentally friendly technologies that can effectively remove toxic heavy metal ions from wastewater. This research showcases a novel approach to electrochemically remove and release cadmium ions (Cd^{2+}) using a cost-effective and water-stable metal-organic framework (MOF) made of copper gallate (CuGA). The MOF was synthesized using naturally abundant gallic acid as a linker in an aqueous solution. The study utilized copper within the CuGA structure as a redox center to attract Cd^{2+} through Cu^{2+}/Cu^{+} reduction, resulting in rapid uptake kinetics and a much higher capacity (> 60 mg g^{-1}) than without electrochemical assistance (~15 mg g^{-1}). Furthermore, the recovery of Cd^{2+} and the regeneration of the electrode were facilitated without the need for regenerants by applying an opposite overpotential to induce the re-oxidation of copper. Physicochemical characterizations, including XPS analysis, were conducted to investigate the electrode's chemical oxidation states and stability after the electrosorption-regeneration process. This work demonstrated the potential of a Cu-based MOF electrode as a reusable platform for efficient Cd^{2+} removal and supports further research and development of new Faradaic electrodes for electrochemical wastewater treatments.

117aa Metal Hexacyanoferrate-Embedded Electrospun PVA/CA Nanofiber for Effective Cesium Ions Remediation

Yonghwan Kim, Hyunjung Kim, Ho Hyeon Eom, Jae Lee

With the increasing production of radioactive waste, efficient and rapid removal of cesium ions has become an important issue for reducing waste volume. This study presents a novel approach to immobilizing potassium copper hexacyanoferrate (KCuHCF) in a composite adsorbent for effective cesium ion separation. Using an electrospinning technique, a nanofiber matrix was created from a polymer blend of poly-vinyl alcohol and sodium alginate, and in situ immobilization of KCuHCF was achieved through the coordination effect of carboxylate functionality. The ratio of polymers was systematically analyzed through electron microscopy and spectroscopic analyses to optimize the content and uniformity of the immobilized KCuHCF nanoadsorbent. The composite adsorbent demonstrated a high adsorption capacity of 114 mg g^{-1} and selectivity for cesium with distribution coefficients of 104 ~ 105 mL g^{-1}, even under seawater conditions where large amounts of competitive cations were present. In a membrane filtration experiments, the composite showed efficient decontamination by removing 180 mg cesium per 1 m^2 with a high flux of about 148 L m^{-2} h^{-1}. Therefore, depositing KCuHCF nanoadsorbent on the surface of the nanofiber and using an adsorption filtration method could be a practical alternative for treating radioactive wastewater.

117ab 2D Cu-Fetcpp MOF Assembled on Znti-LDH to Construct 2D/2D Direct Z-Scheme Heterojunction for Enhanced Photocatalytic CO_2 Reduction

Xiaohong Yin

The artificial photocatalysis to convert CO_2 into renewable fuels has been of great significance in solving climate crisis and depletion of fossil fuels. The development of Z-scheme photocatalyst is crucial to achieve efficient photocatalytic CO_2 reduction. Herein, a 2D/2D Z-scheme ZnTi-LDH/Cu-FeTCPP MOF heterojunction is designed for photocatalytic CO_2 reduction without photosensitizer and organic sacrificial agent. The optimized ZnTi-LDH/Cu-FeTCPP MOF photocatalyst exhibits attractive activity with a CO yield of 37.80 $\mu mol \cdot g^{-1}$, which is 28 and 26 times higher than those of ZnTi-LDH and Cu-FeTCPP MOF, respectively. The improved photocatalytic activity is attributed to the contact interface of Cu-FeTCPP MOF and ZnTi-LDH that generates internal electric field and forms a direct Z-scheme heterostructure to separate photogenerated carriers more efficiently and promote redox reactions. Additionally, cycling tests confirm its favorable stability. This study strategies designing effective LDHs/MOFs Z-scheme heterojunction photocatalysts toward CO_2 photocatalytic reduction.

117ac A Thorough and Evaluative Examination of the Elimination of Heavy Metal Ions from Wastewater

Syed Farzan Ali Shah

The elimination of heavy metal ions from wastewater is paramount for the promotion of a healthy environment and the well-being of humans. Numerous techniques have been developed to remove heavy metal ions from different sources of wastewater. The techniques can be categorized into treatments based on adsorption, membrane filtration, chemical processes, electric separation, and photo catalysis. This study provides a thorough and critical evaluation of these methods, examining the agents and adsorbents used, removal effectiveness, operational parameters, and the advantages and disadvantages of each approach. Additionally, the study presents a summary of the essential discoveries from prior research reported in the literature. In general, the majority of recent research has concentrated on the use of adsorption methods. Greater emphasis should be placed on employing actual wastewater samples, instead of synthetic ones, for investigating the elimination of heavy metals. Future research efforts should prioritize the development of environmentally friendly, cost-efficient, and sustainable materials and techniques.

117ad Hydrothermal Oxidation of H2S Scavenging Wastewater

Marco Maschietti, Jens Muff

Background and Scope of the Research

The removal of hydrogen sulfide (H_2S) from natural gas streams produced offshore is typically carried out by injecting chemicals (H_2S scavengers), which react with H_2S and convert it into by far less harmful species. The most common H_2S scavenger is 1,3,5-tri-(2-hydroxyethyl)-hexahydro-*S*-triazine, which is a water-soluble species known in oil and gas industry as MEA-triazine. It is used as a basic aqueous solution, which is dispersed into the gas stream where it leads to the absorption of H_2S and the consequent aqueous phase scavenging reactions. Downstream of the injection point, the gas is separated from a wastewater, which is called spent H_2S scavengers and contains the scavenging products, as well as any unreacted MEA-triazine. This is a basic wastewater containing a high amount of carbon, nitrogen, and sulfur, which in some cases cannot be disposed by re-injection into subsea geological formations due to its fouling and scaling propensity. In these cases, it is discharged into the sea without any treatment because of lack of feasible alternatives. This is however not unproblematic for the environment as spent H_2S scavengers exhibit a relatively high toxicity. As a matter of fact, it has been reported that the discharge of spent H_2S scavengers in the North Sea accounts for up to 20% of the total Environmental Impact Factor (EIF) associated to the entire water discharge into the sea, even though spent H_2S scavengers are typically less than 0.1% of the total discharge of produced water [1]. Therefore, the development of an efficient and low-footprint process allowing the treatment of this wastewater on topside offshore installations can provide oil and gas operators with a concrete option for further reducing their EIF.

To reach this goal, the project ZeroH2S was started at Aalborg University (AAU) in 2020 (funded by the Danish Offshore Technology Centre), with one of the objectives being the demonstration of the feasibility of hydrothermal oxidation (HTO) as a means for topside offshore treatment of spent H_2S scavengers. This presentation summarizes the results obtained so far at laboratory scale as proof-of-concept [1,2] and discusses the perspectives for the process scale up, which will be executed in 2023-2024.

Materials and Methods

Several samples of North Sea spent H_2S scavengers, collected offshore in the period 2014-2022, were characterized in AAU laboratories. The COD and TOC of the samples were in the range (120 – 320) g/L and (40 – 115) g/L, respectively, with pH in the range 8.9 – 9.6. NMR analysis showed the presence of unreacted MEA-triazine, as well as the scavenging products MEA-dithiazine and monoethanolamine (MEA). In addition, the presence of the intermediate scavenging product MEA-thiadiazine and of two hydrolysis products of MEA-triazine was also evidenced. The concentrations of MEA-triazine, MEA and MEA-dithiazine were in the range (80 – 150) g/L, (20 – 50) g/L, and (10 – 40) g/L. As can be observed, this wastewater is characterized by a high concentration of water-soluble organics, including a large amount of unreacted MEA-triazine. In addition, the ecotoxicity of this wastewater was determined on marine bacteria (*Aliivibrio fischeri*) and algae (*Skleletonema pseudocostatum*). The ecotoxicity analysis revealed that the toxicity of spent H_2S scavengers is dramatically higher than the toxicity of offshore produced water from the same installations. For example, to obtain the same level of inhibition (50% inhibition) of bacterial and algae growth in produced water and spent H_2S scavengers, the former was diluted 9 times for bacteria and 14 times for algae, while the corresponding values for the latter were >4000 and >23000.

The HTO process was tested in a custom-made batch injection 99 mL reactor allowing: (i) water and oxygen to be heated and pressurized to the desired reaction conditions; (ii) the injection of the wastewater into the hot and pressurized water-oxygen system; and (iii) the rapid ejection and quenching of the reaction products. In this way, an accurate control of the reaction time at the desired pressure and temperature conditions can be achieved. The reactor was charged with diluted spent H_2S scavengers, with initial COD in the range 23 to 65 g/kg. The process was studied at three severity levels (200 °C and 70-90 bar; 280 °C and 120-130 bar; 350 °C and 210-250 bar), for reaction times in the range 1 to 360 minutes and two levels of excess of oxygen (106-145% and 190-250% of the COD). Extensive analysis of the aqueous phase produced by the HTO reactions was carried out, including COD, TOC, Total Nitrogen, Total Sulfur, Ion Chromatography, GC-MS, SPME-GC-MS, HPLC, as well as ecotoxicity analysis.

Results and Discussion

The results show a fast degradation of MEA-triazine, MEA-dithiazine, and MEA, since these species were never found in the reaction products. The rate of COD reduction is strongly dependent on the reaction temperature. For example, at 200 °C (higher excess of oxygen and initial COD 30-32 g/kg) the COD reduction was 29% in 3 minutes, 50% in 40 minutes, while a plateau was observed around 84% at long times indicating some persistent species that could not be oxidized. For other things being basically the same, at 350 °C the COD reduction was 71% in 1 minute, 85% in approx. 10 minutes, with the residual COD after 2 hours being only 2% of the initial value. Since the implementation of low footprint solutions is critical in topside offshore operations, the results strongly suggest high severity conditions to minimize the volume of the continuous-flow HTO reactor to be installed offshore.

The analytical results show the presence of pyrazines, pyridines and short chain carboxylic acids as intermediate oxidation products. In most cases, the concentration of these species increases over time, reaches a maximum, and then decreases. The reaction times corresponding to the maximum concentrations depend on the reaction severity and on the species under consideration. For example, focusing on a species of concern due to high toxicity, at 350 °C the mass fraction of pyridine reached a maximum of 333 mg/kg after 10 minutes, while decreasing to 12 mg/kg after 120 minutes. In all cases, at high severity all identified intermediate oxidation products exhibited a maximum concentration within 20 minutes, with a trend toward full carbon conversion to CO_2, nitrogen conversion to ammonium and, to small extent, nitrate, and sulfur conversion to sulfate. Therefore, the analytical results also suggest high severity conditions as favorable to reduce the discharge of water-soluble organics.

The ecotoxicity measurements show higher toxicity reductions at higher temperatures, albeit the reduction of ecotoxicity is in general somewhat lower than the corresponding reduction of COD, the trends of COD and ecotoxicity do not always show similar patterns, and the ecotoxicity data cannot be directly inferred from the concentration trends of individual reaction intermediates.

Overall, the results indicate that substantial ecotoxicity reductions can be obtained by HTO of spent H_2S scavengers (above 90% reduction on marine bacteria at all conditions, up to above 80% reduction on marine algae at 280 °C and 350 °C). The results also indicate that the HTO reactor design should not be based on COD measurements only, but both chemical and ecotoxicity analysis should be considered.

Further Development

In 2023-2024 the process is going to be tested in continuous-flow mode of operation on a scale in the order of 1 L/h, with the aim of confirming the results obtained in batch-mode. In addition, the continuous-flow mode of operation will allow to determine the maximum COD of the feed that can be treated due to potential thermal problems in the reactor, since oxidation reactions are exothermic, which will determine if feed dilution is needed. Moreover, analysis will be carried out on the gas effluent, to verify its suitability for discharge. In case of positive results, the process will be tested onshore at a scale of 10-20 m^3/day, which is the same scale as needed for an offshore installation for treating spent H_2S scavengers in the North Sea.

Literature Cited

[1] N. Montesantos, L.M. Skjolding, A. Baun, J. Muff, M. Maschietti, Reducing the environmental impact of offshore H_2S scavenging wastewater via hydrothermal oxidation, Water Research 230 (2023) 119507.

[2] N. Montesantos, M.N. Fini, J. Muff, M. Maschietti, Proof of concept of hydrothermal oxidation for treatment of triazine-based spent and unspent H_2S scavengers from offshore oil and gas production, Chemical Engineering Journal 427 (2022) 131020.

117ae Zinc Hexacyanoferrate Composite from Temperature-Controlled Pyrolyzed ZIF: Electrochemical Adsorption of Cesium Ions in Aqueous Solutions

Hyunjung Kim, Yonghwan Kim, Ho Hyeon Eom, David Harbottle, Jae Lee

137-Cesium is a significant radioactive isotope that requires removal from the nuclear waste that is produced after nuclear power generation. The aim of this research is to selectively remove cesium ions using a zinc hexacyanoferrate (ZnHCF) composite electrode, which has been designed using ZIF-8 based carbon (ZDC) as support. This study primarily focuses on the influence of the decomposition temperature of ZIF-8 on the formation of the composite and the performance of removing non-radioactive cesium ions selectively. The synthesis of highly conductive and uniformly distributed ZDC was achieved at an optimized temperature of 700 °C, and it exhibited a remarkable cesium adsorption capacity of 204.9 mg g^{-1}. To evaluate the practical performance of the electrode, selectivity tests were conducted in the presence of competing ions. The selectivity of the electrode remained high under Na^+ rich (Na/Cs = 1330, K_d (mL g^{-1}) = 1.04 x 10^5) and K^+ rich (K/Cs = 133, K_d (mL g^{-1}) = 7.24 x 10^4) conditions, with over 95 % of cesium ions removed under groundwater conditions. The adsorption of cesium ions was found to be reversible over five cycles, with complete desorption achieved after the fourth cycle. Moreover, the cesium ion adsorption mechanism was proven to be structural adsorption through oxidation-reduction and phase changes of ZnHCF-ZDC. and it was also shown to be a stable adsorption system. This research is significant as it is the first to apply a composite electric adsorbent using ZIF-based carbon support to remove cesium.

117ag Hydrothermal Oxidation of Spent H2S Scavengers in a Continuous-Flow Reactor: Preliminary Results

Alessandro Perrucci, Maria Nymann Christensen, Sarah Fredung Hückelkamp, Zhuoyan Cai, Peter Mønsted, Jens Muff, Marco Maschietti

Background

The most common process for removing H_2S from natural gas in offshore oil and gas installations consists in: (i) injecting and dispersing an aqueous solution of MEA-triazine into the natural gas stream, allowing H_2S to react with MEA-triazine to form dithiazine (DTZ) and monoethanolamine (MEA), which are by far less harmful than H_2S; (ii) separating the aqueous phase, resulting from the H_2S scavenging process, from the gas. This aqueous phase is a wastewater mainly containing unreacted MEA-triazine (unspent scavenger), DTZ and MEA (spent scavengers). In oil and gas practice this aqueous phase is known as spent H_2S scavengers and it is, in many cases, discharged into the sea without any treatment due to lack of feasible alternatives because of its fouling and scaling potential. The discharge of spent H_2S scavengers into the sea is detrimental to the environment due to the high toxicity of this wastewater [1].

In the research project ZeroH_2S, run at Aalborg University and funded by the Danish Offshore Technology Centre (DOTC) as part of the research program "Produced Water Management", it was proved that unreacted MEA-triazine can be recovered from the spent scavengers by means of nanofiltration [2], while the ecotoxicity of the spent H_2S scavengers can be drastically reduced by applying hydrothermal oxidation (HTO) prior to discharge [1,3]. The two processes were proved at laboratory scale [1-3].

Based on the obtained proof-of-concept, a new project was funded by DOTC, called "Zero H_2S – DEMO", which aims at demonstrating the recovery of MEA-triazine and the feasibility of HTO on the spent H_2S scavengers in continuous-flow mode of operation and on a larger scale. This new project is run by AAU in collaboration with Aquarden Technologies ApS.

This work presents preliminary experimental results of the application of the HTO process in continuous-flow mode of operation. The experimental campaign is starting in April 2023.

Materials and Methods

Spent H_2S scavengers collected in the North Sea from one offshore oil and gas installation will be used as feed for the HTO process. Based on analysis carried out in previous projects at AAU on spent H_2S scavengers collected in the period 2014 – 2022, it is known that this wastewater has an extremely high content of water-soluble organics, with COD and TOC in the range (120 - 320) g/L and (40 - 115) g/L, respectively, and pH values in the range 8.9 - 9.6. The application of NMR analysis confirmed the presence of a large amount of unreacted MEA-triazine, together with the reaction products MEA, MEA-thiadiazine, and MEA-dithiazine, as well as some by-products deriving from the hydrolysis of MEA-triazine.

The HTO tests will be carried out in a continuous-flow reactor available in Aquarden. Two feeds will be used: (i) the spent H_2S scavenger wastewater as it is; and (ii) the spent H_2S scavengers downstream of the membrane separation and recovery of MEA-triazine. The feed flow rate of the continuous-flow reactor will be in the order of 1 L/h. The oxidant will be oxygen-enhanced air. The reactor will be operated at approx. 250 bar. Two reaction temperature levels will be investigated: a lower level (around 280 – 300 °C) and a higher level (around 350 - 360 °C). The values of the wastewater flowrates will be varied to achieve different residence times in the reactor. The oxygen-enriched air flowrate will be set to guarantee the process is run under excess of oxygen, with respect to the COD. An important aspect of the experimental campaign will be the evaluation of the maximum COD level of the feed that can be handled safely, considering that this wastewater has a large amount of organics and oxidation reactions are highly exothermic. This is critical information as it will determine whether dilution is required for the HTO feed in the offshore process.

The aqueous phase produced in the HTO process will be characterized by means of COD and TOC analysis, chemical characterization, and ecotoxicity, in line with the procedures developed in a previous work [1].

Extensive analysis of the aqueous phase produced by the HTO reactions will be carried out, including COD, TOC, Total Nitrogen, Total Sulfur, Ion Chromatography, GC-MS, SPME-GC-MS, HPLC, as well as ecotoxicity analysis, according to procedures developed in previous works [1,3]. Moreover, analysis will be carried out on the gas effluent, to verify that the gas is constituted only by oxygen, nitrogen, and carbon dioxide, i.e., not containing any toxic species which could be a showstopper for the industrial implementation of the technology.

Results and Discussion

Results will be available after the experimental campaign scheduled from April 2023 to September 2023.

Implications of the Results.

The results will be used for making the basic design of a prospective HTO unit to be installed offshore.

Literature Cited:

[1] N. Montesantos, L.M. Skjolding, A. Baun, J. Muff, M. Maschietti, Reducing the environmental impact of offshore H_2S scavenging wastewater via hydrothermal oxidation, Water Research 230 (2023) 119507.

[2] M.N. Fini, N. Montesantos, M. Maschietti, J. Muff, Performance evaluation of membrane filtration for treatment of H2S scavenging wastewater from offshore oil and gas production, Separation and Purification Technology 277 (2021) 119641.

[3] N. Montesantos, M.N. Fini, J. Muff, M. Maschietti, Proof of concept of hydrothermal oxidation for treatment of triazine-based spent and unspent H2S scavengers from offshore oil and gas production, Chemical Engineering Journal 427 (2022) 131020.

117ah Realizing a Resilient Aquaculture Environment Via Biological Processes

Yifan Zhang, Jianzhong He

To realize a healthy ecosystem for aquaculture development, an energy efficient and cost saving aquaculture wastewater treatment system is required to avoid formation of excess nitrate. Here we discovered an innovative technology that applies a microbial community comprising unique denitrifying phosphate accumulating organisms and anammox bacteria to sustain high levels of nutrient removal in low carbon, nitrogen-rich aquaculture wastewater. This culture can remove ammonium, nitrite, and nitrate at rate of 2.85, 1.98, and 2.42 mg-N/L/h, respectively, which can also remove high levels of nitrate present in aquaculture wastewater under limited oxygen and low C/N ratios. Very importantly, the microbial consortia can work efficiently in high salinity (~3% salinity) environment, especially for saline fish farm waters. This leads to the development of a de-nutrient bioreactor that incorporates the anammox reaction to achieve high levels of nutrients (nitrogen and phosphorus) removal without requiring aeration or carbon supplementation for a resilient aquaculture environment.

117ai Removal of Heavy Metals from Aqueous Solution through Biosorption Using Magnetotactic Bacteria

Prabhakar Choudhary, Prabirkumar Saha

Biosorption using magnetotactic bacteria has shown to be an efficient method for the removal of heavy metals from aqueous solutions. These bacteria have unique properties that allow them to adsorb heavy metals from the environment and store them within their cells. When combined with magnetic nanoparticles, magnetotactic bacteria can be easily separated from the solution using a magnetic field, making the process of heavy metal removal faster and more efficient. A comprehensive literature covering published articles on removal of heavy metals from aqueous solution through biosorption have been reviewed. It is found that most of the studies emphasis on heavy metal interaction with the functional groups containing N, O, P and S. The parameters that significantly affect removal of heavy metals are pH, concentration of solute as wells as complexity of solution chemistry. The study here emphasise on biosorption as wells as magnetotaxis behaviour of magnetotactic bacteria (MTB). Currently, nanotechnology is considered comparatively more promising and superior than other methods in handling and removing these toxic pollutants more efficiently. The commonest means of synthesizing biogenic nanoparticles is by using microbes, fungi, yeasts and plants. This is an environment-friendly method but is now facing serious impediments due to the presence of toxic substances in the final effluent and in particular the challenge of effective biomass separation. The idea of this study is to introduce magnetotactic bacteria (MTB) for biomass separation and consider MTB as a novel biosorbent for heavy metal removal and recovery from wastewater. Experimental studies were carried out to culture indigenous MTB from fresh water sources of Guwahati, Assam. Microscopy studies revealed high motile bacteria in environmental sample. Capillary racetrack method (CRT) was used to isolate MTB from rest of non- motile culture. Surface and intracellular characterization were carried out using FETEM and FTIR for MTB. EDX was used for intracellular elemental composition of Fe_3O_4 as well mapping of Fe and O element along the chain of magnetosome was carried out. XRD method was used to determine the forms of iron sources necessary for growth of MTB in local fresh water source. Iron gluconate hydrate and iron lactate hydrate showed better motility behaviour but further investigation is necessary to determine the the necessary iron sources as Fe(III) iron salts have shown high yield in MTB growth compared to Fe(II) iron salts . In later phase, the significance of MTB in heavy metal contaminated freshwater sources of North- East India is being carried out. Soil and sediment sample from IIT Guwahati Lake (Assam, India), Deepor Beel Lake (Assam, India), Margherita coal mines and swamps (Assam, India), and Jharsuguda coal mines and swamp (Odisha, India) were analysed for heavy metals (Arsenic, Cadmium, Chromium, Lead, Nickle and Copper) using AAS. All the analyses were in ppm level. IIT Guwahati Lake (As- 0.02933, Cd- 0.043, Cr -0.151, Pb- 0, Ni- 17.944, Cu- 0.188); Deepor Beel Lake (As- 0.01866, Cd-0.067, Cr- 0.084, Pb- 0.34, Ni- 18.168, Cu- 0.117); Margherita coal mines and swamps (As- 0.023, Cd- 0.122, Cr- 0.460, Pb- 0.26, Ni- 11.466, Cu- 8.546). The incubated samples have confirmed presence of magnetotactic bacteria (MTB). Further extensive study confirms that this method can effectively remove a wide range of heavy metals, including lead, cadmium, and arsenic, with removal rates ranging from 80% to 99%. Overall, biosorption using magnetotactic bacteria offers a promising solution for the remediation of heavy metal-contaminated wastewater.

Comparison on Efficiency of Iron and Aluminum Nanoparticles for Methylene Blue Removal from Aqueous Solutions

Mohamed K. Mostafa[1], Ahmed H. Sadek[2,3]

[1]Faculty of Engineering and Technology, Badr University in Cairo (BUC), Cairo, Egypt (m_khaled@buc.edu.eg)

[2]Environmental Engineering Program, Zewail City of Science, Technology and Innovation, Giza, Egypt

[3]Sanitary and Environmental Engineering Research Institute, Housing and Building National Research Center (HBRC), Giza, Egypt (ahsadek@zewailcity.edu.eg, https://orcid.org/0000-0002-2749-5387)

ABSTRACT

This review demonstrates a comparative study of methylene blue (MB) dye removal from aqueous solutions using zero-valent iron (nZVI) and zero-valent aluminum (nZVAl) nanoparticles. The results indicated that at pH 6, nZVI dosage 10 g/L, initial MB concentration 10 mg/L, temperature 30 °C, and stirring rate 150 rpm, the equilibrium time was 30 min achieving a removal efficiency of approximately 100% for nZVI. While for nZVAl, the results indicated that at an initial MB concentration of 10 mg/L, pH 10, nZVAl dosage 1.0 g/L, stirring rate 200 rpm, and temperature 30 °C, the removal efficiency of approximately 99.0% was achieved within an equilibrium time of 90 min. Moreover, the adsorption data of MB fitted well to the Freundlich isotherm model (R^2: 0.9358) and pseudo-second-order kinetic model (R^2: 0.9976) for adsorption by nZVI. Also, the isotherm adsorption data were best fitted by the Freundlich isotherm model (R^2: 0.9569), while the kinetic data were well described by the pseudo-second-order model (R^2: 0.9999) for adsorption by nZVAl. From the aforementioned results, it is clear that both nZVI and nZVAl showed very high efficiency in removing methylene blue from aqueous solutions with a slight difference in favor of nZVI.

Keywords: Aluminum nanoparticles, aqueous solutions, isotherm models, iron nanoparticles, kinetic models, methylene blue

1. Introduction

A significant quantity of artificial dyes is utilized in various industries, such as textiles, pharmaceuticals, leather, food processing, cosmetics, and paints [1,2]. Each year, around 10-15% (approximately 0.8 million tons) of different dye types produced globally are discharged into the environment by these industries [3]. Due to the increased production and consumption of dyes worldwide, it is estimated that the dye industry will experience an annual growth rate of 2-3% [4,5]. Many developing countries are also witnessing a steady increase in dye production.

However, when these dyes are not properly treated, they can harm the environment, marine life, and pose significant health risks to humans, particularly as they become toxic, resistant to degradation, mutagenic, and carcinogenic [6-9]. Methylene blue (MB) (3,7-bis(Dimethylamino)-phenothiazin-5-iumchloride) ($C_{16}H_{18}ClN_3S.xH_2O$ (x = 2–3)) is a common cationic dye with various applications, including biological staining, coloring paper, hair, cotton, and wool. It is also used for the treatment of methemoglobinemia and urinary tract infections. However, the accumulation of MB in wastewater can have adverse health effects, leading to difficulties in breathing, vomiting, eye burns, diarrhea, and nausea [10-12].

There are numerous processes, both chemical and biological, that have been successfully utilized to treat dye effluents. Some examples include membrane filtration, photocatalytic degradation, irradiation, biological treatment, conventional adsorption, and ultrasonic-assisted adsorption [13-20]. Adsorption is a commonly employed technique due to its ease of operation, availability of adsorbents, selectivity, and potential for scaling up [21, 22].

The use of nano-matrices for adsorption has gained significant attention due to its environmental friendliness, cost-effectiveness, and higher efficiency [23, 24]. Metal and metal oxides are some of the nano-matrices that have been used effectively for dye removal from water, demonstrating high adsorption efficiency [25, 26]. Nanozero valent iron (nZVI) has been recognized as a viable solution for removing cationic pollutants from wastewater [27]. Similarly, the synthesis of various chemical-based nano adsorbent materials is familiar, but they are often expensive and require substantial investment. These materials are primarily sourced from the global market, which directly impacts the cost at an industrial scale [28, 29]. nZVI particles are particularly advantageous compared to other adsorbents due to their large specific surface area, high reducibility, and activity [30, 31]. However, the practical application of nZVI faces several technical challenges. Firstly, the particles tend to agglomerate due to van der Waals forces and magnetic interactions, leading to reduced dispersibility [32]. Additionally, nZVI rapidly undergoes oxidation, diminishing its effectiveness [33, 34]. In recent years, zero-valent aluminum (ZVAl) has shown great promise as a material with significant reduction potential for a wide range of refractory pollutants [35, 36]. ZVAl acts as an excellent electron donor (E_0Al = -1.66 V), which is 3.8 times more negative compared to zero-valent iron (E_0Fe = -0.44 V). Furthermore, ZVAl possesses unique properties, such as being acid-base amphoteric, light weight (ρ_{Al} = 2.70 g/cm^3), having a higher number of electrons per atom (3 electrons per Al), and being abundantly available (as aluminum is the most abundant metal in the earth's crust) [37, 38].

Therefore, the goal of this review was to compare between the efficiency of both nZVI and nZVAl for MB removal from aqueous solutions. For that purpose, the characterization and adsorption performance of both materials were reported. Moreover, a comparison of adsorption kinetic and isotherm studies was established.

2. Material and Methods

Preparation of methylene blue solution

To prepare a concentrated MB solution, 1 g of MB powder was dissolved in 1000 mL of Milli-Q water, resulting in a stock solution with a concentration of 1000 mg/L. The solutions of desired concentrations were prepared by diluting it in a stepwise manner. The pH of the aqueous solutions was adjusted by adding either 0.1 N HCl or 0.1 N NaOH.

Preparation of nZVI and nZVAl

ZVI and nZVAl nanoparticles were prepared following the liquid phase reduction method. The nanoparticles were synthesized in an ethanol and water mediums having a ferric chloride ($FeCl_3$) for nZVI, aluminum sulfate ($Al_2(SO_4)_3$) for nZVAl, respectively, and sodium borohydride ($NaBH_4$) as a reducing agent. Briefly, NaBH4 powder was dissolved in 100 mL of deionized water to form borohydride solution. Further, ferric chloride (for nZVI) and aluminum sulfate (for nZVAl) was added to 100 mL ethanol (for nZVI) and distilled water (nZVAl) to make iron ions and aluminum ions solutions, respectively, followed by efficient stirring to ensure a complete mixing condition. The $NaBH_4$ solution was added drop by drop into both iron solution and aluminum solution, separately, with vigorous stirring. The produced black nanoparticles (nZVI) and white nanoparticles (nZVAl) were separated from the solutions by centrifugation at 2000 rpm for 5 min. Further, the separated particles were washed three times with deionized water and 1:1 (v/v) ethanol/acetone to prevent the rapid oxidation of iron and aluminum nanoparticles. Finally, the nanoparticles were separated by centrifugation and then dried overnight at 110 °C.

Adsorption studies

Equation (1) was used for calculating MB removal efficiency:

$$R(\%) = \frac{C_o - C_f}{C_o} \times 100 \qquad (1)$$

where R is the removal efficiency of MB (%), C_o is the initial MB concentration (mg/L), and C_f is the final MB concentration (mg/L).

Equation (2) was used to estimate the MB adsorption capacity onto nZVI or nZVAl:

$$q = \frac{(C_o - C_f) \times V}{W} \qquad (2)$$

where q is the quantity of MB adsorbed per unit mass of nZVI or nZVAl (mg/g), V is the solution volume (L), and W is the nZVI or nZVAl mass (g).

Adsorption kinetic models

Equation (3) represents the linearized form of the pseudo-first-order model [39-42]:

$$\log(q_e - q_t) = \log(q_e) - \frac{k_1}{2.303}t \quad (3)$$

where k_1 is the pseudo-first-order constant (mg/g. min) and q_t and q_e are the amounts of adsorbed molecules at time t and equilibrium, respectively (mg/g).

Equation (4) represents the linearized form of the pseudo-second-order model:

$$\frac{t}{q_t} = \frac{1}{k_2 \times q_e^2} + \frac{1}{q_e}t \quad (4)$$

where k_2 is pseudo-second-order constant (g/mg.min).

Adsorption isotherm models

The linearized form of the Langmuir isotherm model is presented in Eq. (5) [43-47].

$$\frac{C_e}{q_e} = \left(\frac{1}{Q_m}\right)C_e + \frac{1}{K_L Q_m} \quad (5)$$

where C_e is the MB dye concentration in the aqueous solution at equilibrium (mg/L), Q_m is the maximum adsorption capacity, q_e is the amount of MB molecules adsorbed at equilibrium per gram of nZVI or nZVAl (mg/g), and K_L is the Langmuir constant (L/mg).

Equation (6) represents the linearized form of the Freundlich isotherm model:

$$log(q_e) = \left(\frac{1}{n}\right)log(C_e) + log(K_F) \quad (6)$$

where K_F is the Freundlich constant describing the multilayer adsorption capacity (mg/g).(L/mg)$^{1/n}$, and $1/n$ is the Freundlich constant describes the adsorption intensity/strength.

3. Result and discussions

Characterization of nZVI and nZVAl

According to the TEM analysis shown in Figure 1a, the surface morphology of the freshly synthesized nanoparticles was examined. It was revealed that the nZVI nanoparticles had core-shell structures, indicating that there was an iron oxide layer covering the nanoparticles. The core-shell model of nZVI consists of an inner core primarily composed of Fe^0 and an outer layer derived from oxides/hydroxides surrounding the Fe^0 as a shell.

In Figure 1b, the XRD pattern of nZVI is presented, displaying various peaks in the analysis. The peaks were relatively broadened and of different intensities. The XRD patterns confirmed the presence of iron and iron oxides in the nanoparticles. For instance, the reflection peaks at 2θ = 44.59°, 65.00°, and 82.50° corresponded to α-Fe^0 (110), α-Fe^0 (200), and α-Fe^0 (211) planes, respectively. The sharp peak at 2θ of 44.59° indicated the predominant presence of zero-valent iron (Fe^0) in the prepared nZVI samples.

Examining the TEM image of nZVAl particles in Figure 1c, it can be observed that the particles are semi-spherical, with the majority having a size less than 100 nm. Previous studies have suggested that the outer thin layer (shell) of the nanoparticles is likely composed of aluminum oxide, while the core consists of Al^0.

Figure 1d illustrates the XRD pattern of the nZVAl particles, showing different peak intensities. The peaks at angles 38.43°, 44.60°, 64.88°, and 78.09° corresponded to the α-Al^0 face-centered cubic cell phase, specifically the Al (111), Al (200), Al (220), and Al (311) planes, respectively. These results confirm the presence of aluminum nanoparticles in their zero-valent state (Al^0).

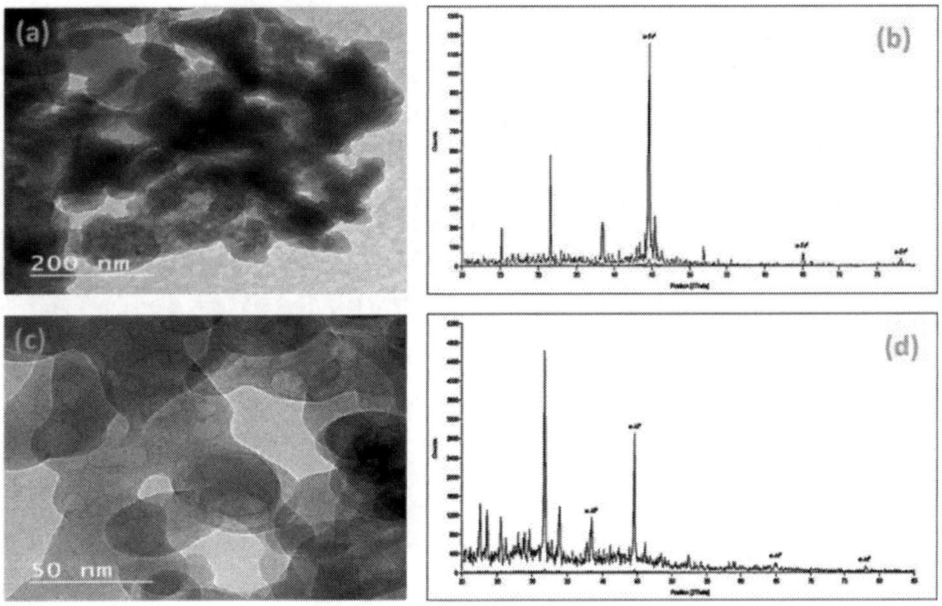

Fig. 1. (a and c) TEM, (b and d) XRD of nZVI and nZVAl, respectively

Adsorption performance

To examine the adsorption of MB onto nZVI, batch experiments were conducted using 250-mL Erlenmeyer flasks. The researchers employed a one-factor at-a-time approach to investigate the impact of various factors on the removal efficiency of MB. These factors include pH values ranging from 2 to 12, nZVI dosage from 2 to 14 g/L, initial dye concentration from 10 to 70 mg/L, temperature from 30 to 50 °C, stirring rate from 100 to 250 rpm, and contact time from 10 to 120 minutes.

Similarly, batch experiments were carried out in 250-mL Erlenmeyer flasks to study the adsorption of MB onto nZVAl. The researchers utilized a one-factor at-a-time technique to analyze the influence of different environmental conditions on the efficiency of MB removal. These conditions included pH values ranging from 2 to 12, nZVAl dosage from 0.1 to 1.0 g/L, MB concentrations from 10 to 80 mg/L, stirring rates from 100 to 300 rpm, and adsorption temperatures from 30 to 60 °C. Table 1 summarizes the removal efficiencies of MB by nZVI compared to nZVAl. It was found that there is imperceptible difference in the removal performance of nZVI and nZVAl towards MB based on the experimental conditions.

Table 1. Comparison of removal efficiencies of MB from aqueous solutions using nZVI and nZVAl

Adsorbent	Experimental factor						Removal efficiency (R%)
	pH	Adsorbent dosage (g L^{-1})	Initial MB (mg L^{-1})	Temperature (°C)	Stirring rate (rpm)	Contact time (min)	
nZVI	6	10	10	30	150	30	100%
nZVAl	10	1	10	30	200	90	99.0%

Kinetic and isotherm studies

To gain a quantitative understanding of the adsorption process, various kinetic models were employed. The kinetic adsorption of MB onto nZVI and nZVAl was determined using the pseudo-first-order and pseudo-second-order models. Isotherm studies were conducted to describe the adsorption equilibrium between the adsorbent and adsorbate molecules. The equilibrium isotherm of MB was determined using Langmuir and Freundlich isotherms. Table 2 represents the kinetic and isotherm studies for adsorption of MB onto nZVI compared with nZVAl.

Table 2. Kinetic and isotherm parameters for adsorption of MB onto both nZVI and nZVAl

Adsorbent	Kinetic parameters					
	Pseudo-first-order kinetic			Pseudo-second-order kinetic		
	k_1 (min^{-1})	q_e (mg g^{-1})	R^2	k_2 (g mg^{-1} min^{-1})	q_e (mg g^{-1})	R^2
nZVI	0.022×10^{-3}	1.052	0.8007	0.056	4.836	0.9997
nZVAl	0.0274	3.22	0.9016	0.0212	26.2	0.9999
	Isotherm parameters					
	Langmuir isotherm			Freundlich isotherm		
	q_m (mg g^{-1})	K_L (L mg^{-1})	R^2	$1/n$ (L g^{-1})	K_F (mg g^{-1})	R^2
nZVI	7.446	1.036	0.6778	0.432	3.431	0.9358
nZVAl	4.24	0.122	0.8253	0.4926	1.45	0.9569

The experimental data for nZVI showed a close fit to the pseudo-second-order model, with a coefficient of determination, R^2, close to unity. The calculated adsorption capacity ($q_{e,cal}$) using the pseudo-second-order model was 4.836 mg/g, which was similar to the experimental value ($q_{e,exp}$ = 4.665 mg/g). This indicates that the rate-limiting step in the adsorption of MB onto nZVI could be chemisorption, involving valence forces and electron sharing between the nanoparticles and MB molecules.

Similarly, for nZVAl, a high R^2 value of 0.9999 suggests a good fit of the pseudo-second-order model to the experimental data. This indicates that chemisorption plays a significant role in the removal of MB dye, involving chemical bonding between MB molecules and the active sites on the surface of the nanoparticles. The adsorption capacity obtained from the experiment ($q_{e,exp}$ = 26.0 mg/g) closely matched the value calculated by the model ($q_{e,cal}$ = 26.2 mg/g).

For nZVI, the Freundlich model provided a good fit to the experimental data, indicated by a coefficient of determination R^2 of 0.9358. This suggests that the adsorption of MB onto nZVI followed the Freundlich isotherm, and the model constants (K_F and $1/n$) were suitable for describing the adsorption capacity. The agreement with the Freundlich isotherm indicates that multilayer adsorption played a significant role in MB removal. Furthermore, the heterogeneous adsorbent surface of nZVI, with different adsorption sites, exhibited a high affinity for MB. The value of the slope $1/n$ (0.432 L/g) indicates favorable adsorption of dye molecules onto nZVI, as it is less than unity.

Similarly, for nZVAl, the Freundlich isotherm also provided a good fit to the experimental data, with a high R^2 value of 0.9569. This suggests that multilayer adsorption played a significant role in the removal of MB dye. The results further indicate that the heterogeneous adsorbent surface of nZVAl exhibited a high affinity for the cationic dye MB. The model constants were $1/n = 0.4926$ L/g and $K_F = 1.45$ (mg/g).(L/mg)$^{1/n}$. The value of $1/n$ being less than unity demonstrates the favorable adsorption of MB molecules onto nZVAl.

4. Conclusions

This review suggests that nZVI can serve as a suitable adsorbent material for the removal of MB dye from aqueous solutions. The adsorption data were accurately described by the Freundlich isotherm model (K_F = 3.431 mg/g, $1/n$ = 0.432 L/g, R^2 value = 0.9358) and the pseudo-second-order kinetic model (k_2 = 0.056 g/mg/min, q_e = 4.836 mg/g, R^2 value = 0.9976). The optimal pH value was determined to be 6, achieving a removal efficiency of approximately 100% with a dosage of 10 g/L of nZVI, an initial MB concentration of 10 mg/L, a temperature of 30 °C, a stirring rate of 150 rpm, and a reaction time of 30 minutes. Moreover, the findings demonstrated that nZVAl particles effectively adsorb MB from aqueous solutions, achieving a removal efficiency of 99% under specific conditions. These conditions include an initial MB concentration of 10 mg/L, a pH of 10, an nZVAl dosage of 1.0 g/L, a stirring rate of 200 rpm, a temperature of 30 °C, and an equilibrium time of 90 minutes. The adsorption data were well described by the Freundlich isotherm model (K_F = 1.45 mg/g, $1/n$ = 0.4926 L/g, R^2 value = 0.9569) and the pseudo-second-order kinetic model (k_2 = 0.0212 g/mg/min, q_e = 26.2 mg/g, R^2 value = 0.9999).

References

1. Fan, J.; Zhang, B.; Zhu, B.; Shen, W.; Chen, Y.; Zeng, F. New Insight into the Mechanism for the Removal of Methylene Blue by Hydrotalcite-Supported Nanoscale Zero-Valent Iron. *Water* **2023**, *15*, 183–183, doi:https://doi.org/10.3390/w15010183.

2. Hamdy, A.; Mostafa, M.K.; Nasr, M. Zero-Valent Iron Nanoparticles for Methylene Blue Removal from Aqueous Solutions and Textile Wastewater Treatment, with Cost Estimation. *Water Science and Technology* **2018**, *78*, 367–378, doi:https://doi.org/10.2166/wst.2018.306.

3. Katowah, D.F.; Ismail, S.H.; Sadek, A.H.; Rahman, M.M. Ultrasensitive QCM Sensor Development for Monitoring of Methyl Orange Dye in Aqueous Phase Based on Novel Cross-Linked Chitosan/PVA/GO/Ce-ZnO Nanocomposite Film. *Materials Science and Engineering: B* **2023**, *297*, 116804, doi:https://doi.org/10.1016/j.mseb.2023.116804.

4. Bayomie, O.S.; Kandeel, H.; Shoeib, T.; Yang, H.; Youssef, N.; El-Sayed, M.M.H. Novel Approach for Effective Removal of Methylene Blue Dye from Water Using Fava Bean Peel Waste. *Scientific Reports* **2020**, *10*, doi:https://doi.org/10.1038/s41598-020-64727-5.

5. Agarwal, J.; Sonia Dyes and Dyeing Processes for Natural Textiles and Their Key Sustainability Issues. **2021**, 439–472, doi:https://doi.org/10.1016/b978-0-12-821483-1.00013-9.

6. Sadek, A.H.; Abdel-Karim, A.; Sajjad Mohsenpour; Ismail, S.H.; Bayoumy, A.M.; Ibrahim, M.; Mohamed, G.G. Polysulfone-Based Mixed Matrix Membranes Loaded with a Multifunctional Hierarchical Porous Ag-Cu Dendrites@SiO_2 Core-Shell Nanostructure for Wastewater Treatment. *Process Safety and Environmental Protection* **2023**, *175*, 677–691, doi:https://doi.org/10.1016/j.psep.2023.05.085.

7. Sudarshan, S.; Harikrishnan, S.; RathiBhuvaneswari, G.; Alamelu, V.; Aanand, S.; Rajasekar, A.; Govarthanan, M. Impact of Textile Dyes on Human Health and Bioremediation of Textile Industry Effluent Using Microorganisms: Current Status and Future Prospects. *Journal of Applied Microbiology* **2022**, *134*, doi:https://doi.org/10.1093/jambio/lxac064.

8. A. Alhadhrami; Mohamed, G.G.; Sadek, A.H.; Ismail, S.H.; Ebnalwaled, A.A.; Abdulraheem Behavior of Silica Nanoparticles Synthesized from Rice Husk Ash by the Sol–Gel Method as a Photocatalytic and Antibacterial Agent. *Materials* **2022**, *15*, 8211–8211, doi:https://doi.org/10.3390/ma15228211.

9. Singha, K.; Pandit, P.; Maity, S.; Sharma, S.R. Harmful Environmental Effects for Textile Chemical Dyeing Practice. *Green Chemistry for Sustainable Textiles* **2021**, 153–164, doi:https://doi.org/10.1016/b978-0-323-85204-3.00005-1.

10. Hamdy, A.; Mostafa, M.K.; Vo, D.-V.N. Regression Analysis and Artificial Intelligence for Removal of Methylene Blue from Aqueous Solutions Using Nanoscale Zero-Valent Iron. **2018**, *16*, 357–372, doi:https://doi.org/10.1007/s13762-018-1677-z.

11. Oladoye, P.O.; Ajiboye, T.O.; Omotola, E.O.; Oyewola, O.J. Methylene Blue Dye: Toxicity and Potential Elimination Technology from Wastewater. *Results in Engineering* **2022**, *16*, 100678, doi:https://doi.org/10.1016/j.rineng.2022.100678.

12. Al-Gethami, W.; Al-Qasmi, N.; Ismail, S.H.; Sadek, A.H. QCM-Based $MgFe_2O_4$@CaAlg Nanocomposite as a Fast Response Nanosensor for Real-Time Detection of Methylene Blue Dye. *Nanomaterials* **2022**, *13*, 97, doi:https://doi.org/10.3390/nano13010097.

13. Tarekegn, M.M.; Balakrishnan, R.M.; Hiruy, A.M.; Dekebo, A.H. Removal of Methylene Blue Dye Using Nano Zerovalent Iron, Nanoclay and Iron Impregnated Nanoclay – a Comparative Study. *RSC Advances* **2021**, *11*, 30109–30131, doi:https://doi.org/10.1039/d1ra03918k.

14. Sadek, A.H.; Asker, M.S.; Abdelhamid, S.A. Bacteriostatic Impact of Nanoscale Zero-Valent Iron against Pathogenic Bacteria in the Municipal Wastewater. *Biologia* **2021**, *76*, 2785–2809, doi:https://doi.org/10.1007/s11756-021-00814-w.

15. Valli Nachiyar, C.; Rakshi, A.D.; Sandhya, S.; Britlin Deva Jebasta, N.; Nellore, J. Developments in Treatment Technologies of Dye-Containing Effluent: A Review. *Case Studies in Chemical and Environmental Engineering* **2023**, *7*, 100339, doi:https://doi.org/10.1016/j.cscee.2023.100339.

16. Ledakowicz, S.; Paździor, K. Recent Achievements in Dyes Removal Focused on Advanced Oxidation Processes Integrated with Biological Methods. *Molecules* **2021**, *26*, 870, doi:https://doi.org/10.3390/molecules26040870.

17. Aragaw, T.A.; Bogale, F.M. Biomass-Based Adsorbents for Removal of Dyes from Wastewater: A Review. *Frontiers in Environmental Science* **2021**, *9*, doi:https://doi.org/10.3389/fenvs.2021.764958.

18. Panchal, D.; Sharma, A.; Pal, S. Novel Photocatalytic Techniques for Organic Dye Degradation in Water. **2021**, 1–22, doi:https://doi.org/10.1016/b978-0-12-823876-9.00013-5.

19. Kamar, M. T., Elattar, H., Mahmoud, A. S., Peters, R. W. & Mostafa, M. K. A critical review of state-of-the-art technologies for electroplating wastewater treatment. Int. J. Environ. Anal. Chem. (2022) doi:10.1080/03067319.2022.2098486.

20. Heberling, J., Adewuyi, Y., Mahmoud, A. S., Mostafa, M. & Peters, R. W. (12e) AOP Performance at Wastewater Treatment Plants | AIChE. AIChE Annual Meeting Proceedings, David L. Lawrence Convention Center, Pittsburgh, PA, October 28 - November 2 https://www.aiche.org/conferences/aiche-annual-meeting/2018/proceeding/paper/12e-aop-performance-wastewater-treatment-plants (2018).

21. Abdelmigeed, M.O.; Sadek, A.H.; Ahmed, T.S. Novel Easily Separable Core–Shell Fe_3O_4/PVP/ZIF-8 Nanostructure Adsorbent: Optimization of Phosphorus Removal from

Fosfomycin Pharmaceutical Wastewater. *RSC Advances* **2022**, *12*, 12823–12842, doi:https://doi.org/10.1039/d2ra00936f.

22. Xie, Q.; Li, Y.; Lv, Z.; Zhou, H.; Yang, X.; Chen, J.; Guo, H. Effective Adsorption and Removal of Phosphate from Aqueous Solutions and Eutrophic Water by Fe-Based MOFs of MIL-101. *Scientific Reports* **2017**, *7*, doi:https://doi.org/10.1038/s41598-017-03526-x.

23. Al-Gethami, W.; Alhashmialameer, D.; Al-Qasmi, N.; Ismail, S.H.; Sadek, A.H. Design of a Novel Nanosensors Based on Green Synthesized $CoFe_2O_4$/Ca-Alginate Nanocomposite-Coated QCM for Rapid Detection of Pb(II) Ions. *Nanomaterials* **2022**, *12*, 3620, doi:https://doi.org/10.3390/nano12203620.

24. Aghababai Beni, A.; Jabbari, H. Nanomaterials for Environmental Applications. *Results in Engineering* **2022**, *15*, 100467, doi:https://doi.org/10.1016/j.rineng.2022.100467.

25. Chavali, M.S.; Nikolova, M.P. Metal Oxide Nanoparticles and Their Applications in Nanotechnology. *SN Applied Sciences* **2019**, *1*, doi:https://doi.org/10.1007/s42452-019-0592-3.

26. Hamdy, A.; Ismail, S.H.; Ebnalwaled, A.A.; Mohamed, G.G. Characterization of Superparamagnetic/Monodisperse PEG-Coated Magnetite Nanoparticles Sonochemically Prepared from the Hematite Ore for Cd(II) Removal from Aqueous Solutions. *Journal of Inorganic and Organometallic Polymers and Materials* **2020**, *31*, 397–414, doi:https://doi.org/10.1007/s10904-020-01741-0.

27. Tarekegn, M.M.; Hiruy, A.M.; Dekebo, A.H. Nano Zero Valent Iron (NZVI) Particles for the Removal of Heavy Metals (Cd^{2+}, Cu^{2+} and Pb^{2+}) from Aqueous Solutions. *RSC Advances* **2021**, *11*, 18539–18551, doi:https://doi.org/10.1039/d1ra01427g.

28. Baig, N.; Kammakakam, I.; Falath, W. Nanomaterials: A Review of Synthesis Methods, Properties, Recent Progress, and Challenges. *Materials Advances* **2021**, *2*, 1821–1871, doi:https://doi.org/10.1039/d0ma00807a.

29. Khajeh, M.; Laurent, S.; Dastafkan, K. Nanoadsorbents: Classification, Preparation, and Applications (with Emphasis on Aqueous Media). *Chemical Reviews* **2013**, *113*, 7728–7768, doi:https://doi.org/10.1021/cr400086v.

30. Hamdy, A. Experimental Study of the Relationship between Dissolved Iron, Turbidity, and Removal of Cu(II) Ion from Aqueous Solutions Using Zero-Valent Iron Nanoparticles. *Arabian Journal for Science and Engineering* **2020**, *46*, 5543–5565, doi:https://doi.org/10.1007/s13369-020-05079-0.

31. Li, S.; Li, L.; Zhang, W. Nanoscale Zero-Valent Iron (NZVI) for Heavy Metal Wastewater Treatment: A Perspective. *Engineering* **2023**, doi:https://doi.org/10.1016/j.eng.2023.08.012.

32. Hamdy, A.; Mostafa, M.K.; Nasr, M. Techno-Economic Estimation of Electroplating Wastewater Treatment Using Zero-Valent Iron Nanoparticles: Batch Optimization, Continuous

Feed, and Scaling up Studies. *Environmental Science and Pollution Research International* **2019**, *26*, 25372–25385, doi:https://doi.org/10.1007/s11356-019-05850-3.

33. Ye, J.; Luo, Y.; Sun, J.; Shi, J. Nanoscale Zero-Valent Iron Modified by Bentonite with Enhanced Cr(VI) Removal Efficiency, Improved Mobility, and Reduced Toxicity. *Nanomaterials* **2021**, *11*, 2580–2580, doi:https://doi.org/10.3390/nano11102580.

34. Dong, H.; Deng, J.; Xie, Y.; Zhang, C.; Jiang, Z.; Cheng, Y.; Hou, K.; Zeng, G. Stabilization of Nanoscale Zero-Valent Iron (NZVI) with Modified Biochar for Cr(VI) Removal from Aqueous Solution. *Journal of Hazardous Materials* **2017**, *332*, 79–86, doi:https://doi.org/10.1016/j.jhazmat.2017.03.002.

35. Li, Y.; Bao, X.; Yang, S.; Li, Q.; Fan, D.; Wang, H.; Zhao, D. Application Potential of Zero-Valent Aluminum in Nitrophenols Wastewater Decontamination: Enhanced Reactivity, Electron Selectivity and Anti-Passivation Capability. *Journal of Hazardous Materials* **2023**, *452*, 131313–131313, doi:https://doi.org/10.1016/j.jhazmat.2023.131313.

36. Sadek, A.H.; Mostafa, M.K. Preparation of Nano Zero-Valent Aluminum for One-Step Removal of Methylene Blue from Aqueous Solutions: Cost Analysis for Scaling-up and Artificial Intelligence. *Applied Water Science* **2022**, *13*, doi:https://doi.org/10.1007/s13201-022-01837-7.

37. Li, S.; Yang, S.; Bao, X.; Li, Y.; Wang, M.; Zhao, D. Carbon-Modified/Embedded Zero-Valent Aluminum Microparticles Will Control Electron Release for Efficient Adsorption and Degradation of Aqueous Pollutants. *Journal of Cleaner Production* **2022**, *366*, 133013–133013, doi:https://doi.org/10.1016/j.jclepro.2022.133013.

38. Sadek, A.H.; Fahmy, O.M.; Mahmoud Nasr; Mostafa, M.K. Predicting Cu(II) Adsorption from Aqueous Solutions onto Nano Zero-Valent Aluminum (NZVAl) by Machine Learning and Artificial Intelligence Techniques. *Sustainability* **2023**, *15*, 2081–2081, doi:https://doi.org/10.3390/su15032081.

39. Farag, R. S., Elshfai, M. M., Mahmoud, A. S., Mostafa, M. K. & Peters, R. W. (592d) Green Synthesis of Nano Iron Carbide: Preparation, Characterization and Application for Removal of Phosphate from Aqueous Solutions. Annual AIChE Meeting Proceedings; Pittsburgh, PA, October 28 - November 2 https://www.aiche.org/conferences/aiche-annual-meeting/2018/proceeding/paper/592d-green-synthesis-nano-iron-carbide-preparation-characterization-and-application-removal (2018).

40. Mahmoud, A. S., Mostafa, M. K. & Peters, R. W. A prototype of textile wastewater treatment using coagulation and adsorption by Fe/Cu nanoparticles: Techno-economic and scaling-up studies. Nanomater. Nanotechnol. 11, (2021).

41. Mahmoud, A. S., Mohamed, N. Y., Mostafa, M. K. & Mahmoud, M. S. Effective Chromium Adsorption From Aqueous Solutions and Tannery Wastewater Using Bimetallic Fe/Cu

Nanoparticles: Response Surface Methodology and Artificial Neural Network. Air, Soil Water Res. 14, (2021).

42. Mahmoud, M. S., Mahmoud, A. S., El-Said, M. A. & Mostafa, M. K. Comparison of aluminum and iron nanoparticles for chromium removal from aqueous solutions and tannery wastewater, empirical modeling and prediction. Emergent Mater. 5, 1729–1744 (2022).

43. Mostafa, M. K., Mahmoud, A. S., Mahmoud, M. S. & Nasr, M. Computational-Based Approaches for Predicting Biochemical Oxygen Demand (BOD) Removal in Adsorption Process. Adsorpt. Sci. Technol. 2022, (2022).

44. Mahmoud, A. S., Mahmoud, M. S., Yousri, N. & Mostafa, M. K. Advanced Wastewater Treatment by Using Entrapped nZVI into Alginate (Ag) Biopolymer: Adsorption Isotherm, Kinetic Models, and Statistical Analysis. Key Eng. Mater. 921, 147–160 (2022).

45. Mahmoud, A. S., SaryEl-deen, R. A., Mostafa, M. K. & Peters, R. W. (655d) Artificial Intelligence for Organochlorine Pesticides Removal from Aqueous Solutions Using Entrapped Nzvi in Alginate Biopolymer | AIChE. AIChE Annual Meeting Proceedings https://www.aiche.org/conferences/aiche-annual-meeting/2017/proceeding/paper/655d-artificial-intelligence-organochlorine-pesticides-removal-aqueous-solutions-using-entrapped (2017).

46. El-Shafei, M. M., Mahmoud, A. S., Mostafa, M. K. & Peters, R. W. Effects of entrapped nZVI in alginate polymer on BTEX removal. Environmental Division 2016 - Core Programming Area at the 2016 AIChE Annual Meeting 15–22 https://www.aiche.org/conferences/aiche-annual-meeting/2016/proceeding/paper/60b-effects-entrapped-nzvi-alginate-polymer-on-btex-removal (2016).

47. Farag, R. S. et al. (670d) Study the degradation and adsorption processes of organic matters from domestic wastewater using chemically prepared and green synthesized nano zero-valent iron. AIChE Annual Meeting, Conference Proceedings vols 2019-Novem https://www.aiche.org/conferences/aiche-annual-meeting/2019/proceeding/paper/670d-study-degradation-and-adsorption-processes-organic-matters-domestic-wastewater-using-chemically (2019).

117ak Combining Ion Exchange and Deammonification for Sustainable Ammonium Removal from Mainstream Wastewater

Sheyla Chero-Osorio, Susieanna Persaud, Tione Grant, John Kuhn, Sarina Ergas

Introduction

Excess reactive nitrogen discharges severely impact surface and groundwater, resulting in environmental, economic, and public health problems. Domestic wastewater is considered a major point source of nitrogen pollution. Conventional biological nitrogen removal (BNR) processes are widely used to remove inorganic nitrogen from domestic wastewater through the processes of nitrification and denitrification. Conventional BNR processes typically require a high energy demands to provide oxygen for nitrification and high chemical demands to provide organic carbon for denitrification. Biological deammonification is a novel BNR process that combines partial nitritation (NH_4^+ => NO_2^-) with anaerobic ammonium oxidation (anammox; $NH_4^+ + NO_2^-$ => N_2) (Ergas and Aponte-Morales, 2014). The combined partial nitritation/anammox (PN/A) process considerably reduces the oxygen and organic carbon requirements at full-scale wastewater treatment plants (WWTPs). Therefore, deammonification is promising as a more sustainable BNR alternative with lower energy and chemical requirements than conventional BNR (Cao et al., 2017).

Implementation of deammonification has been successful for treatment of high-ammonium strength wastewaters, such as sidestreams from anaerobic digestion process. Full-scale sidestream PN/A plants are in operation in the US, Europe, and Australia (Bailey et al., 2018). Elevated ammonium concentrations result in high free ammonia concentrations, which are suitable to inhibit competitor microbes that otherwise interfere with the PN/A process. However, typical domestic (mainstream) wastewater has a much lower ammonium (and free ammonia) content, making the PN/A process challenging to implement and control (Anthonisen et al., 1976). To date, there are no reported full-scale mainstream PN/A plants operating worldwide (Izadi et al., 2021).

Our research project has implemented a strategy of concentrating ammonium on chabazite, a natural zeolite mineral with a high ion exchange (IX) capacity and selectivity for ammonium ions. The fundamental basis of this process is to pass large volumes of mainstream wastewater through the chabazite media bed until saturation is reached. Subsequently, the mineral is bioregenerated via PN/A by microbial biofilms attached to the chabazite surface that consume ammonium. The bioregeneration process allows chabazite to be reused for many treatment cycles without having to add fresh zeolite or regenerant brines (Aponte-Morales et al., 2016, 2018).

The ***overall goal*** of the proposed project is to implement IX-PN/A process in a single-stage for ammonium removal from mainstream wastewater. Specific objectives are to: a) evaluate whether chabazite is effective in concentrating ammonium and providing the environmental conditions needed for PN/A, b) demonstrate that the microbes involved in deammonification can bioregenerate chabazite for reuse over many IX-PN/A cycles, and c) identify the diversity among the microbial community governing the IX-PN/A process.

Methodology

A 2-L sequencing batch biofilm reactor (SBBR) filled with an adsorptive medium containing chabazite was used to conduct IX-PN/A as a single process. Once the IX process reached a complete breakthrough curve ($C_{effluent}/C_{influent} \approx 100\%$) (Figure 1, left), the reactor was inoculated with biofilms from PN/A enriched cultures, and bioregeneration was initiated by applying recirculation to provide aeration and aid mass transfer. The steps in a typical IX-PN/A treatment cycle consisted of IX, sidestream addition, and bioregeneration (by PN/A). During IX, the IX-PN/A bioreactor was fed with mainstream wastewater (~40 mg/L ammonium as nitrogen) using variable feeding rates (22-34 mL/min) to reach a target breakpoint. Sodium and ammonium concentrations were measured using Standard Method 3500 (APHA) by Metrohm

881 Compact IC Pro Ion Chromatography. Subsequently, synthetic sidestream (~500 mg/L ammonium as nitrogen) was added to ensure a high free ammonia content in the liquid phase to favor conditions for PN/A. During bioregeneration, samples were taken periodically to measure ammonium, nitrite, and nitrate concentrations over time and determine the nitrogen removal and zeolite bioregeneration. Measurements of nitrogen species (NH_4^+, NO_2^-, and NO_3^-) and other ion species (Na^+, SO_4^{2-}, and PO_4^{3-}) concentrations were done by Chromatography. The pH, temperature, turbidity, and concentrations of dissolved oxygen, sulfide, and chemical oxygen demand (COD) were measured by the end of each IX-PN/A cycle following APHA Standard Methods.

Results and Discussion

Results show that, during the IX stage, chabazite successfully removed large ammonium loads from low ammonia-strength wastewater. During a typical IX cycle, ammonium was removed from mainstream wastewater until a breakpoint of $C_{effluent}/C_{influent} \approx$ 30-40% was achieved (Figure 1, right). During the bioregeneration stage, the ammonium concentration in the effluent typically reached <1 mg/L (as N) due to the activity of PN/A microorganisms. At the end of the bioregeneration stage, the chabazite was ready to start the next IX-PN/A treatment cycle (Figure 2). The single-stage IX-PN/A was operated for twelve treatment cycles (to March 2023). Sodium and ammonium profiles over a typical cycle are shown in Figure 3. During IX, ammonium capture was accompanied by sodium release, which slowed as the chabazite reached saturation. During bioregeneration, microbial ammonium uptake from the chabazite surface was accompanied by sodium adsorption from the liquid phase onto the surface, providing exchangeable sodium ions for the following IX-PN/A cycle.

The microbial basis on which the IX-PN/A process relied was that ammonium concentration progressively decreased due to the activity of ammonia-oxidizing microorganisms (AOM) and anammox bacteria, with suppression of competitor nitrite-oxidizing bacteria (NOB). Some NOB activity and nitrate accumulation was observed in some treatment cycles (Figure 2) with high recirculation rates provided for microbial detachment and sampling. Microbial analysis by polymerase chain reaction (PCR) also confirmed the existence of AOM and anammox bacteria through the IX-PN/A column.

Conclusions

Deammonification, or partial nitritation/anammox (PN/A), is a novel BNR technology with lower energy and chemical requirements than conventional BNR. Bioregeneration of chabazite with PN/A eliminates the need for chemical regenerant inputs, replacement with fresh media, or brine waste production. As the natural chabazite is reused for many treatment cycles, this process offers a sustainable strategy for a circular economy. Finally, coupling IX with PN/A is an innovative, effective BNR for mainstream wastewater, with lower costs, environmental impacts, and carbon footprint. Future work will focus on a) identifying the effect of operating conditions on PN/A performance through experimental and modeling studies and b) compare system performance with different strategies for biofilm formation: naturally formed PN/A biofilm, chabazite-coated biocarriers for attached biofilm growth, or biocoatings encapsulating PN/A microbes.

References

Anthonisen, A. C., Loehr, R. C., Prakasam, T. B. S., and Srinath, E. G. (1976). Inhibition of nitrification by ammonia and nitrous acid. *Journal (Water Pollution Control Federation)*, 835-852.

Aponte-Morales, V. E., Payne, K. A., Cunningham, J. A., & Ergas, S. J. (2018). Bioregeneration of chabazite during nitrification of centrate from anaerobically digested livestock waste: Experimental and modeling studies. *Environmental science & technology*, *52*(7), 4090-4098.

Aponte-Morales, V. E., Tong, S., & Ergas, S. J. (2016). Nitrogen removal from anaerobically digested swine waste centrate using a laboratory-scale chabazite-sequencing batch reactor. *Environmental Engineering Science*, *33*(5), 324-332.

Bailey, E. L., Bilyk, K., Hanna, A., & Wankmuller, D. (2018). Evaluation of Sidestream Deammonification Process Enhancements for Treating High Strength Filtrate at the City of Raleigh's Neuse River Resource Recovery Facility. *Proc. Water Environ. Fed., 2018*, 5286-5307.

Cao, Y., van Loosdrecht, M., and Daigger, G. T. (2017). Mainstream partial nitritation–anammox in municipal wastewater treatment: status, bottlenecks, and further studies. *Applied microbiology and biotechnology*, 101(4), 1365-1383.

Ergas, S. J., and Aponte-Morales, V. (2014). Biological Nitrogen Removal. *Comprehensive Water Quality and Purification: Vol. 3 Remediation of Polluted Waters*, S. Sengupta Ed., Elsevier.

Izadi, P., Izadi, P., & Eldyasti, A. (2021). Towards mainstream deammonification: Comprehensive review on potential mainstream applications and developed sidestream technologies. *Journal of Environmental Management*, *279*, 111615.

117al Enhancing Lead Removal from Wastewater Using Waste-Derived Boron-Doped Carbon

hafiz Muhammad Aamir, Matthew Alexander

Lead is a common hazardous heavy metal found in agricultural soil and industrial wastewater. It is a dire need for society to remove lead from wastewater via a facile and cost-effective route, which is needed to improve water quality. Here we obtain boron (B) doped porous carbon (BC) via two-step thermal pyrolysis of waste tires, as it is a promising raw material. The B doping changes the coordination chemistry, physicochemical properties, and surface structure of carbon. The degree of graphitization, boron content, and pore size distribution of BC was controlled by thermal treatment and $NaBH_4$ content. The surface topography, elemental composition, degree of graphitization, chemical structure, and lead (Pb) removal efficiency of BC was thoroughly analyzed by SEM, EDS, Raman, XRD, XPS, and ICP-OES analysis. Benefiting from the structural disorder, high specific surface area, and least coordinated exposed carbon, the BC revealed high adsorption efficiency (98.02%) for lead removal from wastewater.

117am A Silk Fibroin-Based Wastewater Treatment System By Automated Air-Delivery

Fengjie He, Hui Zhao, Shengjie Zhai

As our personal drinking water sources become increasingly contaminated, a large population on earth live without access to clean and safety water and almost every two minutes a child dies from a water-related disease. Thus, treating polluted water sources like rivers, lakes and streams becomes a matter of great urgency to alleviate water scarcity and supply safe drinking water. Nevertheless, existing commercial adsorbents such as activated carbon consume a significant amount of energy during mass production and have low removal efficiency when operating in harsh environments. Herein, a sustainable silk fibroin (SF) film is fabricated as an efficient bioadsorbent for wastewater remediation. The experimental results exhibited the fabricated SF films showed high adsorption capacities towards different types of water-soluble dye contaminants. The maximum adsorption capacity of SF films can reach up to 174.48 mg/g. Moreover, the SF films also demonstrated strong acid and alkaline resistance, allowing them to meet the harsh real wastewater treatment requirements (pH 1~13). On the other hand, in order to fulfill the dynamic and intricate treatment requirements of natural water sources, we integrated internet of things (IoT)-based real-time water quality monitoring part and the SF film based sustainable bioadsorbent into a complete treatment system. During the operation of the integrated system, the water quality data are real-time captured, processed and uploaded by the Raspberry Pi to Google Cloud. Simultaneously, the cloud server automatically analyzes data to develop an optimized treatment strategy. For the delivery of the treatment system to target water body (including the hard-to-reach sites), it is achieved by the Unmanned Aerial Vehicles autonomously. In conclusion, we provide a promising and closed-loop strategy for high-performance remediation and management of polluted natural water sources.

117an Photocatalytical Degradation of Congo Red (CR) Dye By Nano Titanium Dioxide Coated Glass Bead Under UV Light

Asad Khan

Titanium dioxide (TiO_2) nano-scale particles were synthesised and coated firmly on soda lime glass beads were used as the catalyst for degradation of Congo red (CR) Dye. Sol-gel method was used for the for synthesising fine and uniformly dispersed nano-sized TiO_2 (n-TiO_2) particles followed by imobilising on the soda lime glass beads by dry mixing and heating. These imoblised TiO_2-glass bead catalyst was packed in glass columns to form a reactor for evaluating the performance of the catalyst for photo-catalytic degradation of Congo-red (CR) dye in the presence of ultraviolet radiation. The effect of calcinations temperature on the TiO_2 Anatase and Rutile phase transformation; preparation of Glass surface (etching) and effect of glass support on the catalyst performance; the CR dye concentration used and pH of solution; the dye degradation in dark condition and the impact of U.V light exposure; the duration of U.V. light irradiation and its impact on the dye degradation; the catalyst (n-TiO_2) loading used with the catalyst regeneration were some of the parameters studied to evaluate and optimise the performance of the n-TiO_2 coated glass beads. It was observed that increasing the calcination temperature resulted in the phase transformation of anatase to rutile phase with complete rutile phase transformation at 800° C calcination temperature. The U.V. spectroscopy studies signified that photo-catalytic degradation is optimised in acidic zone as lower pH favored the formation of more soluble CR ions. A reduction in the n-TiO_2 desecration was observed with profound catalyst regeneration cycle up to five times for the same catalyst used due to the employed glass bead support. The U.V. light has optimised the CR dye degradation with 66% photo-catalytic dye degradation in just 10 minutes of the U.V. exposure. The TiO_2 spherical structure and nano-sized particles formation was confirmed by SEM analysis and recorded in the range of 130 to 270 nm. The elemental analysis was evaluated by EDX analysis and XRD analysis detected no characterization peaks other than TiO_2, signifying the synthesis of pure TiO_2 in the presented study. The surface area was measured using BET technique and was found 26.18 and 5.32 m^2/g for Titania samples prepared at 600° C and 800° C calcinations temperature respectively. U.V. spectrophotometer analysis appraised the CR dye degradation.

117ap Analyzing the Rationale and Assembly for Combined Oxidation Processes: Photocatalytic and Peroxide Treatment Coupled with Cavitation for Effluent Treatment

Varaha Prasad Sarvothaman, William Roberts

The production of wastewater is inevitable as a result of industrial and agricultural activities. The introduction of organic molecules in the form of solvents, pharmaceuticals, pesticides lead to generation of wastewaters necessitating its treatment. The treatment of these wastewaters requires multiple processing steps. Conventional treatment methods are not effective in treating wastewaters containing recalcitrant pollutants [1]. Advanced oxidation processes (AOPs) are processes which can generate reactive oxygen species, which can efficiently breakdown organic pollutants in wastewater. In recent years, there has been an avid interest in the use of semiconductor photocatalysis and hydrodynamic cavitation based techniques for effluent treatment [2].

Photocatalysis is a technique of generating reactive oxidant species without the catalyst being consumed. Titania and zinc oxide are common photocatalysts employed for semiconductor photocatalysis. UV light or sunlight can be used to illuminate photocatalysts and drive effluent breakdown. Research of coupling photocatalysis with hydrodynamic cavitation has been attempted in fewer than 10 studies in the recent years, with the objective of achieving required scalability with cavitation reactors and (chemical) oxidant effects with photocatalysis. This coupling is promising and has shown synergy and is effective in breakdown of phenolics [3]. Studies which investigate the breakdown of organic pollutants in these coupled studies use concentrations typically less than 100 ppm. However, in order to effectively treat effluents and be relevant to industrial processing, higher initial concentrations need to be considered and additional oxidants such as hydrogen peroxide or ozonation are required. The use of hydrogen peroxide in conjunction with photocatalysis does not lead to proper utilization of oxidant species, as hydrogen peroxide gets consumed in scavenger type of reactions. Thus, it is important to demarcate the phase at which these (hydrogen peroxide and photocatalysis) are present in the reaction mixture, in order to efficiently utilize these oxidants.

A hydrodynamic cavitation loop was developed to investigate the coupling of cavitation with hydrogen peroxide and photocatalysis. A pool of organic pollutants comprising – chlorocresol, dichloroaniline and paracetamol was prepared to simulate wastewater. The characterization of individual processes of i) cavitation and peroxide, and ii) cavitation and photocatalysis was first carried out. After determining the optimal parameters of the individual processes, a matrix of tests where photocatalysis was introduced at different stages of the reaction was performed. Similarly, peroxide was introduced at different stages of the reaction. The concentration of water matrix was quantified using TOC based techniques. An initial result of the coupling of peroxide and photocatalysis (shown below) does show a monotonic decrease with increase in peroxide concentration, which was on expected lines for the system. A complete characterization of the system with the test matrices is underway and will provide useful information for the strategy whilst coupling this powerful strategy of AOPs to treat effluents.

References

[1] V.P. Sarvothaman, Hydrodynamic Cavitation for Effluent Treatment: Using Vortex-based Cavitation Devices, PhD Thesis, Queen's University Belfast, Queen's University Belfast, 2020.

[2] V.V. Ranade, Bhandari, V.M., Nagarajan, S., Sarvothaman, V.P., and Simpson, A.T., Hydrodynamic Cavitation: Devices, Design and Applications, Wiley-VCH2022.

[3] M.V. Bagal, P.R. Gogate, Degradation of diclofenac sodium using combined processes based on hydrodynamic cavitation and heterogeneous photocatalysis, Ultrasonics sonochemistry 21(3) (2014) 1035-1043.

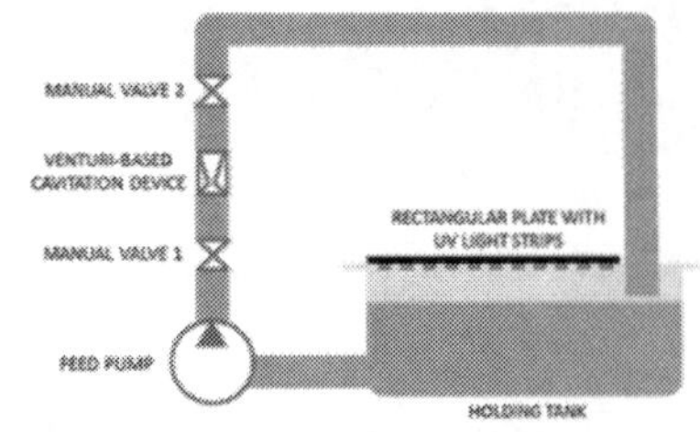

Schematic of cavitation based oxidation of pollutants

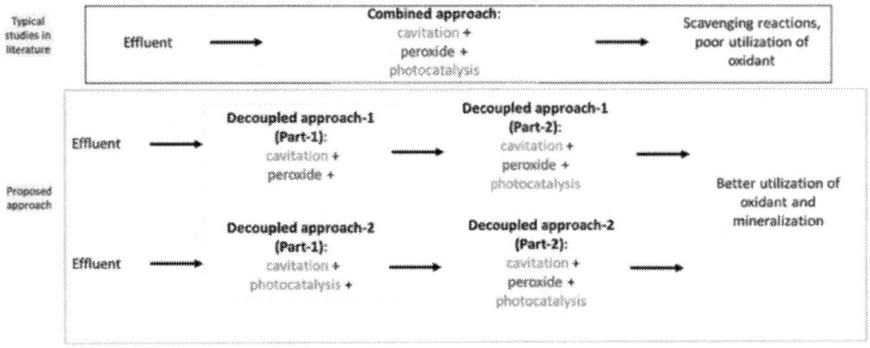

Methodology of proposed study

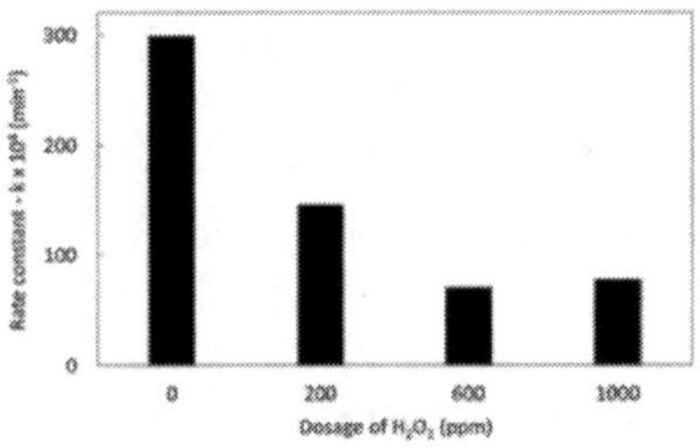

Influence of peroxide concentration with photocatalysis

117aq Sustainable and ECO-Friendly Algae-Based Produced Water Treatment

Fares Almomani, Malak Alsarayreh

Produced water (PW) is the largest waste stream generated in oil and gas industries, it contains a mixture of different organic and inorganic components. Qatar is considered one of the top ten oil and gas production countries, so a massive amount of produced water (PW) is expected. Discharging produced water without proper treatment will pollute surface and underground water and soil. One of the identified priority challenges for Qatar vision 2030 is the water security. The present work develop, for the first time, a sustainable and eco-friendly integrated algae-based treatment system. That algae system utilizes produced water, natural solar light, and CO_2 in an engineered or closed system (photo-bioreactor), to produced treated water with a quality suitable for reuse in different applications such as cooling system . On other hand, the produced algae biomass can be used as a source of animal feed, bio-fertilizer, and alternative feedstock for biofuel. The work was carried out at pilot scale under natural conditions of temperature and solar irradiation. The results identified the types and concentrations of the potentially toxic compounds and micro and macro nutrients present in the produced water for proper algae growth. Key parameters of light, CO_2 loading, and temperature for the selected microalgae species to establish optimal growth conditions were optimized. Chemical composition of the produced algal biomass were tested and the potential for value-added product development or as a source of animal feed, bio-fertilizer and alternative feedstock for biofuel was evaluated. Results showed that native algae species *Mychonastes, Chlorella, Chlorophyta, Desmodesmus, and Scenedesmus* can be used successfully for the treatment of PW, organic matter and nutrients removals in the range 88-93% and 80-89% were achieved. Pre-treatment process for the removal of suspended solids and oil content was essential to produce treated water that can be used as in cooling application. Algal feedstock showed high oil content and a potential as a sustainable and eco-friendly fuel source

117ar Removal of Heavy Metals from Wastewater Using Dead Algae Cells

Fares Almomani

The presence of heavy metals in the systems of water supplies become a global issue faces the most developed and developing countries worldwide. These metals are extremely toxic, non-biodegradable, and once they enter the food chain, they have the ability to accumulate at low concentration in the living organisms in general and human bodies in specific. Under specific environmental conditions, heavy metals may be accumulated to toxic levels and cause many serious diseases such as kidney failures, nervous system damage, cancer, and it can be deadly at high concentrations. During the last years, many different treatment processes have been used in order to remove heavy metal from industrial wastewater. This includes chemical precipitation, cementation, ion exchange, and membrane filtration. However, these treatment processes have some limitations such as the production of a huge quantity of heavy metals rich sludge, high operation cost, and some processes inapplicable on large scales (Metcalf et al. 2003). On the other hand, adsorption as a removal process is widely used because of its low cost and it is applicable to large-scale applications. Furthermore, the adsorption process by using low-cost adsorbents (i.e. fungi, yeast, bacteria, and algae) provides many advantages over the conventional treatment methods to heavy metals removal from industrial wastewater. For instance, it has a high removal efficiency with cost-effectiveness. Moreover, the use of bioprocesses technologies minimizes the production of chemical/biological sludge which makes this process environmentally friendly (Abbas et al. 2014). Algal cultures are considered the most suitable adsorbent which is more cost-effective and require fewer limitations such as high pressures and temperatures. Furthermore, it is widely available all over the world. Hence, there is growing attention in testing the ability of different lgae species for heavy metals removal from industrial wastewater. Finding suitable operating conditions is very important to address the concerns of metal-contaminated wastewater. This paper aims to test the ability of non-living algae in removing Copper, Nickel, and Aluminum from industrial wastewater samples and compare the results with other conventional treatment processes. The performance of non-living algae (spirulina) in removing aluminum from industrial wastewater samples will be studied. Additionally, the optimal operating conditions (T, pH, contact time) will be determined.

117as Development of Nanoengineered Photocatalytic Membranes for the Removal of Contaminants of Emerging Concern in Water: An Integrated System

vimbainashe Mercy Chakachaka, Charmaine S. Tshangana, Bhekie B. Mamba, Adolph Anga Muleja

The removal of emerging contaminants of concern such as endocrine-disrupting compounds (EDCs) is an important consideration to produce safe drinking water and to discharge environmentally friendly wastewater. In this study, an integrated system was developed to simultaneously degrade and filter selected emerging contaminants of concern in water. Key parameters such as time, pH, water matrix, mixing speeds, flow rate, and light intensity showed an undeniable effect on photocatalytic and membrane separation processes. For the graphene oxide quantum dots-based membranes (GQDs-Ms): the incorporating graphene oxide quantum dots (GQDs) into the polymer matrix led to enhanced hydrophilicity, pore size, porosity, improved flux as well as superior inhibition of bacteria cells. Whereas the addition of ferrites improved mechanical and rejection properties of membranes, inorganic salts and EDCs through the electrostatic exclusion and size exclusion mechanism. The possible EDCs degradation mechanism was presented based on intermediates detected by Quadrupole Time-of-Flight (QTOF) Mass Spectrometry. The overall integrated system achieved 99 % of EDCs removal efficiency. The system optimization was based on improving irradiation to generate more reactive species and mass transfer to increase the reaction rate. The simulated flow field revealed the presence of zones with different velocities in the reactor.

117at Piney Point: Breakthrough Solution in the Face of an Environmental Emergency

Don Luke

In April of 2021, a former phosphate facility located near the shores of Tampa Bay and the Gulf of Mexico, developed a leak in one of their lined containment structures. The leak threatened the integrity of the structure and had the potential to catastrophically release over 700 million gallons of slightly acidic wastewater containing very high levels of phosphorus and ammonia. If released this water had the potential to not only threaten neighboring communities but also posed a disastrous scenario for Tampa Bay where this water would serve to stimulate algae blooms and exacerbate seasonal red tide. This paper chronicles the emergency response with a focus on the efforts to contain and treat this water to discharge standards to mitigate any potential release. For over 9 months, our company, under rigorous supervision and monitoring by State and Federal environmental regulators, has been engaged to ensure that this water was treated to be suitable for discharge.

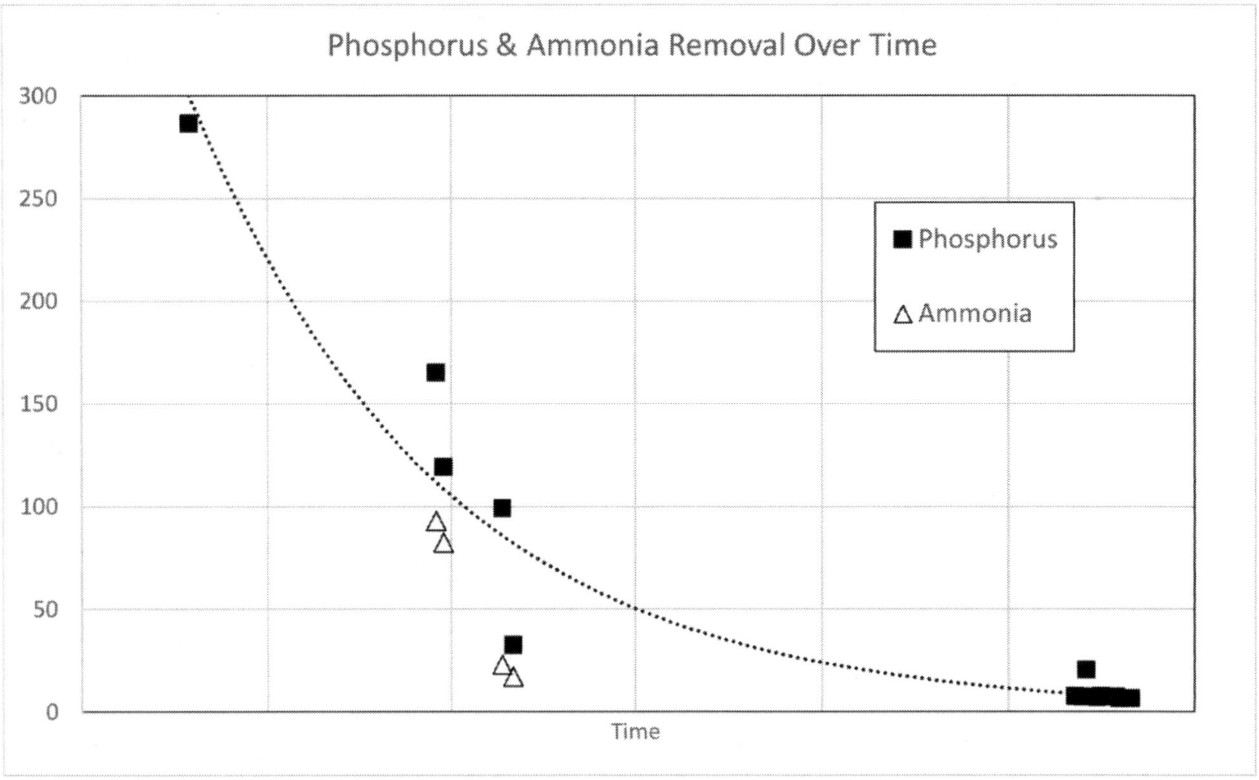

117au Extractive Crystallization of Calcium Acetate Deicer from Water Plant Residuals

Samuel Degife

There are a large number of water treatment plants that need to use chemical softening of the water to bring hardness of the water to acceptable levels. The lime-soda softening process produces significant amounts solid residuals containing calcium carbonate, magnesium hydroxide and unreacted base that need to be disposed of. Discharge of the residuals to surface waters is prohibited. Hence much of this material is applied to agricultural lands or landfilled at considerable cost due to the high dewatering and transportation costs. In this study, the use of the residuals to produce calcium acetate by reacting the sludge with acetic acid for use as a chemical deicer during winter for snow and ice control. Currently, the industrial process for the production of calcium-magnesium-acetate (CMA) deicer is conducted by reacting acetic acid and dolomite lime in the aqueous phase. The reaction product solution containing CMA would then be subjected to evaporation process to produce dry powder as product. This concentration step via evaporation of water is costly as it consumes a great deal of energy. In this project, an extractive crystallization process using ethyl acetate as a solvent was studied to reduce energy costs associated with the synthesis of calcium acetate from water plant residuals.

Extractive crystallization was used to synthesize calcium acetate from water plant residuals or calcium carbonate and acetic acid using two immiscible phases. The aqueous phase containing calcium carbonate was used as the calcium source while the ethyl acetate organic phase containing acetic acid was used as the acetic acid source. The neutralization reaction takes place at the interphase of the two phases and then the produced calcium acetate migrates to the aqueous phase. By adjusting the phase ratios, an aqueous phase that is saturated with respect to calcium acetate can be produced. This paper will present data on the optimization of the water to organic phase ratios in the effective conversion of the reactants to the desired deicer product. Data will also be presented on the effects of mixing the two phases on the reaction and conversion rate.

117av Physiochemical Treatment of Wastewater for Marine Applications

Christian Seaward, Jess Fike, Ramalingam Subramaniam

In this study, the design of a novel wastewater treatment method utilizing direct flocculation is explored. This novel design hopes to reduce the time, cost, and space requirements of existing wastewater treatment methods. Direct flocculation involves turbid mixing of the wastewater with an emulsion polymer flocculant to eliminate most of the total suspended solids (TSS) residing in the wastewater by coagulating the solids into larger coliforms, so they can be physically separated. The resulting solution is then treated with sodium hypochlorite as a disinfectant, and sodium meta bisulfate to dechlorinate the solution by way of precipitation. Our study showed success with substantial removal of TSS, as well as significant reduction of chemical oxygen demand (COD) and 5-day biological oxygen demand without nitrification (BOD_5). The final levels of TSS, COD, and BOD_5 were compared to the Marine Environmental Protection Committee's standards for wastewater discharge and met the listed specifications (TSS<35 mg/L, COD<125 mg/L, BOD_5<25 mg/L).This shows potential for this treatment method for commercial use, and further studies should be pursued.

117aw Quantifying Carbon Sequestered from Symbiotic Plant-Microbe Interactions for Sustainable Agriculture

Kong Wong, Keith E. Duncan, Sergey Komarov, Dierdra Daniels, Clara Lebow, Daniela S. Floss, Yuan-Chuan Tai, Christopher N. Topp

Plants engage in symbiosis with arbuscular mycorrhizal fungi (AMF) to exchange plant-derived carbon for essential plant inorganic nutrients, such as nitrogen and phosphorus. This symbiosis may improve plant tolerance to abiotic and biotic stresses and sequester carbon in the form of root and microbial biomass. The carbon cost of this nutrient exchange and the partitioning of plant-derived carbon to fungal storage organs likely depends on the AMF species or strain. A better understanding of the fate of carbon within various plant-AMF systems will enable the optimization of AMF consortia for carbon sequestration and plant growth. However, quantification of AMF effects on plant carbon allocation is limited, and most methods require destructive sampling. Here, we combine X-ray computed tomography (XCT) and positron emission tomography (PET) to observe and quantify *in situ* the flow of 11-carbon from leaves to roots to hyphae. Preliminary results in a maize-*Rhizophagus irregularis* system suggest that *R. irregularis* induces an increase in carbon allocation to the maize root system compared to uninoculated controls. Co-registered XCT and PET images suggest the increase in carbon correlates with increased lateral root growth near the inoculation site. While the increased number of lateral roots account for some of the ^{11}C signal increase, the remaining carbon is likely allocated to AMF. Our *in situ* method for quantifying carbon allocation has deepened our understanding of maize-AMF symbiosis and will enable future studies on the impact of plant and fungal genetics on carbon allocation in plant-AMF symbioses.

117ay A Dynamic Material Flow Analysis of the Global Demand of Polymers

Yunhu Gao, André Cabrera Serrenho

Introduction

Polymers are ubiquitous thanks to their low-cost and versatile properties. The production of plastics has increased from 1.7 Mt in 1950 to 368 Mt in 2019[1]. The wide application of polymers has led to a series of environmental issues, including greenhouse gas (GHG) emissions and ocean pollutions. The life cycle of polymers accounts for 4.5% of the global GHG emissions[2].

In order to reduce the environmental impact of polymers, capping the production and consumption of polymers is proposed by researchers[1,3–5]. Circular economy is also proposed to reduce the waste of polymers[6,7] and reduce production. However, a global model to quantify the practical limitation of circular economy and demand reduction is not available yet.

Material flow analysis is widely adopted to investigate the global and regional mass flow of polymers[8]. However, a holistic breakdown of polymer's mass flow in each world region is missing. Few studies forecasted the future demand of polymers by extrapolating the historical consumption[8–10], but the accuracy of regional demand of polymers is disabled or compromised by the lack of regional granular data to fit the model, e.g., production and trade of polymer raw materials and products with embedded polymers.

Method

To bridge the gap, the study adopts the dynamic material flow analysis model to quantify the productions, trades, and consumptions of 11 groups of polymers (LDPE, LLDPE, HDPE, PP, PS, PVC, PET, PU, rubber, fibres, and other plastics) and 1205 products made of polymers in eight world regions from 1978 to 2020, and fits the historical trend in order to project the future regional and global demands of polymers in different scenarios.

Results and discussion

Among the 8.86 Gt cumulative consumption of polymers, 2.41 Gt polymers are still in use in 2020. Among all the regions, Northeast Asia ranks the first, covering 28% of the total stock in-use, and North America, Europe and South Asia are responsible for 20%, 19% and 10% of the total stock in-use, respectively.

In terms of application sectors, 51% of the in-use stocks takes place in building and construction, due to the long average residence time of 35 years. 19% plastics are used in the transportation sector. The other sectors are responsible for less than 10% of the total in-use stocks.

The stocks in-use per capita in Oceania and North America are 1.28 t/cap and 1.27 t/cap, respectively. The average stock in-use in Europe is 840 kg/cap, and all other regions have less than 400 kg/cap. As the regional GDP per capita increases, the regional stock in-use per capita also increases, which is the basis for the projection of future demands of polymers in different scenarios.

The results will show how much GHG emissions can be reduced globally by different changes in recycling rate across eight different world regions modelled.

References

1. Bauer, F. *et al. One Earth* **5**, 361–376 (2022).

2. Cabernard, L., *et al. Nat Sustain* **5**, 139–148 (2022).

3. Simon, N. *et al. Science* **373**, 43–47 (2021).

4. Bergmann, M. *et al. Science* **376**, 469–470 (2022).

5. Syberg, K. *Nature* **611**, S6 (2022).

6. Kakadellis, S., *et al. Science* **373**, 49–50 (2021).

7. Meys, R., *et al. Science* **76**, 71–76 (2021).

8. Geyer, R., *et al. Sci Adv* **3**, 25–29 (2017).

9. OECD. *Global Plastics Outlook: Policy Scenarios to 2060.* (2022)

10. Stegmann, P., *et al. Nature* **612**, 272–276 (2022).

117az Physicochemical Characterization and Water Quality Assessment of the Cane Creek Lake: A Potential Freshwater Resource Located in Cookeville, TN

Shafieh Karami

Cane Creek Lake is a freshwater reservoir located in Cookeville, TN-USA. The lake covers an area of approximately 270 acres, and it is at the heart of the Cane Creek Recreational Park. The lake collects water from the surrounding hills located near the City of Cookeville. In general, the freshwater's fraction on the earth's surface is 2.5% of which only 1% is suitable for drinking purposes. Lakes account for 0.3% of the total surface waterbody sources, and have been historically utilized to meet human water demands. However, water quality of lakes is constantly affected by geographical, climate, and human factors including, for example, hydrology, geology, terrestrial cover, human and animal activities, and climate changes, among others. Therefore, lakes' water quality must be assessed by using various physicochemical parameters in order to determine the feasibility of the water for city consumption. To assess the possibility of providing water to the City of Cookeville from this source, samples were taken from upstream and downstream and parameters including pH, conductivity, TSS, DO, COD, turbidity, total alkalinity, and hardness were measured in order to evaluate lake's water quality. One of the goals of this study is to determine current water quality and conditions to identify possible treatments needed for the water to become suitable for consumption in the City of Cookeville. The results showed that COD, turbidity, and alkalinity quantities exceed the maximum limits, and this outcome suggests that further treatments are needed.

Keywords: Water Quality Assessment, Cane Creek Lake, Freshwater, COD

188g Removal of Allergens from Pistachio Shell By Soxhlet Method; An Experimental Study

Leila Fereidooni, Lakshmiprasad Gurrala, Ana Colaco Morais, Mark B. Shiflett

Agricultural wastes are accounted as cheap and available enriched materials, which have been widely used in different applications such as fertilizers, composite structures. One of the most available agricultural wastes are pistachio shells, which are being disposed more than 500000 tons per year, in the United States. These wastes would be proper sources of animal feed. Allergens are found in different shell powder containing nuts, and could make some problems for animal and people dealing with the shells powder, if not extracted from the shell powder. By adhering to the surface of an immune cell with allergens, the secretion of Immunoglobulin E (IgE) occurs. This event triggers a cell signaling cascade, leading to an allergic response. Based on the literature, the allergens include the 2S albumin, 7S vicilin, and 11S legumin proteins, conglutin, and glycinin. In our experimental study, we identified the allergens of pistachio shell. Different methods of allergen removal, including rinsing with solvent, acid treatment, lights radiation processing, and Soxhlet extraction were investigated. The Soxhlet method was chosen (because it showed a selective removal), and pistachio proteins were extracted and separated by SDS-PAGE. The 7S vicilin-like protein was identified and extracted, as the equal code of pistachio shell Pis v 3.

48e Evaluation of Green Hydrogen Production from Enterobacter Sp. Using Waste Effluents from Biodiesel

Gisele Góes, Fernando Pessoa, Ana Lucia Barbosa

The demand and production of biodiesel for energy generation has been increasing, being one of the renewable sources, using resources such as biomass. In addition, it is a raw material with high availability and promotes the reduction of greenhouse gases to the planet. However, its production generates waste accumulation, and procreating an adverse effect to the objective: to be a fuel with less environmental impact, considering an adequate life cycle.

Residual effluents from biodiesel, composed of a large amount of crude glycerol, which, with improper disposal, cause environmental, economic, and social problems. However, combined with fermentation, it can generate the production of monomers, energy and other biofuels, such as biohydrogen. Hydrogen is considered a clean, highly reactive fuel and provides a higher value-added destination for biodiesel production effluents.

The development of technologies for the biological production of hydrogen is very promising. The processes for obtaining hydrogen by microbial means have been gaining prominence, due to the possibility of using renewable energy sources, as well as the reuse of waste materials, thus knowing the amount of by-products stored in energy sources. Furthermore, microorganisms such as Enterobacter sp. are capable of converting glycerol into a series of metabolites for the production of biohydrogen. This process is possible, because this bacterium presents characteristics such as anaerobiosis and the possibility of co-production of biogas with hydrogen. According to Correia et al, (2009) the production by biological route contributes with 1% of the total production. However, with the development of increasingly emerging technologies, this percentage rises, thus promoting more efficient and sustainable production.

Thus, the present work aims to evaluate the technical viability of biohydrogen production from Biodiesel effluents making use of computer simulation using SuperPro Designer®, software that allows planning, programming and minimizing production bottlenecks to carry out the modeling of the processes, in addition to assisting in the cost analysis of the processes. The data used were obtained from bibliographic consultation, equipment companies and professionals in the area.

The results from the behavior of the reactions, comparison of the flowcharts obtained with data in the literature and economic analysis of the process, indicate that the bioconversion of effluents from biodiesel, using Enterobacter sp. have advantages and technical feasibility of hydrogen production.

188a Use of an Adsorbent Produced from Residual Sludge from the Aluminum Industry for the Removal of Chlorpyrifos

Juan F. Saldarriaga, Eva Epelde, Zuria Tabernilla, Julián E. López

The aluminum industry produces a high number of wastes that are concentrated in the sludge of their water treatment plants. The industry is one of the most prosperous in recent years, being aluminum one of the most widely used non-ferrous metals worldwide. Its production has increased in the last decade from 37,606 thousand metric tons in 2009 to 63,697 thousand metric tons in 2019, showing a growth of 59% for this period [1]. The main applications include transportation, packaging, construction, and electrical engineering [2]. This sludge is rich in aluminum, which could be extracted and purified. This work aims to evaluate, by different extraction methods, the recovery of the aluminum contained in a wastewater sludge derived from a metallurgical industry. For this purpose, Bayer, HCl, and isopropanol extraction methods are compared for different reagent ratios. Besides, the extracted product has been evaluated for its application as adsorbent for the removal of an organophosphate pesticide (chlorpyrifos, CPS) from contaminated waters. In this work, three extraction methods have been evaluated to recover the aluminum from the sludge samples provided by a metallurgical industry from Medellín, Colombia. The sludge has been characterized to confirm its high aluminum content. The Bayer extraction method has been carried out, using 1:10, 1:15 and 1:20 sludge/solution mass ratios, and HCl extraction, by using 1:15 and 1:20 ratios. The extraction with isopropanol has also been evaluated. The extraction agent for this method was analytical grade isopropanol. The tests were carried out in accordance with the procedure proposed by Saiz et al. [3]. The goal of this process is to produce aluminum monohydroxide by reacting sludge with isopropanol to produce aluminum isopropoxide. Because of its widespread use as a pesticide in agriculture, CPS has been used as a reference water pollutant in adsorption tests, particularly in developing countries such as Colombia. Adsorption tests were performed with different weights (0-5 g) of the precipitate obtained from the Bayer extraction process in a 1:15 ratio. All samples were made in quadruplicate to ensure reproducibility. The vials were filled with 100 mL of the standard solution, the flasks were shaken, and samples were taken at 5, 30, 60, 90, 120, 150, 180, 210, and 240 minutes. The equilibrium parameters for the Freundlich and Langmuir models, have been optimized by minimizing an objective function, *OF*, defined as the sum of the squares of the differences between the experimentally measured adsorbate concentration in the liquid phase, and the values calculated by the model. In MATLAB, a calculation algorithm based on the Nelder-Mead algorithm has been implemented, which uses the ode subroutine to solve the mass balance equationsand the *fminsearch* subroutine to calculate the without restrictions minimum of the objective function. According to XRF results, Al (68.52%) stands out among other species, with concentration levels comparable to those found in other studies. As a result, this sludge contains a high concentration of Al, which has a high potential for extraction and little interference from other metallic species. The high concentration of S (28 wt%) observed could be attributed to neutralization treatments performed in the company at its origin. The amounts of Al, Fe, Cu, and Ni present in the aqueous phase have been measured to determine the recovery of the metals, as well as the selectivity of the extraction method. The highest metal recovery (99.3%) is achieved by HCl extraction for a sludge/solution ratio of 1:20, while the Bayer method (1:15 ratio) has shown the best selectivity. On the other hand, the extraction with isopropanol has not been successful for any of the ratios evaluated. The extracted material has been tested as an adsorbent for the removal of chlorpyrifos (200 mg/l), where the removal percentages have been higher than 95% for all the weight ratios studied. Hence, this Al-enriched material shows good prospects to be used as an adsorbent in the treatment of polluted water.

Among the various Al extraction methods investigated, the Bayer method proved to be the most practical, particularly for 1:15 and 1:20 sludge/NaOH ratios, where an Al recovery of 80% and 100%, respectively, was obtained. However, the selectivity for the 1:20 ratio was significantly lower than that for the 1:15 ratio. Furthermore, with a 1:20 ratio, an average recovery of 60% iron, 40% copper, and 2.5% nickel was obtained. Based on these findings, the Al samples extracted by the Bayer method with a ratio of 1:15 (BES15) were tested for CPS adsorption, and a removal of 99% was achieved for 5 to 240 mins, with no evidence of desorption. According to the Langmuir isotherm, a good fit has been established for the adsorption process, in which a single layer is formed at the start of the process.

References

1. IAI, Primary aluminum production, Int. IAI Inst. (2019). http://www.world-aluminium.org/statistics/primary-aluminium-production/.
2. V. Marques, R.L. Barcelos, G.O.C. Parma, E. Girotto, A.C. Junior, N.C. Pereira, RF. Magnago, Recycled polyethylene terephthalate and aluminum anodizing sludge-based boards with flame resistance, Waste Manag. 92 (2019) 1–14. https://doi.org/10.1016/j.wasman.2019.05.013.
3. S. Saiz, G.C. Vargas, J.C.M. Piraján, Obtención de pseudoboehmita mediante el método de sol-gel empleando dos catalizadores diferentes, Av. Investig. En Ing. 1 (2010) 35–44.

175a Contribution to the Circular Economy from the Use of Tannery Hair Biochar for the Removal of Emerging Compounds in Polluted Water

Katherine Herrera, Luisa F. Morales, Julián E. López, Juan F. Saldarriaga

The presence and destination of pharmaceutical compounds in water constitutes one of the main emerging events in current environmental chemistry. The most significant effects that have been reported are a variation in the chemical composition of discharges, their identification as possible organic pollutants and the effect on some biological treatment processes due to their presence in high concentrations [1]. The pharmaceutical pollutants that generate the greatest concern are antibiotics, since their consumption reaches approximately 100 tons per year, surpassing any other types of medications [2]. Hair is found within the waste from livestock and tannery activities, which in the process is discarded mixed with other waste and compounds such as hydroxides and chromium, making its later use complex, due to the lack of specialized technology to decrease its dangerousness. During the pyrolysis process of hair, the proteins that are part of the keratin when they reach ~250 ºC are melted, then at 300 ºC the formation of aromatic compounds occurs, at 350 ºC the formation of cyclic amines and finally between 350 and 600 ºC it produces the degradation of cyclic and aromatic framework. However, the use of adsorbents from waste hair valorization is interesting because of its composition, such as the presence of keratin. The treatment temperature for the tannery hair waste was 450 °C, and the material was allowed to remain there for 20 minutes in both batch and continuous feed. According to Rodríguez et al. [3] and Herrera et al. [4], this temperature range is ideal for producing carbonaceous materials with superior surface, textural, and chemical properties for adsorption. The biochar was washed to remove any impurities that might interfere with the analysis results. A 1:3 water-biochar ratio was used, centrifuged for 5 minutes, and allowed to settle for 5 minutes before removing the supernatant and repeating the process until a clear water-biochar mixture was obtained. Tannery hair biochar obtained through the pyrolysis process at 450 °C was mixed in a 1:1 weight ratio with KOH (EMSURE, London, England) and mixed for 1 hour until a homogeneous mixture was obtained before being left in contact for 72 hours. For the tests on AMOX, DFC, AZM, and ERY, a 100 $mg \cdot L^{-1}$ aqueous solution was used. A sample of the aqueous solution containing 100 mL was used for an adsorbable ingredient concentration of 100 $mg \cdot L^{-1}$. However, there were reading mistakes when assessing the pharmacological mixture using concentrations of 100 $mg \cdot L^{-1}$ of each one in the same solution due to interference produced by the wavelength associated with each one.

For the removal of drugs, two different tests were carried out: one for 28 days and the other for 4 hours, in order to evaluate the efficiency and the percentage of removal. It was found that the biochar obtained at 450 ºC is the one that presents the best efficiencies and that after being activated, the efficiency increases considerably. With the biochar at 450 ºC, removals of 90% of AMOX and more than 80% of DFC were achieved.

By examining it, biochar made from waste tannery hair has these qualities and is an excellent adsorbent. As removal percentages of more than 60% were achieved in this example for each medication, it is possible to claim that good removal percentages are provided when examining medicines separately or in mixes. The findings of this investigation allow us to draw the conclusion that pyrolysis of tannery hair at 450 °C results in a carbon with physicochemical properties acceptable for use as an absorbent. This carbon enables the removal of newly developing contaminants like antibiotics. Because biochar may partially eliminate the developing pollutants found in traditional treatment systems, its application would ensure improvements in the process. This could be accomplished by adding filtration columns or fixed bed filters after the standard filtration procedure or after the coagulation, flocculation, and sedimentation procedures. On the other hand, by forgoing a typical treatment for a waste that is deemed dangerous, pyrolysis creates a new product and a new stage within the leather industry's value chain by adding value to it. Similar to how tannery hair waste can be fully utilized by undergoing an alternative value-added treatment, the manufacturing of leather can be improved by making it more environmentally friendly.

References

1. Alvarez, S., Sheepdog Escudero, G., Garcia, Panorama., J. (2017). Elimination of Emerging Contaminants from Wastewater. *Research Science*. 2017, 485, 10–16.
2. Tejada, C.; Quiñonez, E.; Peña, M. (2014). Contaminantes Emergentes en Aguas: Metabolitos de Fármacos. Una Revisión. Fac.Rev. Cienc. Básicas, 10 (1), 80–101. https://doi.org/10.18359/rfcb.341.
3. Rodríguez, F., Montoya-Ruiz, C., Estiati, I. & Saldarriaga, J. F. Removal of Drugs in Polluted Waters with Char Obtained by Pyrolysis of Hair Waste from the Tannery Process. ACS Omega 5, 24389–24402 (2020).
4. Herrera, K., Morales, L. F., Tarazona, N. A., Aguado, R. & Saldarriaga, J. F. Use of Biochar from Rice Husk Pyrolysis: Part A: Recovery as an Adsorbent in the Removal of Emerging Compounds. ACS Omega 7, 7625–7637 (2022).

226e *Moving Past Resilience in Progressive Supply Chains from the Perspective of Large-Scale Shipping of Future Energy Carriers*

Elizabeth Abraham, Dhabia Al-Mohannadi

With the projected shift towards more sustainable forms of energy accompanied by an increase in overall global energy consumption, energy supply chains will need to become progressive. To this end, there is a need to ensure that they will not only be able to accommodate the future energy mix and its associated carriers but also any disruptions that may occur during and after the transition such as climatic or geopolitical upheavals. The transport of these carriers is an intermediate and crucial step of the energy supply chain that relies on various modes of transportation depending on the quantity of energy transported and the distance it must traverse from production to distribution. Recently, more focus has been cast on expanding the possibilities of obtaining energy from more localized sources that employ ground transportation such as pipelines and trucks, to more extensive supply chains that span across borders and require marine transportation. However, a significant transformation is expected in present-day shipping fleets to become sustainable in their operations [1] and further accommodate these more renewable carriers that are primarily comprised of chemicals that can easily store energy. Moreover, these chemicals have established operational flexibility in terms of the quantity and location of their production and distribution that make them ideal candidates to transport energy across large distances. In this work, an analysis on moving towards guaranteed resilience is explored through the incorporation of risk mitigation strategies to supply chain optimization for the large-scale shipping of future energy carriers [2]. Through this integrated approach, the optimal production and distribution locations can be determined along with the suitable energy mix that meets specified economic and environmental constraints while being resilient to disruptions. The analysis provides decision-makers with a preliminary understanding of the tradeoffs between costs, operability, and resilience in the design of future supply chains. The application of this approach is illustrated through an example where energy trades that are set to achieve guaranteed resilience are optimized for varying economic and environmental targets.

1. IMO, 2013. Concept of a Sustainable Maritime Transportation System.
2. Chrisandina, N.J., Vedant, S., Iakovou, E., Pistikopoulos, E.N., El-Halwagi, M.M., 2022. Multi-scale integration for enhanced resilience of sustainable energy supply chains: Perspectives and challenges. Computers & Chemical Engineering. 164, 107891.

245b Rate Limiting Factors for Pfas Degradation on Carbonaceous Sorbents Using UV/Sulfite Generated Hydrated Electrons

Hosea Santiago, Zimo Lou, Jinghao Li, Wan Zhang, Susie Dai, Gregory V. Lowry

Per- and polyfluoroalkyl substances (PFAS) are a large class of persistent synthetic contaminants detected in water globally, threatening human and environmental health. Carbon-based sorbents used to remove PFAS from water can be regenerated through thermal treatments that simultaneously destroy PFAS. However, this process is energy intensive and can emit harmful byproducts back into the environment. Advanced reduction processes (ARP) can mineralize PFAS in homogeneous aqueous solutions; however, reactivity in heterogeneous systems is poorly understood. We explore an alternative regeneration treatment for carbonaceous sorbents with ARP using hydrated electrons (e_{aq}^-) generated by UV/sulfite. Three carbon sorbents, granular activated carbon (GAC), carbon nanotubes (CNT), and polyethyleneimine-modified lignin (Lignin), were chosen based on their varying sorption strength and surface chemistry. Reaction rates and extent of defluorination were determined by tracking the amount of parent compound perfluorooctanoic acid (PFOA) in water and on the sorbent; and measuring fluoride and target perfluorocarboxylic acid (PFCA) transformation products, respectively. Results indicate that compounds in the aqueous phase (aq) reacted rapidly compared to sorbed (s) ones. This outcome was expected due to the highly reactive and aqueous nature of e_{aq}^-. Once desorbed, $PFCA_{aq}$ reaction rates are negatively compromised by the carbonaceous particles. Results suggest that reaction rates are hindered by the sorbents' capabilities of scavenging e_{aq}^-, decreasing sulfite (SO_3^{-2}) sensitizer concentration, and decreasing the incident light available for the sensitizer to generate e_{aq}^-. Because of these limiting factors, we propose that a regeneration scheme using UV/SO_3^{-2} generated e_{aq}^- requires a desorption step and removal of sorbents before initiating the reaction of the concentrated liquid. This study encourages using sorbents that can efficiently capture and then release PFAS on command to promote efficient degradation in a concentrated aqueous media.

245c Evaluating Process Costs and Performance Metrics of Granular Activated Carbon for the Removal of Pfas Under Model Uncertainty

Hunter Barber, Kurban Sitterley, Alexander Dudchenko, Adam Atia, Fernando Lima

After longstanding emissions of per- and polyfluoroalkyl substances (PFAS), concentrated efforts to implement systems for the removal of these harmful compounds are growing throughout the United States (U.S. Department of Energy, 2022). Granular activated carbon (GAC) adsorption has been demonstrated to be generally effective for the removal of PFAS (Verma et al., 2021). However, the estimated performance and cost of the system are not well understood at the design phase due to the infancy of the application of PFAS removal processes and lack of accurate and validated models.

Bench- and pilot-scale data characterizing PFAS adsorption onto GAC are available in limited quantities for a range of real and idealized feed water compositions (Hwang et al., 2021). However, breakthrough profiles as a function of absorbent properties, PFAS concentrations, background component concentrations, and column designs are unavailable. Under these limitations, the mass transfer mechanisms by which PFAS adsorb onto GAC and other factors that inhibit it are unknown, making it challenging to develop predictive models for PFAS adsorption.

The first objective of this work is to use a simplified GAC model to estimate the performance of PFAS adsorption onto GAC. WaterTAP, a framework for simulating water treatment processes, was used as a simulation platform for this purpose with integrated supplemental tools (Beattie et al., 2021). The constant pattern homogeneous surface diffusion model (CPHSDM) for GAC adsorption was used to predict PFAS removal (Hand et al., 1984). Adsorption isotherm and diffusion coefficient parameters were regressed to fit available data to describe the removal performance. The complexities of the true adsorption mechanisms and how they affect performance may be captured through the variability of regressed model parameters across diverse experimental data sets. Results of this work provide a distribution of the parameters to describe the uncertainty of the model.

The second objective is to evaluate cost and other performance metrics (such as system size and energy consumption) across the distribution of model parameters. This analysis was used to describe low- and high-end system costs and provide trends of the metrics that are codependent on model parameters. Cost mapping for variations in case-specific aspects such as water quality (captured by uncertainty in model parameters) and system size (captured by simulation of varying designs) gives an informed approximation of system performance and design for PFAS removal by GAC.

References:

Beattie, K., Gunter, D., Ben, K., Lee, A., Ladshaw, A., Drouven, M., Bartholomew, T., Bi, X., Bianchi, L., Arnold, T., Atia, A., Wang, C., Miara, A., Sitterley, K., Srikanth, A., Dudchenko, A., Amusat, O., Kinshuk, P., & Young, E. (2021). *WaterTAP* (v1.0.0). Lawrence Berkeley National Lab. (LBNL), Berkeley, CA (United States); National Energy Technology Laboratory (NETL), Pittsburgh, PA, Morgantown, WV, and Albany, OR (United States); Oak Ridge National Lab. (ORNL), Oak Ridge, TN (United States); National Renewable Energy Lab. (NREL), Golden, CO (United States). https://www.osti.gov//servlets/purl/1785311

Hand, D. W., Crittenden, J. C., & Thacker, W. E. (1984). Simplified Models for Design of Fixed-Bed Adsorption Systems. *Journal of Environmental Engineering*, *110*(2), Article 2. https://doi.org/10.1061/(ASCE)0733-9372(1984)110:2(440)

Hwang, M., Grieco, S., Chang, J., Wille, A., Lozier, J., Plumlee, M., Pannu, M., Medina, R., Pham, C., Safarik, J., Huang, A., Dadakis, J., & Olsen, C. (2021). *PFAS Treatment Testing Study* (No. PPS0219211302SCO). https://www.ocwd.com/wp-content/uploads/ocwd_pfas_treatment_testing_study_final_report.pdf

U.S. Department of Energy. (2022). *PFAS Strategic Roadmap: DOE Commitments to Action 2022-2025*. U.S. Department of Energy.

Verma, S., Varma, R. S., & Nadagouda, M. N. (2021). Remediation and mineralization processes for per- and polyfluoroalkyl substances (PFAS) in water: A review. *Science of The Total Environment*, *794*, 148987. https://doi.org/10.1016/j.scitotenv.2021.148987

<u>Disclaimer:</u>

245d Uptake and Release of Perfluorinated Surfactants from Macro and Microplastics

Philip J. Brahana, Ahmed Al Harraq, Bhuvnesh Bharti

Microplastics and perfluorinated (PFAS) surfactants are two of the most notable emerging contaminants reported in the environment. Micron and nanoscale plastics possess a high surface area-to-volume ratio, which increases their potential to adsorb other common environmental pollutants such as PFAS surfactants. The PFAS surfactants are widely considered as 'forever chemicals' and a potential threat to human health. However, it is their amphiphilic properties which is often overlooked. This lack of consideration has resulted in a diminished understanding of the environmental mobility of PFAS surfactants, as well as their interactions with environmental media along with the change in their behavior after enduring environmental stresses. Here, we report an intricate relationship between the two anthropogenic pollutants, namely microplastics and PFAS surfactants. Through a series of adsorption-desorption experiments, we delineate the role of the fluoroalkyl tail in PFAS surfactants on their binding affinity to micro and macroplastics. As the number of carbon atoms in the fluoroalkyl chain increases, there is a corresponding increase in the adsorption of PFAS surfactants onto microplastics. This relationship can become modified by environmental weathering, where the PFAS surfactants are released from the macro and microplastic surface after exposure to simulated sunlight. This study identifies a synergistic relationship between PFAS surfactants and plastic pollutants, where they can mutually impact their thermodynamic and transport properties.

245e Investigation into Cationic Surfactant and Polyelectrolyte Coated β-Zeolites for Rapid and High-Capacity Adsorption of Short- and Long-Chain Pfas

Dwyer Stuart, Md Shahadat Hossain, Deepak Kumar, Mario Wriedt, Bandaru V. Ramarao

Per- and polyfluoroalkyl substances (PFAS) have been implicated as carcinogens and have emerged as contaminants of serious health and safety concerns. The objective of this work is to investigate the potential use of β-zeolites for the efficient and rapid removal of short- and long-chain PFAS from water. PFAS adsorption capacities of β-zeolites (CP811C and CP814E) with varying silica and alumina (Si/Al) ratios were evaluated. The zeolites were coated with either (cetyltrimethylammonium bromide - CTAB) or (poly(diallyldimethylammonium chloride) - PDADMAC) to enhance adsorbent affinity for highly soluble PFAS. Perfluorooctanoic acid (PFOA) and heptafluorobutyric acid (PFBA) were selected as model solutes, representing long- and short-chain PFAS, respectively, and were adsorbed onto unmodified and modified β-zeolites in ambient conditions. Characterization via scanning electron microscopy (SEM), Fourier-transform infrared spectroscopy (FTIR), and zeta potential analysis showed successful surface modification. All studied β-zeolites rapidly adsorbed PFAS, reaching equilibrium within 1-2 hours. Among them β-zeolite CP811C was found to have higher PFOA uptake than CP814E (91.87 vs. 57.77 mg/g). However, CTAB coatings reduced PFOA adsorption capacity on β-zeolites by up to 58.7%. Similarly, PDADMAC coating also reduced PFOA adsorption capacity by up to 15.4%, but increased PFBA adsorption capacity by up to 51.3%. Such batch adsorption kinetics data was well-described by the pseudo-second order (PSSO) rate model. Additionally, PFOA and PFBA adsorption equilibrium onto β-zeolite CP811C was fitted by the Langmuir isotherm model, with maximum adsorption capacity of 181.87 and 27.89 mg/g respectively. Such faster adsorption rates and higher adsorption capacities make β-zeolites promising for PFAS remediation compared to conventional Granular activated carbon and ion exchange resins PFAS adsorbents.

245f Relation between Organic Matter Optical Parameters and Photodegradation of Emerging Contaminants

Emad Sanei, Maxwell Hornak, Itzel Marquez

Emerging contaminants are susceptible to degradation by reaction with Reactive Intermediates (RIs) - such as triplet-state organic matter ($^3DOM^*$), singlet oxygen ($^1O_2$), and hydroxyl radical ($^•OH$) - produced when organic matter present in water is exposed to light. It has been shown that contaminant degradation kinetics strongly depend on the organic matter composition and characteristics.

Organic matter present in water bodies may be naturally derived - natural organic matter (NOM) - or come from wastewater treatment effluent - effluent organic matter (EfOM) -. The generation of RIs has been shown to differ between EfOM and NOM, affecting the photodegradation of emerging contaminants in the aquatic environment, especially when effluent is released to the environment. Therefore, it is important to understand the differences between EfOM and NOM in terms of photodegradation of contaminants. However, characterizing organic matter can be expensive and complex.

Several optical parameters have been defined to extract information from UV-Vis and fluorescence spectroscopy about organic matter's composition and properties. For instance, $SUVA_{254}$, HIX, FI, $S_{275-295}$, and E2:E3 have been used as indicators of DOM characteristics such as molecular weight, aromaticity, and light absorbance. Because of this, optical parameters of organic matter may provide insight on the reactivity of organic matter. In this work, we aim to correlate optical properties of EfOM and NOM to degradation kinetics of emerging contaminants.

Here, *p*-cresol was used as a surrogate for other phenolic emerging contaminants, which are persistent in nature and among the most toxic chemicals. *p*-Cresol photodegradation experiments were carried out in seven different EfOM samples and four different NOM samples to investigate contaminant photodegradation kinetics. EfOM samples consist of wastewater effluent obtained from a variety of wastewater treatment processes. NOM samples were standards obtained from the International Humic Substances Society. Fluorescence and UV-Vis spectroscopy was performed to characterize different fractions in the various organic matter samples. Optical parameters such as $SUVA_{254}$, HIX, FI, $S_{275-295}$, and E2:E3 were determined to indicate organic matter characteristics such as molecular weight and aromaticity. EfOM and NOM solutions were spiked with *p*-cresol and exposed to light in an AMETEK SUNTEST XLS+ sunlight simulator. The concentration of *p*-cresol was measured by an Alliance HPLC instrument every 30 minutes for four hours. Degradation kinetic parameters were determined, including quantum yields, RIs steady-state concentrations, and *p*-cresol degradation rate constants. Finally, Pearson correlations were determined among optical parameters and *p*-cresol degradation rate in both EfOM and NOM samples.

This work will provide a report on trends in *p*-cresol photodegradation in both NOM and EfOM samples. Moreover, the relation of *p*-cresol photodegradation kinetic parameters to optical parameters will be evaluated. It will also discuss the *p*-cresol degradation mechanism work in different aquatic matrices. The effect of organic matter composition on emerging contaminants degradation will be presented. These findings will benefit future studies by providing correlations to predict the degradation of emerging contaminants in NOM and EfOM.

245g Separation and Destruction of per and Polyfluoroalkyl Substances through Galvanic Reaction

Alan Hanley

Per and polyfluoroalkyl substances, or PFAs for short, are a rapidly growing concern due to recent studies connecting PFA exposure to cancer and immune system problems. With over 9000 PFA variants identified to date, research into the removal and destruction of these molecules is vital to the health and safety of the world population. Two common PFA molecules are perfluorooctane sulfonic acid (PFOS) and perfluorooctanoic acid (PFOA); the EPA has published health advisory limits of 0.02 ppt and 0.004 ppt for drinking water, respectively, and maximum contaminant levels of 4.0 ppt for both contaminants. These low concentration requirements make remediation methods such as filtration and/or incineration ineffective, cost prohibitive, or environmentally damaging. Our process combines a continuous flow reactor with an electroless electrochemical reaction, requiring no external energy. We've determined that PFOA and PFOS can be separated, concentrated, and destroyed from water sources ranging from drinking water to landfill leachate using our continuous galvanic reactor. Our experiments show a 95% reduction in PFA molecules after a single treatment stage; higher removal can be achieved using multiple stages. Experiments show that residence time can be reduced to minutes instead of hours, with waste concentration ratios exceeding 500:1. Once concentrated, we can destroy these contaminants using electrolysis, chemical oxidation, or thermal decomposition, resulting in complete mineralization of the original contaminants. This process provides a simple, scalable, and inexpensive resolution to an emerging global environmental issue.

245h Multifunctional Zwitterionic Hydrogels for Rapid and Simultaneous Elimination of Organic and Inorganic Micropollutants from Water

Devashish Gokhale, Andre F. Hamelberg, Patrick Doyle

The removal of micropollutants from water is made difficult by their chemical diversity, low concentrations, and slow uptake by industrial adsorbents. Traditional adsorbents, like activated carbon and ion-exchange resins, are effective in eliminating only specific classes of micropollutants, necessitating the use of multiple unit operations that sequentially treat water. At the same time, slow kinetics require the unit operations to have long residence times and be much larger to effectively eliminate micropollutants. Here we present a new multifunctional zwitterionic porous hydrogel platform for the single-step bulk absorption of both organic and inorganic micropollutants from water. Shifting from traditional surface adsorbents to bulk absorbents greatly increases the effective functional area, while allowing us to retain the same process-level equipment that has been well studied in the context of conventional adsorbents. A facile synthesis allows easy functionalization of the hydrogels for eliminating multiple micropollutant classes, while also allowing a measure of selectivity. The hydrogels can be synthesized in several form factors for specific applications, including microparticles for use in packed beds, or tablets for treating water in a glass or canteen. The zwitterionic hydrogels are shown to remove six chemically diverse micropollutants (including BPA, xenoestrogens, and other organics, and lead and other metals), at least 10 times more rapidly than a commercial activated carbon/ion exchange resin mixture, and 100 times faster than other multifunctional adsorbents reported in literature. Zwitterionic hydrogels treat complex micropollutant mixtures, including in the presence of background hardness. Lower compressibility, extremely rapid mass transport, and high stability make our materials highly scalable for real world applications. We demonstrate the use of a prototype packed bed filter with a practical residence time and containing zwitterionic hydrogel microparticles to clean highly contaminated water to below regulatory limits for a period of two weeks.

245i Thin-Film Nanocomposite Membranes Functionalized with Fe-Based Metal-Organic Frameworks for Enhanced Removal of Small and Uncharged Urea from Water

Tin Le, Milad Esfahani

Urea is widely used in fertilizer production for agricultural purposes, which risks runoff into soil and water sources. Leaching of urea into water resources poses a concern due to algal bloom formation which can kill local wildlife and cause environmental eutrophication. Thin-film composite (TFC) membranes are among the best-performing membranes for water filtration; however, they do not show acceptable rejection for small and uncharged molecules such as urea. To combat this issue, adsorptive membranes combine the power of membrane filtration with adsorbent materials to provide enhanced selective removal, specifically for small and uncharged molecules. Metal-organic frameworks (MOF) are a new generation of effective adsorbents with excellent tunability in structure and compatibility with polymeric membranes. In this study, we synthesized MIL-100 (Fe), an iron-based MOF, using a mild room temperature method and tested for urea adsorption. The urea adsorption capacity of MIL-100 (Fe) was examined under varying experimental conditions such as pH, temperature, MOF concentration, and urea concentration. The results revealed the superfast adsorption (reached almost 90% removal within 2 min) of neutrally-charged urea molecules with a high adsorption capacity of 3321 mg/g and a BET surface area of 1174 m^2/g. To fabricate the adsorptive membranes, the MIL-100 (Fe) was grafted on the surface of the polyamide layer of thin-film composite membranes via covalent binding for enhanced removal of urea from water. The MIL-100 (Fe)-TFC membrane showed enhanced permeability (14.10 vs 0.87 LMH/bar) and no significant decrease (47% vs 41%) in Mg salt rejection, indicating successful grafting without damaging the PA layer. In addition, although both bare TFC and MIL-100(Fe)@TFC showed similar rejection of 63% of urea, the MIL-100(Fe)@TFC adsorbed a portion of the urea compared to the bare TFC, which rejected it. The quantification of adsorption versus rejection is currently under investigation.

273a Estimation of the Environmental Impacts of [P2228][2-CNPyr] and its Monoethanolamine Based Hybrid Solvent Using Life Cycle Impact Assessment Methodology

Adhish Chandra Saketh Madugula, Tracy Benson

Ionic liquids (IL) containing phosphonium cations paired with aprotic heterocyclic anions (AHAs) have shown to be effective in absorbing CO_2, particularly for post combustion processes with significant water concentration. In particular the ionic liquid, triethyl(octyl)phosphine 2-cyanopyrrolide ([P2228][2-CNPyr]), was found to be an ideal candidate to be blended with aqueous monoethanolamine (MEA) as the resulting hybrid solvent showed a significant reduction in the viscosity (compared to neat [P2228][2-CNPyr]) and exhibited a significantly lower the enthalpy of CO_2 desorption (compared to aqueous MEA). Therefore, based on these characteristics, the phosphonium based IL and its corresponding hybrid solvent can be considered as a greener alternative to MEA. However, the available literature on the Lifecycle Assessment (LCA) of phosphonium based ILs suggest the contrary. These studies find that the synthesis route currently being employed to produce the phosphonium cation utilize phosphine and phosgene as precursors which greatly diminish the environmental feasibility of phosphonium based ionic liquids. Therefore, this work suggests an alternate phosphine and phosgene free synthesis route and studies the environmental impacts for producing 1kg of [P2228][2-CNPyr] and its corresponding hybrid solvent mixture with aqueous MEA through this synthesis route. The impacts were estimated based on the Tool for Reduction and Assessment of Chemicals and other environmental Impacts 2.1 (TRACI 2.1) methodology using GaBi LCA Software by Sphera®. This study takes into consideration the various precursors/intermediates in the synthesis route for the production of the two solvents. The TRACI 2.1 methodology was chosen as it calculates the impact of the over ten environmental categories for a holistic environmental impact assessment. This study found that the altered synthesis route could produce the IL and its corresponding hybrid solvent with a lower overall environmental impact than that of MEA. In fact, in this study, both the solvents exhibited lower environmental impacts in 6 and 7 of the 10 environmental categories respectfully when compared to MEA. The study also finds that despite the overall lower environmental impacts of the two solvents, the phosphonium cation synthesis route still remained as a major environmental hotspot. The majority of the environmental impacts was due to the electricity requirement in the IL synthesis route and the material precursors were found to have a lower impact on the environment than that of the electricity requirement in this route. This trend was found to be similar for the production route of both the solvents. Overall, this study found that adopting phosgene and phosphine free synthesis route to produce the phosphonium cation made both the solvents more environmentally sustainable than MEA. Therefore, the authors of this study promote the exploration of alternative, green chemistry synthesis routes for economical and environmentally sustainable IL based CO_2 capture solvents.

273b Fuzzy Decision-Making for Sustainability Performance Improvement of Complex Systems

Abdurrafay Siddiqui, Yinlun Huang

Industrial sustainability problems are always complex because of the existence of complicated interrelated factors that affect system's economic, environmental, and social performance and, more critically, the pervasive presence of various types of uncertainty that is associated with accessible data and information, and is reflected in problem understanding and possessed knowledge. These have made decision making for sustainability performance assessment and improvement very challenging, especially for predictive solution derivation. In theory, uncertainty can be classified into two types: aleatory and epistemic. While the former refers to the inherent variation associated with physical systems and the environment and is objective and irreducible, the latter is caused by the lack of knowledge and information and is subjective and reducible. Thus, sustainability under uncertainty has been a very active research area in sustainability science and engineering. Methodologically, decision making by resorting to fuzzy systems theory may offer a unique perspective in systematically handling epistemic uncertainty.

In this paper, we introduce a predictive fuzzy decision-making methodology for deriving optimal solutions under uncertainty for multistage sustainability performance improvement. By this methodology, a large-scale industrial system is decomposed into a number of subsystems. For each subsystem, fuzzy decision making is made to satisfy its set sustainability goal for a number of development stages. The decisions derived in this lower layer of the decision hierarchy will then be analyzed and coordinated at the higher layer for the entire system's goal achievement in multiple stages. The coordinated decisions will be transmitted back to the lower layer for each subsystem to adjust its previous decision. This recursive process will be terminated when the sustainability goal for the entire system is achievable. The methodology is systematic, computationally efficient, and suitable for analyzing sustainability status of any system, and predicting its short-to-long-term behavior. The efficacy of the methodology will be demonstrated through a case study on the short-to-long-term sustainability of an industrial system involving chemical manufacturing, metal finishing, and automotive manufacturing.

273c Aggregation Methods for Life Cycle Sustainability Assessment: Decision Making Guidance

Natalia Cano-Londoño, Erasmo Cadena, Sue Ellen Taelman, Jo Dewulf

Methods based on life cycle thinking have been developed to comprehensively assess the impacts (burdens) and benefits of products (goods and services) and business systems to determine qualitative and quantitative environmental, social, and economic indicators across the value chain. ***However, an established comprehensive sustainability assessment method for products still needs to be included that coherently integrates all pillars of sustainability.*** Methods such as Multi-Criteria Decision Analysis (MCDA), outranking models (ELECTRE, PROMETHEE, VIKOR), Multi-Objective Decision Making (MODM), Data Envelopment Analysis (DEA), Product Environmental Footprint (PEF) normalization and weighting, distance to target weighting methods, and monetary weighting methods, addressing this challenge with the aim to both simplifying the decision process and identifying the optimal case out of a set of alternatives based on certain decision criteria (environmental, social and economic), the majority of them providing a single indicator to facilitate decision-making. However, each method has its own methodological complexity and uses different decision criteria, making its application by beginners and practitioners across the academic and SME sectors unpractical. Our research aim is to provide practical decision tree guidance in selecting and applying aggregation methods for LCSA based on the objective of LCSA integration: 1) obtaining (an) aggregated score (s), 2) Only ranking alternatives, or 3) only visualization of the three sustainability dimensions; as well as, some criteria of importance: a) Potential to generate an aggregate score for each of the three sustainability domains or some domains, b) compensatory sustainability, c) potential to use for a single product and d) requirements of semi-quantitative or quantitative inputs, e) comparative analysis among others.

To achieve the main objective, a literature search was conducted via scientific search tools (e.g., web of science) to identify integration methods applied across the sustainability domains, where an initial examination and additional pre-screenings limit the number of relevant papers retained to 23. Based on these literature reviews, a selection of 14 integration methods or sub-procedures leaders in the multiple criteria-based decision aiding were done, which can be categorized into three groups: Group 1 - Integration methods across sustainability domains (for LCSA), Group 2 – Integration methods with a promising visualization approach for LCSA, Group 3 – Other aggregation methods or sub-procedures commonly used within the environmental and social domains. The integration methods selected were analyzed based on the research aim of LCSA integration as well as some criteria of importance (decision criteria). A structured pathway (foundational decision tree guidance) was proposed to easily select the most appropriate integration method for assessing a set of alternatives under multiple criteria or objectives.

273d Holistic Assessment of Decarbonization Potential and Their Related Policies for Introducing Anaerobic Digestion in Food Waste Management

Borja Hernández Blázquez, Alexandros N. Karaiskakis, Marianthi Ierapetritou

According to United Nations Environment Program, 17% of the global food production, corresponding to 931 million tons of food, was wasted in 2019. Among them, 11% is generated in households, which amounts to 74 kg per year.[1] In the United States, 40% of the food produced is thrown away every year, which translates into 99 kg of waste per person and $218 billion worth of lost opportunity.[2,3] Besides its economic significance, food waste has a carbon footprint of approximately 170 million $MTCO_2e$ (excluding landfill emissions and food surplus). Current food waste management involves primarily landfilling, composting, land application, and incineration. Among them, landfills and incineration are the two most popular technologies receiving 75% and 18% of the overall food waste in the US.[4] These technologies provide low value to the waste generated, and they generate higher emissions in direct (pollutants generated in combustion) or indirect forms (landfill gas). Reducing their contribution is then necessary for mitigating CO_2 emissions. In this context, the Environmental Protection Agency has recommended more sustainable technologies such as composting and anaerobic digestion (AD). However, the products generated by these technologies are not competitive enough (e.g. the production cost of biomethane from AD of food waste is around $22/MMBTU, and the cost of natural gas is ~$7/MMBTU). Thus, developing policies that incentivize their introduction are necessary.

In this work, we compare incentive mechanisms based on their decarbonization potential to promote the introduction of AD technology. Introducing these incentives requires developing a framework that performs a holistic life cycle assessment (LCA) that integrates the emissions of the AD process with the specific geographical conditions. The proposed holistic Life Cycle Analysis (LCA) framework is composed of the following parts:

1. <u>Data collection with specific geographical information.</u> The specific emission factors, the amount of waste generated and the type of management (percentage sent to landfill and type for substitution in system expansion) is collected from open data sources.[5]
2. <u>LCA of AD process.</u> The amount of waste generated, the treatment technologies used for each of them, and the emission factors are substituted into a flexible LCA of the AD process following a system expansion approach where the products and raw materials are substituted. The inventory of the LCA has utilized data from the collaborating company regarding the material and energy flows. Two alternatives for digestate management (landfill disposal and hydrothermal carbonization) and three uses for biogas (upgrading to biomethane, production of electricity and production of hydrogen) are considered.
3. <u>Incentives quantification:</u> The emissions determined by this framework for a plant are scaled up to the entire state with modular plants. The emissions in each of the states have been quantified as Carbon Intensity (CI) score (emission per unit of energy) to be used in the assessment of the incentive's mechanism. The Low Carbon Fuel Standard (LCFS) mechanism is used for this purpose.

The holistic LCA, apart from being utilized in the incentives' development, it estimates the maximum decarbonization potential of AD in food waste treatment. An avoidance potential of ~21.8 MMtons CO_{2e}/y (or 345 gCO_2/kg_{waste}) and the potential of generating 161 TBTU of natural gas is estimated for the entire US with an average decarbonization cost of ~$128/ $tonne_{CO2e}$. The states that benefit more with the new LCFS mechanism are those with the lowest CI Score represented in Figure 1. The incentives introduced with the LCFS mechanism based on LCA with a system expansion approach, reduce the average cost of treating the waste by nearly a half, from $79/$tonne_{waste}$ to $31/$tonne_{waste}$. A comparison of the policy mechanism presented in this work has been performed against the recent Federal Energy Regulatory Order (FERC) 2222 policy, which has shown to favor the introduction of AD in farm-waste treatment facilities.[6]

This FERC-2222 is obtained to have a cost of ~\$27/tonne$_{waste}$, lower than the one obtained with LCFS mechanism. However, FERC-2222 favors states with less decarbonization potential, which are also the states with more technological development. Thus, an indirect mechanism like LCFS can provide similar economic benefits to FERC-2222 and it can also provide a fairer distribution of the incentives incentivizing the installation of AD facilities is less technological developed regions.

REFERENCES

1. United Nations. Food Waste Index Report, **2021**.
2. US Environmental Protection Agency. International Efforts on Waste Food Recovery. 2022.
3. US Department of Agriculture. Food Waste FAQs. **2022**. *Available in: https://www.usda.gov/foodwaste/faqs*
4. Badgett, A. Milbrandt, A. Food waste disposal and utilization in the United States: A spatial cost benefit analysis. **2021**. Journal of Cleaner Production, 314, 128057.
5. Explore solutions to food waste. 2022. Available in: *https://insights-engine.refed.org/food-waste-monitor?view=overview&year=2019*
6. Erickson, E.D. Tominac, P.A. Zavala, V.M. Biogas production in United States dairy farms incentive by electricity policy changes. **2023**. Nature Sustainability, 1, 12, 1-9.

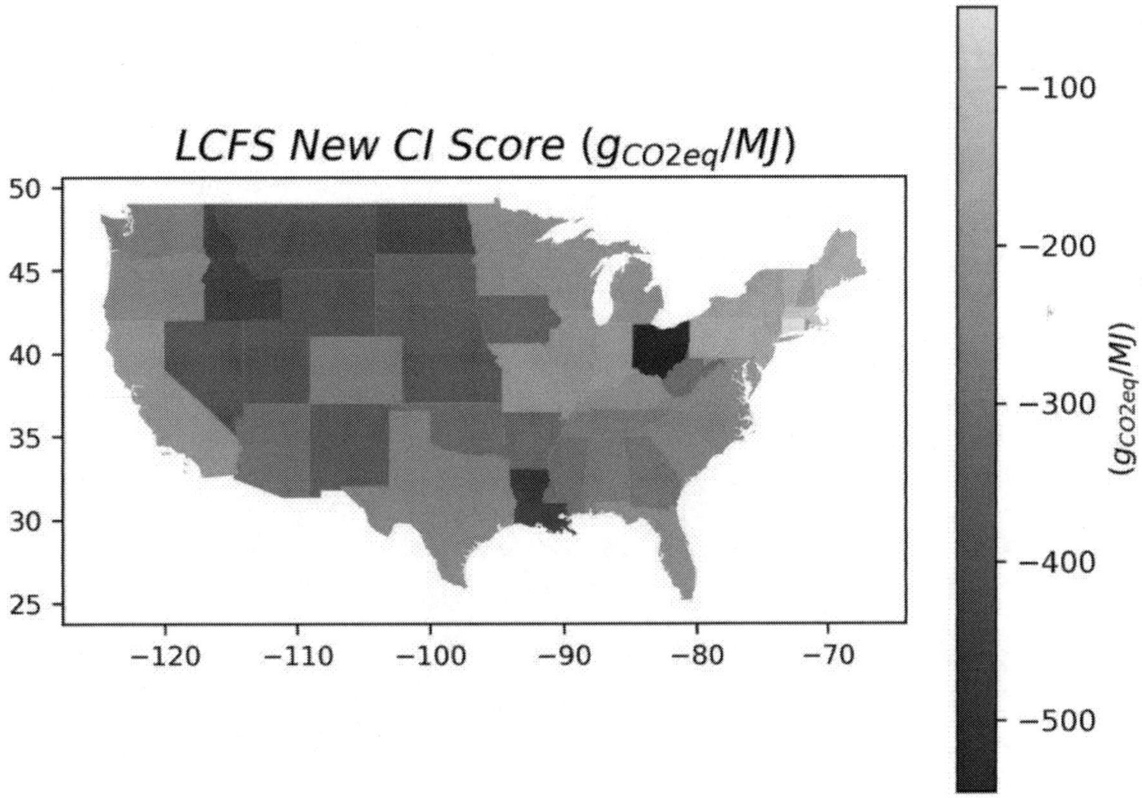

273e A Novel Data-Driven Approach to Reducing the Environmental Footprint of an Integrated Gas-Oil Separation Network

Abdullah Bahamdan, Ehecatl Antonio del Rio Chanona, Nilay Shah

As the pressure to meet net-zero targets builds, the energy industry, particularly the oil and gas sector, is investigating various opportunities spanning the supply chain. Accordingly, companies are considering the adoption of advanced analytic techniques to monitor operations performance and reduce their environmental footprint. However, achieving this objective while maintaining a solid financial performance is critical, thus, requiring innovated and integrated solutions.

This work has focused on the upstream business since it accounts for a substitutional percentage of profit margin and emissions. In large oil reservoirs, a cluster of producing wells supply a network of gas-oil separation facilities interconnected via swing pipelines. The primary function of the facilities is to separate the three phases and remove contaminants. Nonetheless, they play a vital role in eliminating flaring and greenhouse gas emissions by recovering associated gases.

For the past few years, scientists and practitioners have devoted several efforts to optimizing network operations. However, it was focused on achieving higher savings in operating expenditures (OPEX) while overlooking sustainability. Furthermore, previous contributions had several limitations in the proposed method—for example, they utilized a physics-based model with expensive computation time.

Due to these limitations, we have developed a novel data-driven method that addresses them. Our approach uses artificial neural networks in a mixed integer linear programming (MILP) framework to quantify the entire network's environmental footprint and operating expenditures, then optimize it by manipulating the feed allocation and equipment utilization. The surrogate models are trained using a high-fidelity simulation model to capture possible feed uncertainties. Furthermore, the proposed method has been tested using several production scenarios obtained from actual production rates and demonstrated outstanding capabilities. As a result, it has significantly reduced greenhouse gas emissions while maximizing profitability.

273f Advancing Environmental Sustainability through Green Chemistry and Engineering Principles and Matrices.

Tapas K. Das

Speaker: Prof. Tapas K. Das, PhD, PE, BCEE, Fellow AIChE & ICS

School of Engineering, Saint Martin's University, Lacey, Washington, USA

Prepared for Presentation at the 2023 AIChE Annual Meeting in Orlando, Florida

Abstract:

The goal of this presentation is to forge a pathway for environmental sustainability by incorporating basic principles and concepts of green chemistry, green engineering, recycle, reuse, and recovery; and zero discharge. Bearing in mind, the design and implementation of completely green products and processes is an enormous challenge. Because the number of chemical synthesis pathways is enormous, there is no one systematic, fail-safe method for ensuring that the chemistry being implemented is green. Indeed, it is more nearly correct to inquire if a proposed chemical manufacturing process is simply "greener" than other alternatives. Thus, green chemistry recognizes the importance of incremental improvements applying a set of 12 principles to define and guide the scope of green chemistry. Sustainable engineering involves the design, commercialization, and use of processes and products that are feasible and economical while minimizing both the generation of pollution at the source and the risk to human health and the environment. The discipline embraces the concept that decisions to protect human health and the environment can have the greatest impact and cost effectiveness when applied early in the "design and development phase of a process or product." Sustainable engineering transforms existing engineering disciplines and practices to those that promote environmental sustainability. A few mini-case studies are presented.

Key References:

Anastas, P.T. and Warner, J.C. (1998). *Green Chemistry Theory and Practice*, New York, NY: Oxford University Press.

Anastas, P.T. and Zimmerman, J.B. (2003). Design through the 12 Principles of Green Engineering, *Engineering Science and Technology*, **37**(5), 94A-101A.

Das, T.K. (2005). *Toward Zero Discharge: Innovative Methodology and Technologies for Process Pollution Prevention,* John Wiley and Sons, Hoboken, NJ.

Rahimi, A. and Garcia, J.M. (2017). "Chemical Recycling of Waste Plastics for New Materials Production", *Nature Reviews Chemistry,* 1, 1-11.

Das, T.K. (2020). *Industrial Environmental Management: Engineering, Science and Policy*, 1st ed., John Wiley and Sons, Hoboken, NJ.

273g Absolute Environmental Sustainability Assessment of Chemicals and Bridging Gaps to Becoming 'Nature Positive'

Ying Xue, Bhavik Bakshi

'Nature Positive by 2030', 'Net-Zero by 2050'...have been widely accepted as global goals for guiding decisions towards a healthier planet [1]. The chemical industry is one of the largest and most pervasive industries, with chemicals used in numerous applications including agriculture, medicine, food, and manufacturing. However, this industry has also caused significant and irreversible environmental impacts that have led to the urgent need for sustainability assessment and action. To this end, conventional sustainability assessment methods, such as life cycle assessment (LCA), provide relative sustainability metrics by comparing one system to another. However, these metrics are sensitive to the selection of the reference group. Alternatively, absolute environmental sustainability assessment (AESA) defines an absolute value - nature's carrying capacity, against which the environmental impacts of a system are compared [2].

In this work, we extended the multiscale method based on the Techno-ecological synergy (TES) framework for AESA [3]. We selected four major chemicals - ammonia, methanol, ethylene, and benzene - and focused on carbon dioxide emissions to assess their AES metrics at the country and global levels. We compared different sharing principles, including emissions, gross value added (GVA), and historical accumulative emissions, to gain insights into the responsibility that each stakeholder should take in achieving sustainable global goals. We also compared our multiscale AESA method with the benchmark method based on the planetary boundary (PB) framework, highlighting the limitations of the latter. The AES metric is quantified by comparing a system's or product's environmental impacts versus ecological thresholds. We used Ecoinvent 3.8 life cycle inventory database to estimate the life cycle emission of these four chemicals in nine major producer countries. While most AESA research is conducted at the global scale, we argue that results should be conducted at smaller scales to guide stakeholders towards sustainability goals and we explored how to do it. After quantifying gaps for these chemicals to becoming 'nature positive', we looked into different 'green pathways' to explore the potential opportunities in low carbon transition of the chemical industry. For instance, using renewable energies from biomass, solar, wind; installing carbon capture and storage units; reforestations and restore ecosystems, etc.

The planetary boundary framework defines nine key earth system processes [4]. Each earth system process has a quantitative boundary at global or subglobal scale which can be referred to ecological threshold. In PB-based AESA methods, boundaries will be directly downscaled to a process or product system [5-7] which makes the result highly dependent on the choice of sharing principles. Issues above could provide misleading results as spatial heterogeneity is ignored. Our TES framework avoids these issues by quantifying ecosystem services (ES) through biophysical models, providing high geographical resolutions for ecological threshold quantification. Due to the nested character of ecosystems, ESs are quantified from small to large scales. We applied a 2-scale TES framework, including national and global scales, to reflect stakeholder capabilities to take in CO2 emissions sustainably. Public and private ownership of ESs are considered during allocation to avoid introducing subjectiveness and also to encourage stakeholders to restore ecosystems. Ecological thresholds at global scale will be partially allocated to specific products by certain sharing principles such as gross value added, population, etc. We compared different sharing principles that are commonly used. Considering the fact that countries began industrialization at different times and at different levels of development, countries like united states and European union countries have emitted considerable amount of CO2 for decades while countries like China, India have a shorter high emission time range. From the egalitarianism aspect, accumulative historical emission should be considered and we explored a novel criteria based on it. Applying the same method, we extended the scope to whole the chemical industry using chemicals and materials industry model.

Results show large variation of transgression levels among different countries, some stay within ecological thresholds while some not which are not reflected from other studies in this topic. When these regional transgression levels be scaled up to global level, the variations will be averaged. In conclusion, this work contributes to the urgent need for sustainability assessment and action. The multiscale TES-based method used in this work provides targeted insights for stakeholders to transition towards a low-carbon chemical industry, avoiding the limitations of the PB framework.

1. Global Goal for nature. https://www.naturepositive.org/. Accessed: 2022-12-13.

2. Bjørn, Anders, Katherine Richardson, and Michael Zwicky Hauschild. 'A framework for development and communication of absolute environmental sustainability assessment methods.' Journal of Industrial Ecology 23.4 (2019): 838-854.

3. Xue, Ying, and Bhavik R. Bakshi. 'Metrics for a nature-positive world: A multiscale approach for absolute environmental sustainability assessment.' Science of The Total Environment 846 (2022): 157373.

4. Rockström, Johan, et al. 'A safe operating space for humanity.' nature 461.7263 (2009): 472-475.

5. D'Angelo, Sebastiano Carlo, et al. 'Planetary boundaries analysis of low-carbon ammonia production routes.' ACS Sustainable Chemistry & Engineering 9.29 (2021): 9740-9749.

6. González-Garay, Andrés, et al. 'Plant-to-planet analysis of CO 2-based methanol processes.' Energy & Environmental Science 12.12 (2019): 3425-3436.

7. Ryberg, Morten W., et al. 'How to bring absolute sustainability into decision-making: an industry case study using a planetary boundary-based methodology.' Science of the Total Environment 634 (2018): 1406-1416.

243a Implications of Performance Variables on Sustainable Aviation Fuel (SAF) Production and Deployment

Arpit Bhatt, Yimin Zhang, Ling Tao

Jet fuel consumption is expected to double from 2018 to 2050 while accounting for 2.8% of total greenhouse gas (GHG) emissions globally. This increase in consumption and emissions associated with jet fuels have led to need for alternate technologies for producing sustainable aviation fuels (SAF) from bio-based sources that has the potential to meet the SAF Grand Challenge goals of producing 35 billion gallons of SAF production by 2050 and reducing up to 50% of GHG emissions compared to conventional jet fuels by 2030. While industries, government, and airlines are committed to achieving net zero carbon emissions, SAF is a promising near-tern option for meeting these goals for the transportation sector.

While previous studies have individually looked into analyzing feedstock types or varying the reaction chemistry to boost jet fuel slate in the hydrocarbon fuel mixture, performed resource assessment, techno-economics, or lifecycle GHG emissions analysis, or investigated policies impacting SAF production in general, none of them looked into synergistic impacts of all these performance variables affecting the production of SAF. This knowledge gap on how the combination of different parameters affects the SAF industry is important to identify key hotspots that can help accelerate the deployment of SAF for meeting the decarbonization goals and thus provides a bigger picture of issues at a regional scale.

Because the availability of feedstocks and use of jet fuel vary across locations, we consider Chicago O'Hare International Airport (ORD) in Illinois as a case study based on the passenger data, jet fuel consumption, and feedstock availability, and assess the implications of key performance variables to understand the costs associated with production and delivery of SAF, policy incentives that can help reduce costs, evaluate feedstocks available around Chicago airport (50 to 200-mile radius), determine the reduction in lifecycle GHG emissions for SAF production compared to conventional jet fuel, and discuss infrastructure logistics around ORD for blending SAF with Jet A. We focus our analysis on three ASTM approved SAF conversion technologies including Fischer-Tropsch Synthetic Paraffinic Kerosene (FT-SPK) utilizing woody biomass feedstock, Hydroprocessed Esters and Fatty Acids SPK (HEFA-SPK) utilizing fats, oils, and grease (FOG) feedstock, and Alcohol-to-Jet SPK (ATJ-SPK) utilizing crop residues (iso-butanol intermediate). We utilize Biofuel Atlas, Environmental Protection Agency data, and Census Bureau population data for feedstock resource analysis, Aspen Plus process simulation model based discounted cash flow rate of return analysis for techno-economics, and Greenhouse gases, Regulated Emissions, and Energy use in Technologies (GREET) model, SimaPro, and existing literature to estimate emissions associated with various life cycle stages. Policy incentives and infrastructure logistics data were obtained from literature.

Our results show that feedstocks are available within 200-mile of ORD to displace up to 55% of airport's current conventional jet fuel with lifecycle GHG emissions reductions potentially reaching up to 86% by using SAF rather than Jet A. Also, FT-SPK has the lowest SAF selling price although high upfront capital costs and low jet fuel yield may limit its commercialization. HEFA and ATJ still need technological improvements to beat the market price of Jet A. We find that feedstock price and renewable fuel incentives are key variables affecting the production cost. Moreover, infrastructure is available at ORD to blend SAF with Jet A sourced both locally and from the gulf coast through terminals directly connected via pipelines.

Insights from our analysis can help various stakeholder groups, including aircraft manufacturers, airlines, policymakers, fuel blending/refining companies, farmers, researchers, and agricultural companies and also meet the targets set by SAF Grand Challenge Report. While resources are available to produce SAF and replace a significant portion of Jet A fuel, a detailed supply-chain analysis may need to be performed to allocate resources that compete for production of other hydrocarbon fuels, and common and specialty chemicals.

243b Low-Carbon Methanol Production Routes Via Hydrogen Electrolysis: Techno-Economic and Life Cycle Analyses

Jonathan Martin, Eric Tan, Jennifer King, Daniel Ruddy

The most common process for methanol (MeOH) production is currently steam methane reforming (SMR) using natural gas (NG), a fossil fuel, as the feedstock. Even with carbon capture and sequestration (CCS) of the tail gas CO2 from natural gas SMR MeOH plants, the cradle-to-gate carbon intensity (CI) of this methanol is still approximately 1 kg CO2e / kg MeOH. Several lower-carbon alternatives have been proposed which do not use NG as the feedstock and instead combine captured CO2 and renewably produced H2. These alternatives are more expensive to produce than conventional SMR methanol, mainly due to the high production costs of renewable H2. One alternative MeOH production route is the CO2 hydrogenation process, which currently has plants producing MeOH at up to 110,000 t/y [1]. This route generally utilizes CO2 from CCS of stack gasses from existing industrial plants, such as natural gas combined cycle (NGCC) power plants. A combined techno-economic analysis (TEA) and life cycle analysis (LCA) of these two processes (natural gas SMR and CO2 hydrogenation) was conducted to determine the conditions necessary for the methanol produced from CO2 hydrogenation to become economically competitive with conventional SMR-produced methanol. The TEA/LCA utilized several renewable technology assessment tools developed at the National Renewable Energy Laboratory (NREL), particularly:

- Hybrid Optimization and Performance Platform (HOPP) – used to design optimal hybrid wind/solar/electrolysis plants to minimize the cost of green hydrogen
- The Annual Technology Baseline (ATB) – an annually-updated set of projections of capital costs of different electricity production technologies
- H2A Hydrogen Production Analysis models
- Cambium – a model of marginal grid electricity costs/emissions for several scenarios, including:
  - "Mid-case": Central estimates for technology costs, fuel prices, and demand growth
  - "High Natural Gas Prices": the same set of base assumptions as the first scenario, but where natural gas prices are drawn from the AEO2022 Low Oil and Gas supply scenario.

The result of this TEA/LCA, shown in Figure 1 below, found that the CO2 hydrogenation route would reduce methanol CI by approximately half, but increase production costs by more than 150% in the "mid-case" scenario for plants coming online between 2030 and 2050. To be economically competitive with SMR-produced methanol would require a CO2 emissions price of more than \$500/t CO2e. This is well above the Biden administration's value for the social cost of carbon of \$51/t CO2, as well as the EPA's recent proposal of \$190/t CO2. For CO2 hydrogenation-based MeOH to be economically competitive by 2050, with a more reasonable CO2 emissions price would require both the "High National Gas Prices" scenario as well as higher-than-expected technology development for green hydrogen, enough to bring green H2 production costs down to \$1/kg H2 by 2050.

Another MeOH production route, currently under development by NREL, would instead utilize direct reactive capture and conversion (RCC) of CO2 from flue gas in a series of pressure swing reactors (PSRs) utilizing dual functional materials (DFM) catalysts. This process is being developed at the NREL lab scale and is being modeled in ASPEN at the same scale as the CO2 hydrogenation process. A comparative TEA/LCA of this RCC process will be presented at AiCHE.

Figure 1. Methanol TEA and LCA, comparing the conventional SMR and CO2 hydrogenation routes. Line color indicates the methanol production technology, and line style indicates the feedstock price/technology development scenario simulated. Top left: cost of producing methanol (both routes); top right: cost of producing renewable hydrogen for CO2 hydrogenation-based methanol, bottom left: methanol carbon intensity (both routes); bottom right: CO2 emissions price required for CO2 hydrogenation-based methanol to overcome the production cost gap with SMR-based methanol.

[1] carbonrecycling.is/projects#projects-shunli

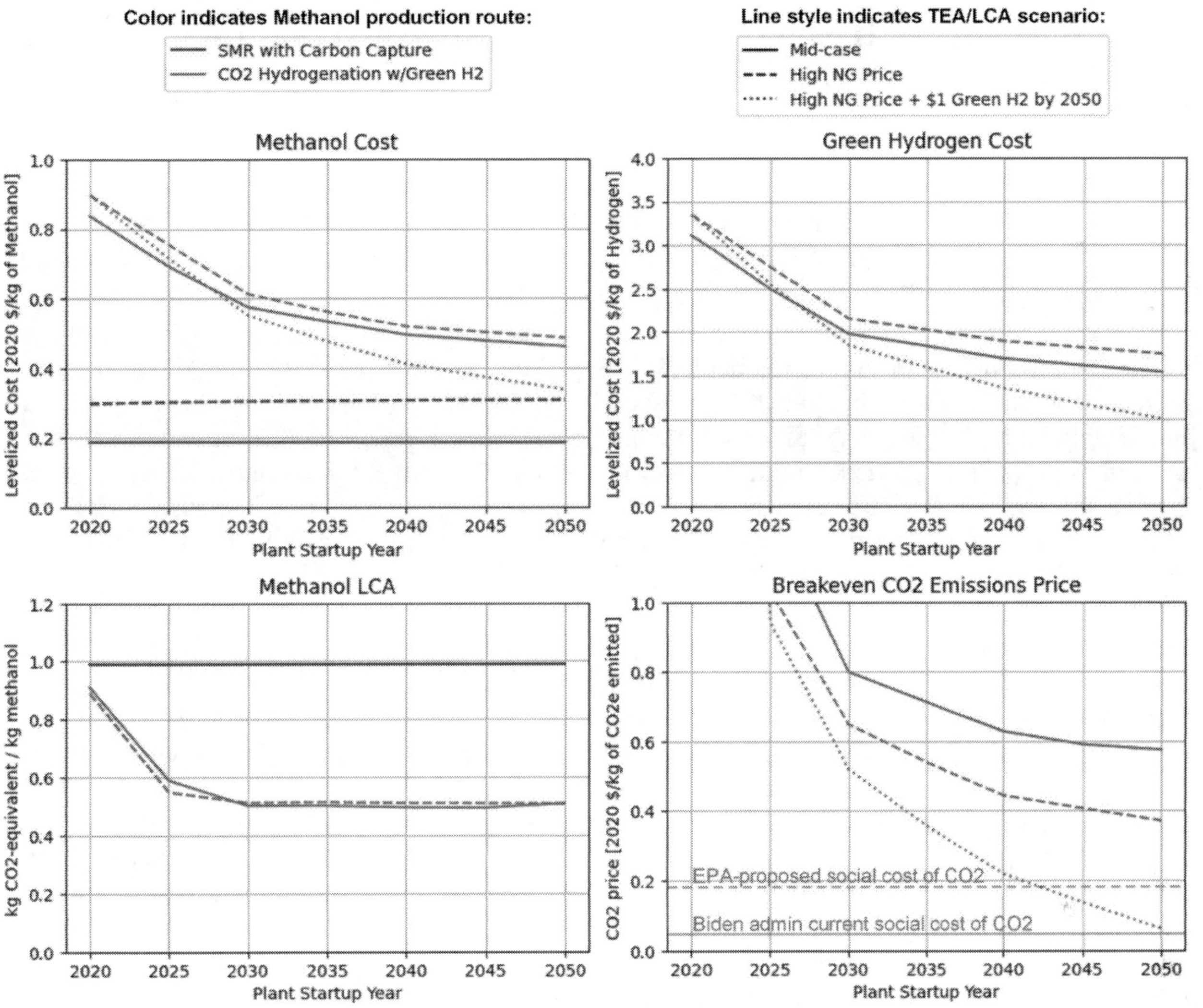

243d Techno-Economic Analysis Towards Revalorization of Biomass through Pellet Production As a Heating Substitute of Methane and Natural Gas: Spain Case

Guillermo Galán Iglesias, Roberto Cifuentes, Mariano Martin

The extensive use of fossil fuels such as coal and oil, burned for energy in the industry, transportation sector, and other human activities, has resulted in huge greenhouse gas emissions, composed mainly of CO_2, reaching levels above over 32Gt CO_2 per year (EPA, 2022). Moreover, the continued utilization of these fuels also contributes to air pollution and poses risks to human health. Over the last years, natural gas was positioned as the better fossil fuel to obtain energy, reducing emissions and improving efficiencies in combined cycle and cogeneration plants, boilers, and heating systems, as well as in domestic use. However, the current international situation with the Ukrainian War has exposed the European dependency on the supply of natural gas from Russia. Alternatives are to be developed and deployed towards energy security , using local resources to obtain energy and chemicals. Thus, cleaner energy sources and technologies, such as renewable energy and energy-efficient systems must be developed. Focusing on local resources, biomass has become more important in the last years. This may be used to produce biogas and biomethane through anaerobic digestion (Martín-Hernández et al., 2020), (Taifouris & Martin, 2018). Methane could be also produced from water electrolysis and CO_2 capture (De la Cruz & Martin, 2016), but that technology is still under development and it's out of the scope of the study. The utilization of these processes to reduce energy dependence is not enough, so the use of biomass for the production of pellets has become an increasingly important new alternative to exploit and revalue local biomass, achieving a consumption to heat production in the EU around of 17 Mt In the year 2018 (European Pellet Council, 2019), and 23.1 Mt in the year 2021 (USDA, 2022). Pelletization is an increasingly widespread process, generally using lignocellulosic wastes from conditioning and pruning forests, agricultural wastes, etc., and submitting them to the set of operations that allow increasing their density and improving the physicochemical properties, minimizing the problems associated with transportation and storage, low density and moisture content. It consists of a set of operations based on previous drying, reduction, and homogenization of particle size, adjusting the composition of moisture, densification of the residues, cooling, and storage.

The problem is addressed as a multiscale facility location problem, which will allow deciding the better emplacement of the areas to obtain biomass, its transport, and treatment process. The problem is solved following the methodology presented by Heras & Martín (2021). The process yield, product, and investment cost of pellets are estimated based on a techno-economic analysis performed. The equation-based approach is used to model all of the processes. The dry biomass undergoes a crushing process to reduce the size of the particles. Next, part of the biomass will be conducted in a boiler, where it is burned to obtain energy, submitting the rest of the biomass to the following drying process, reducing moisture content up to 12%. Finally, the biomass will be subjected to a pressure granulation process. Depending on the energy needs of the plant, part of the pellets could be reintroduced into the process. Several biomasses are considered. Concerning the macro scale, the KPI of the process for each biomass will be used to formulate a network optimization problem to decide the place to obtain the residues and the best location to install the facility. The problem is formulated as NLP, which includes the social benefits in form of direct and indirect job generated, wealth, and CO_2 emissions in the location chosen to install the facilities and to transport the residues.

For the case study, Spain is selected, taking into account only the peninsula. The Balearic and Canary

Islands, the autonomous cities of Ceuta and Melilla, and the rest of the Spanish enclaves in the Mediterranean Sea are not considered. On the one hand, the study is focused on biomass selection, thus lodgepole pine, eucalyptus, hardwood, switchgrass, and corn stover are chosen as the more appropriate lignocellulosic biomasses (Yancey et al, 2013). Later, a systematic comparison of product and investment costs, and income is analyzed to determine the more appropriate biomass for developing the process. On the other hand, the approach to the problem is focused on the facility location, selecting a total of 345 regions to place the facilities (Taifouris & Martin, 2022) depending on biomass availability. Generally, the central and south of the peninsula are the areas with higher probabilities to install the locations, due to the presence of more solar irradiation, huge water reserves to produce cereals, and the pine and eucalyptus growing.

References:

De la Cruz V, Martin M. Characterization and optimal site matching of wind turbines: Effects on the economics of synthetic methane production. 2016;133:1302-1311.

European Pellet Council. European Wood Pellet Consumption. https://epc.bioenergyeurope.org/about-pellets/pellets-statistics/european-consumption/; 2019.

Heras J, Martin M. Multiscale analysis for power-to-gas-to-power facilities based on energy storage. Computers and Chemical Engineering. 2021;144:107147.

Martín.Hernández E, Guerras LS, Martin M. Optimal technology selection for the biogas upgrading to biomethane. Journal of Cleaner Production. 2022;267:122032.

Taifouris M, Martin M. Multiscale scheme for the optimal use of residues for the production of biogas across Castile and Leon. Journal of Cleaner Production. 2018;185:239-251.

Taifouris M, Martin M. Integrating intensive livestock and cropping systems: Sustainable design and location. Agricultural Systems. 2022; 203:103517.

USDA. EU Wood Pellet Annual. https://www.fas.usda.gov/data/european-union-eu-wood-pellet-annual; 2022.

US EPA. Overview of Greenhouse Gases. https://www.epa.gov/ghgemissions/overview-greenhouse-gases; 2022.

Yancey NA, Tumuluru JS, Wright CT. Drying, Grinding and Pelletization Studies on Raw and Formulated Biomass Feedstock's for Bioenergy Applications. Journal of Biobased Materials and Bioenergy. 2013;10:1-10.

Efficient Capture of Carbon Dioxide from Thin Air for Synthesis of Fuels and Chemicals (invited Undergraduate Talk)

Tanishka Kucheria, Erin Heeschen, Magda Barecka

The ability to synthesize sustainable fuels and chemicals will largely rely on the access to biogenic sources of carbon, that could be used in the chemical manufacturing processes instead of petroleum-derived feedstocks. Using renewable energy to electrochemically separate carbon dioxide (CO_2) is a promising approach for source biogenic carbon on a large scale. One of the main challenges limiting the performance of the existing electrochemical CO2 capture systems is related to the slow mass transfer in flow reactors. In this project, we propose a novel design of electrochemical reactors that allows to significantly limit the distance between the electrodes and facilitates mass transfer, leading potentially to reduced energy consumption and extended system lifespan. This undergraduate research project focuses on modelling and experimental investigation of intensified electrochemical systems. This advancement not only ensures efficient CO2 capture, translating to reduced costs and environmental impact, but also bolsters efforts against climate change.

The Presence of Waste Plastics and Microplastics in Lakes and Rivers, and their Impact on the Marine Environment

Mohamed K. Mostafa [1], Ahmed S. Mahmoud [2], Djihan Hassan [3], and Robert W. Peters [4]

[1] Faculty of Engineering and Technology, Badr University in Cairo (BUC), Cairo, Egypt
[2] Institute of Environmental studies, Arish University, North Sinai, Egypt
[3] Environmental Consultant, Freelance
[4]Department of Civil, Construction and Environmental Engineering, University of Alabama at Birmingham, Birmingham, AL, USA

Abstract

Waste plastics (>5 mm) and microplastics are tiny pieces of plastics (<5 mm) that have been manufactured, shed from textiles or formed as degradation products of macroplastics. These materials can be taken up by aquatic organisms, which can be taken-up and incorporated into the food chain. Humans can consume microplastics from fish as well as other impacted sources such as bottled water and tap water. Microplastics can pose risks to marine organisms, and also act as vectors carrying additional adsorbed chemical pollutants and pathogens. Microplastics pose significant risks to the biota and fauna. This presentation describes the presence and impacts of waste plastics and microplastics in freshwater systems such as lakes and rivers. Microplastics are ubiquitous worldwide in surface waters. Microplastics and waste plastics have been detected virtually everywhere on earth. For example, microplastics have been detected in China, Brazil, Tanzania, the Great Lakes in the U.S., Spain, Argentina, Italy, Mongolia, and Canada. Microplastics have even been detected in remote mountain lakes. Global rivers and effluents act as a dominant transport pathway for land-based plastic debris for the marine environment. This is particularly true for urban environments. Combined sewer overflows likely contribute likely contribute additional plastic loads during high-flow conditions. Waste plastics can degrade due to the mechanical degradation by wave action and sand abrasion. It is estimated that more than 1,000 rivers account for more than 80% of global annual emissions into the oceans ranging from 0.8 million to 2.7 million tons per year, with smaller urban rivers among the most polluting rivers. The plastics can remain in the aquatic environment for years, decades, and potentially centuries. Hydrometeorological variables such as wind, runoff, and river discharge, mobilize, transport, and deposit plastics in different river components (e.g., riverbanks, floodplains, lakes, and estuaries); these conditions are especially impacted by storms and floods. Flooding conditions exacerbate the concentration of microplastics in rivers globally. Additionally, microplastics can impact marine species in rivers; Nile tilapia and catfish have microplastics detected in their gastrointestinal tracts, present as fibers and films. Plastics detected include polyethylene, polypropylene, and polyethylene terephthalate. Water birds are also impacted by microplastics. This technical presentation will summarize results of various research studies related to microplastics present in the oceans and rivers, and their impacts on the ecosystem and fauna.

Keywords:

Microplastics, Waste, Impacts, Environmental Health, Marine Ecology.

1. Introduction

The global plastic market size was valued at 609 B USD in 2022 and is anticipated to reach 824.46 B USD by 2030. This results from increasing consumer demand for higher performance plastics as well as the rising trend and need for replacing metals and ferrous alloys with plastics across different industries including consumer goods, medical, automotive and industrial machinery [1]. Moreover, most plastics in use today are virgin/primary plastics, made from petroleum. Global production of plastics from recycled/secondary plastics has more than increased fourfold between 2000 and 2019 from 6.8 Million Metric tons (MMT) to 29.1 MMT, but still account for about 6% of the size of total plastics production [2].

This increase in plastic production comes hand in hand with an increase in environmental concern and need for sustainable solutions to manage and reduce consequential plastic waste. Currently, only around 9% of plastic waste is recycled, 69% is treated and disposed of through incineration and landfilling; leaving 22% without appropriate waste management [2, 3]. With regards to the reduction in the quantities of generated plastic waste, there has been an increasing focus on research and market demand for more sustainable plastic materials e.g. biodegradable plastics, as well as advancements in environmental regulations. Furthermore, many initiatives have been implemented by companies to reduce their environmental impact from plastic production by utilizing recycled content within products and collecting and recycling post-consumer plastic waste [4].

Microplastics are small pieces of plastic that are less than five millimeters in size. They are produced either from primary sources such as from commercial product development e.g. cosmetic products; or from secondary sources that result from the degradation of larger plastics (macroplastics) [5,6]. Macroplastics, pieces of plastic that are more than five millimeters in size, undergo weathering once exposed to environmental conditions and result in the formation of microplastics [4,7].

Microplastics are considered pollutants both for the aquatic and terrestrial ecosystems due to their ecotoxicological impacts that pose great threats to human, animal and plant health [5,6]. Due to their small size, microplastics can be readily ingestion by animals (e.g. fish and other marine organisms) that can be later ingested by others and by humans resulting in the bioaccumulation and potential biomagnification along the various stages of the food chain. Additionally, microplastics act as carrier vectors for pathogens, heavy metals and other pollutants such as polycyclic aromatic hydrocarbons (PAHs) and polychlorinated biphenyls (PCBs) [8, 9, 10].

This review aimes to provide a comprehensive understanding of microplastics, their types, pathways and their impacts on the ecosystem.

2. Sources and Types of Microplastics

Plastics are composed of a wide variety of materials that have been developed over the years. Fragmented plastics are being addressed based on their origin, process of fragmentation, chemical composition and density as well as their size and shape [11].

Potential primary sources of microplastics include plastic pellets, scrubbers, commercial cleaning abrasives, plastic resin flakes, personal care products containing microbeads, paint, wastewater treatment, sewage sludge and motor vehicle tire wear [4, 11, 12, 13, 14].

Secondary microplastic result from the breakdown of larger plastics due to their exposure to environmental effects. Environmental weathering involves the breakdown of long-chain plastic polymers into smaller ones [15] and can be triggered by one of two notable mechanisms; abiotic or biotic degradation. Abiotic degradation involves wind abrasion, wave action, photodegradation, ultra-violet radiation, heat, and chemical reactions that lead to plastic breakdown or fragmentation via mechanical stress or chemical reactions [16] or biotic degradation caused by microorganisms such as enzymes and by biological disintegration [4, 11]. Some common sources of secondary microplastics include wastewater treatment plant effluent (where they are found in the secondary treatment process) [8]; effluent from laundry activities where microplastics in clothing products are released [12]; the use of fishing equipment including nets and ropes [17] and the tearing of car tires [13].

Microplastics can also be classified based on their chemical composition and density. There are six major types [5, 18, 19]:

- Polystyrene (PS): Commonly found in cosmetic products (exfoliators and scrubs), plastic furniture, kitchenware, packaging material.
- Polyethylene is sub-categorized into Low-density polyethylene (LDPE) and High-density polyethylene (HDPE). LDPE is commonly found in plastic films and wraps, juice boxes, disposable shopping bags whereas, HDPE can be found in shampoo bottles, recycle bins, pipe systems, toys.
- Polypropylene (PP): Commonly found in fishing gear, carpets, camping equipment.
- Polyvinyl chloride (PVC): Commonly found in pipe accessories and PVC fittings.
- Polyethylene terephthalate (PET): Food packaging and textiles.
- Polyurethane (PUR): Commonly found home furnishing (foam), coatings, adhesive and synthetic skins.
- Others such as nylon, polyester

PE and PP hold first and second places in the market followed by PET [20]. The chemical composition of these plastics and microplastics. Plastics can vary in size, and as such they have been classified into four main groups [11, 21, 22]:

- Greater than 25 mm, >> Macroplastics
- Between 5 to 25mm, >> Meso-plastics
- Between 1 to 5mm >> Microplastics
- Between 1nm to 1μm >> Nano-plastics

Microplastics can take the form of pellets, microbeads (sphere), foams, fibers, films, fragments, and microfibers depending on their origin [5, 23].

3. Microplastics in the environment

It is well documented that microplastics can be found in the hydrosphere, pedosphere and atmosphere. The hydrosphere is the main sink for microplastics, where the disposal of treated wastewater for example results in transmitting microplastics to aquatic habitats [5]. It is reported that everyday seven million microplastic items are released into the aquatic habitat from wastewater treatment plant effluents [24,25]. Human activities in the form of agricultural practices involving the use of fertilizers, compost, soil conditioners, wastewater treatment plant sludge, and additives on agricultural lands to enhance crop growth and yield can also result in the introduction of microplastics to the soil [11, 24, 26].

Microplastics are transported through aquatic and/or atmospheric currents. Microplastics can be transported from land to water and vise versa. Lighter and smaller microplastics can be carried by the wind as airborne microplastics and consequently be transported over large distances and possibly to remote areas such as glacier and mountains [5,11]. Heavier microplastics on the other hand, can accumulate and/or be buried in the soil (pedosphere). Rainfall and surface runoff from agricultural lands can transport microplastics to surface waters (the hydrosphere). Stormwater runoff transports microplastics resulting from the normal wear of tires on roads to surface water [5]. As such, atmospheric microplastics eventually enter the hydrosphere or pedosphere.

The means for understanding the fate and distribution of microplastics in the hydrosphere are not well-defined due to the plethora of variables it depends on; such as the variation in the different degradation pathways as well as the water and climate dynamics [5]. However, two probable pathways for travel have been proposed, namely; horizontal and vertical distributions.

Plastic and microplastics in the hydrosphere experience many phenomena including ocean currents, abrasion and abiotic or biotic degradation resulting in the formation of different shapes, sizes, densities, and compositions. In general, microplastics with density greater than that of the surrounding water will sink (e.g. PET, PS and cellulose acetate), whereas the lighter ones will float or remain suspended (e.g. PP and PE) [27]. Vertical distribution pathways also depend on turbulent mixing, movement of aquatic organisms and cluster formations. A study [28] suggests that deep-sea sediments are hot spots for microplastic accumulation. Other studies have reported microplastics in sediments to range between 8 to 600,000 pieces/kg sediment [29,30].

In terms of horizontal distribution pathways of microplastics, this depends on the velocity and flow direction of water currents, and wind which can be very complex to calculate and predict. Moreover, in the case of rivers, dam and other constructions can affect the horizontal travel and the location of microplastics [31, 32]. As a result, suspended microplastics can either return to coasts/beaches or be transported to remote regions [7, 24].

4. Effect of Microplastics on the Hydrosphere and Marine Ecology

Due to the small size of microplastics, they are easily consumed by aquatic species such as sea turtles, penguins, fish and crustaceans as well as coastal birds and animals [33] and result in a range of adverse health effects. The extent of these effects increases with the increases in ingested quantities; which vary widely. Many microplastics also contain additives that are used to enhance material properties during the plastic manufacturing process such as plasticizer, anti- oxidant, flame retardant, etc. These chemicals once leached can cause toxic effects. Further, microplastics can amass and release hazardous organic pollutants into the aquatic ecosystem such as polybrominated diphenyl ethers, and other additives that are incorporate during their production [34].

According to the literature, there seem to be three forms of negative effects that can result from the aquatic contamination by microplastics. These effects, either cause physical, chemical and/or biological changes in the species affected by the microplastics as follows:

- Physiological effects resulting from the ingestion of the microplastics: Microplastics can be excreted quickly by many marine species however, their ingestion reduces food intake, causes bowl obstructions and results in development disorders and in behavioural change such as feeding habits [35, 36, 37].
- Death resulting from the release of hazardous pollutants into the aquatic environment: This is a result of the leaching of microplastic chemicals into species' tissue altering body chemistry and/or causing bioaccumulation. The toxicity can also differ according to the ratio of additives needed for each plastic [38].
- Harmful reaction resulting from the absorption of pollutants by microplastics while in the environment: Microplastics can become carriers for contaminants such polycyclic aromatic hydrocarbons, PCB, DDT, organo halogenated pesticides, hexachlorocyclohexanes, and chlorinated benzenes [39]. These can cause reproductive disorders.

Since humans are regular consumers of sea food [40] which are highly affected by microplastics, there is a high possibility of microplastic contamination being transfer to them [41]. Microplastics have also been found in tap and bottled water as well as sea salt. Once contaminated food, water and/or species are ingested by human beings, they can be affect in an number of ways that include, skin allergies, impaired vision, respiratory failure, headaches and dizziness, liver disfunction, gastrointestinal issues, infertility, hormonal changes and cancer [34].

5. Microplastics in Africa: An Example

Microplastic abundance varies greatly across Africa, with concentrations as low as 40 items/kg in Ethiopia's Lake Ziway and as high as 7,960 items/kg in Tunisia's Lake Bizerte [42]. Research showed that pellets constitute the majority of the microplastics in the lake ecosystems accounting for about 70% followed by fragments at 17%. Moreover, evidence showed that microplastics had been ingested by 20% of locally caught fish, such as Nile perch, and tilapia in Tanzania.

In a quantitative survey of microplastic contamination across South Africa's southeast coast, the density of microplastics in sediments ranged between 700 and 3,000 particles/m^2, and ranged between 250 and 1,200 particles/m^3 in the water column [43]. Furthermore, [44] showed that more microplastics were present in sediments closer to the surface layer and that microplastics profiles across time-dated cores were compatible with the development of plastics.

In a study [45] the relevance of microplastic contamination on African fishing resources was evaluated. The study showed that freshwater environmental pollution in Africa is the main focus of contemporary microplastic pollution research, whereas fish health in the Mediterranean Sea and plastic contamination in South Africa are the main areas of emphasis. Additionally, fish from northwest African seas appear to be less polluted according to the obtained data, but the river Nile nations, particularly Egyptian waters, and the Gulf of Guinea, are hotspots of fish microplastic pollution.

Finally, [46] highlights that more empirical research on freshwater microplastics across species and is needed, particularly in Africa where such data has been substantially underreported. Studies on edible animals, such as crabs, snails, and fish, via which microplastics may spread readily to people, are crucial. Such research' findings may help policymakers develop measures to reduce the load of microplastic pollution and its effects on aquatic life, human and environmental health [47].

6. Conclusions

The quantities of microplastics are continuously rising, due to the increasing production of virgin plastics from primary sources and the disintegration of larger plastics. This results in ever-increasing microplastic waste that is plaguing the global ecosystem. Microplastics are transported through aquatic and/or atmospheric currents largely ending up in the hydrosphere.

In the aquatic environment, microplastics as mediums for the absorption of harmful chemicals, causing adverse health effects when ingested and disbursing toxic chemicals into the environment when weathered/broken-down. Microplastics, due to their small sizes, can be readily ingestion by animals (e.g. fish and other marine organisms) that can be later ingested by others and by humans resulting adverse biological, chemical and physiological health effects along the various stages of the food chain.

In order to address the critical impacts of microplastics pollution on the environment, international cooperation is necessary. The production of virgin plastic must be reduced through innovation, enhanced product design and developing more sustainable material alternatives. This can include the use of bio-based feedstock for plastic and additive production. Low-cost and sustainable plastic waste management is also essential. This includes, waste minimization, reuse, recycling, appropriate treatment and disposal.

Currently, bans and taxes on single-use plastics exist in more than 120 countries. Despite this being a needed initiative, it falls short in reducing overall plastic and microplastic pollution. Most regulations address specific items such as plastic bags, which do not constitute a significant portion of plastic waste, and do not address reducing plastic consumption. Disposal taxes for landfills and incinerators that incentivize plastic recycling and reuse are too low or aren't available in many countries and should be mainstreamed. Additionally, schemes such as extended producer responsibility for plastic packaging can also be employed. Finally, efforts need to be directed at creating a well-functioning market for recycled plastics. This can be achieved by setting recycled content targets and investing in improving recycling technologies.

7. References

[1] Global Plastic Market Report 2023: Shift In Trend Toward Replacement of Glass & Metals Drives Growth. **2023.**

[2] OECD. Plastic Pollution Is Growing Relentlessly as Waste Management and Recycling Fall Short, Says OECD. 2022. Available online: https://www.oecd.org/environment/plastic-pollution-is-growing-relentlessly-as-waste-management-and-recycling-fall-short.htm

[3]United Nations Environment Programme (UNEP). Plastic Planet: How Tiny Plastic Particles are Polluting Our Soil. **2021**.

[4] Evode, N.; Qamar, S.A.; Bilal, M.; Barceló, D.; Iqbal, H.M. Plastic waste and its management strategies for environmental sustainability. *Case Stud. Chem. Environ. Eng.* **2021**, *4*, 100142.

[5] Haque, F.; Fan, C. Fate and Impacts of Microplastics in the Environment: Hydrosphere, Pedosphere, and Atmosphere. Environments **2023**, 10, 70. https://doi.org/10.3390/environments10050070.

[6] Andrady, A.L. The plastic in microplastics: A review. *Mar. Pollut. Bull.* **2017**, *119*, 12–22.

[7] Zhang, K.; Hamidian, A.H.; Tubić, A.; Zhang, Y.; Fang, J.K.; Wu, C.; Lam, P.K. Understanding Plastic Degradation and Microplastic Formation in the Environment: A review. *Environ. Pollut.* **2021**, *274*, 116554.

[8] Cole, M.; Lindeque, P.; Halsband, C.; Galloway, T.S. Microplastics as Contaminants in the Marine Environment: A Review. *Mar. Pollut. Bull.* **2011**, *62*, 2588–2597.

[9] Wang, J.; Li, Y.; Lu, L.; Zheng, M.; Zhang, X.; Tian, H.; Wang, W.; Ru, S. Polystyrene Microplastics Cause Tissue Damages, Sex-Specific Reproductive Disruption and Transgenerational Effects in Marine Medaka (*Oryzias melastigma*). *Environ. Pollut.* **2019**, *254*, 113024.

[10] Yang, H.; Chen, G.; Wang, J. Microplastics in the Marine Environment: Sources, Fates, Impacts and Microbial Degradation. *Toxics* **2021**, *9*, 41.

[11] Lamichhane, G., Acharya, A., Marahatha, R. *et al.* Microplastics in Environment: Global Concern, Challenges, and Controlling Measures. *Int. J. Environ. Sci. Technol.* **2023,** 20, 4673–4694. https://doi.org/10.1007/s13762-022-04261-1

[12] Galvão, A.; Aleixo, M.; De Pablo, H.; Lopes, C.; Raimundo, J. Microplastics in Wastewater: Microfiber Emissions from Common Household Laundry. *Environ. Sci. Pollut. Res.* **2020**, *27*, 26643–26649.

[13] Kole, P.J.; Löhr, A.J.; Van Belleghem, F.G.A.J.; Ragas, A.M.J. Wear and Tear of Tyres: A Stealthy Source of Microplastics in the Environment. *Int. J. Environ. Res. Public Health* **2017**, *14*, 1265.

[14] An, L., Liu, Q., Deng, Y., Wu, W., Gao, Y., Ling, W. (2020). Sources of Microplastic in the Environment. In: He, D., Luo, Y. (eds) Microplastics in Terrestrial Environments. The Handbook of Environmental Chemistry, vol 95. Springer, Cham. https://doi.org/10.1007/698_2020_449

[15] Peterson, J.D.; Vyazovkin, S.; Wight, C.A. Kinetics of the Thermal and Thermo-Oxidative Degradation of Poly-styrene, Polyethylene and Poly(propylene). *Macromol. Chem. Phys.* **2001**, *202*, 775–784.

[16] Aragaw, T.A. Surgical Face Masks as a Potential Source for Microplastic Pollution in the COVID-19 Scenario. *Mar. Pollut. Bull.* **2020**, *159*, 111517.

[17] Xue, B.; Zhang, L.; Li, R.; Wang, Y.; Guo, J.; Yu, K.; Wang, S. Underestimated Microplastic Pollution Derived from Fishery Activities and "Hidden" in Deep Sediment. *Environ. Sci. Technol.* **2020**, *54*, 2210–2217.

[18]Gündoğdu, S. Polymer Types of Microplastic in Coastal Areas. In *Microplastic Pollution: Environmental Occurrence and Treatment Technologies*; Springer: Berlin/Heidelberg, Germany, **2022**, pp. 77–88.

[19] Alabi, O.A.; Ologbonjaye, K.I.; Awosolu, O.; Alalade, O.E. Public and Environmental Health Effects of Plastic Wastes Disposal: A Review. *J. Toxicol. Risk Assess.* **2019**, *5*, 21.

[20] Leng, Z; Padhan, R.K.; Sreeram, A. Production of a Sustainable Paving Material Through Chemical Recycling of Waste PET into Crumb Rubber Modified Asphalt. J Clean Prod. **2018**, 180:682–688. https://doi.org/

[21] Lee, J; Lee, J.S.; Jang, Y.C.; et al. Distribution and Size Relationships of Plastic Marine Debris on Beaches in South Korea. Arch Environ Contam Toxicol. **2015**, 69:288–298. https://doi.org/10.1007/s00244-015-0208-x

[22] Gigault, J.; Halle, A.; Baudrimont, M.; et al. Current Opinion: What is a Nanoplastic? Environ Pollut. **2018**, 235:1030–1034. https://doi.org/10.1016/j.envpol.2018.01.024

[23] Cowger, W.; Gray, A.; Christiansen, S.H.; et al. Critical Review of Processing and Classification Techniques for Images and Spectra in Microplastic Research. Appl Spectrosc. **2020**, 74:989–1010. https://doi.org/10.1177/0003702820929064

[24] Auta, H.; Emenike, C.; Fauziah, S. Distribution and Importance of Microplastics in the Marine Environment: A review of the Sources, Fate, Effects, and Potential Solutions. *Environ. Int.* **2017**, *102*, 165–176.

[25] Sutton, R.; Mason, S.A.; Stanek, S.K.; Willis-Norton, E.; Wren, I.F.; Box, C. Microplastic Contamination in the San Francisco Bay, California, USA. *Mar. Pollut. Bull.* **2016**, *109*, 230–235.

[26] Rodríguez-Seijo, A.; Lourenço, J.; Rocha-Santos, T.A.P.; Da Costa, J.; Duarte, A.C.; Vala, H.; Pereira, R. Histopathological and Molecular Effects of Microplastics in *Eisenia andrei* Bouché. *Environ. Pollut.* **2017**, *220*, 495–503.

[27] Coyle, R.; Hardiman, G.; Driscoll, K.O. Microplastics in the Marine Environment: A Review of Their Sources, Distribution Processes, Uptake and Exchange in Ecosystems. *Case Stud. Chem. Environ. Eng.* **2020**, 2, 100010.

[28] Van Cauwenberghe, L.; Vanreusel, A.; Mees, J.; Janssen, C.R. Microplastic Pollution in Deep-Sea Sediments. *Environ. Pollut.* **2013**, *182*, 495–499.

[29] Thompson, R.C.; Olsen, Y.; Mitchell, R.P.; Davis, A.; Rowland, S.J.; John, A.W.G.; McGonigle, D.; Russell, A.E. Lost at Sea: Where Is All the Plastic? *Science* **2004**, *304*, 838.

[30] Doyle, M.J.; Watson, W.; Bowlin, N.M.; Sheavly, S.B. Plastic Particles in Coastal Pelagic Ecosystems of the Northeast Pacific Ocean. *Mar. Environ. Res.* **2011**, *71*, 41–52.

[31] Yulan, Z.; Tanguang, G.; Shichang, Kang.; Mika S. Importance of Atmospheric Transport for Microplastics Deposited in Remote Areas, Environmental Pollution. **2019**, 254, Part A, 112953, ISSN 0269-7491, https://doi.org/10.1016/j.envpol.2019.07.121.

[32] Kumar, R.; Sharma, P.; Verma, A.; Jha, P.K.; Singh, P.; Gupta, P.K.; Chandra, R.; Prasad, P.V.V. Effect of Physical Characteristics and Hydrodynamic Conditions on Transport and Deposition of Microplastics in Riverine Ecosystem. *Water* **2021**, *13*, 2710.

[33] Marn, N.; Jusup, M.; Kooijman, S.; Klanjscek, T. Quantifying Impacts of Plastic Debris on Marine Wildlife Identifies Ecological Breakpoints. Ecol Lett 23. **2020**, 1479–1487. https://doi.org/10.1111/ele.13574

[34] Issac, M.N.; Kandasubramanian, B. Effect of Microplastics in Water and Aquatic Systems. *Environ Sci Pollut Res* **2021**, 28, 19544–19562. https://doi.org/10.1007/s11356-021-13184-2

[35] Cozar, A.; Echevarria, F.; Gonzalez-Gordillo, J.I. et al. Plastic Debris in the Open Ocean. Proc Natl Acad Sci. **2014** 111:10239–10244. https://doi.org/10.1073/pnas.1314705111

[36] Udayakumar, K.V.; Gore, P.M.; Kandasubramanian, B. Foamed Materials for Oil-Water Separation. Chem Eng J Adv. **2021**, 5:100076. https:// doi.org/10.1016/j.ceja.2020.100076

[37] Horton, A. A.; Jürgens, M. D.; Lahive, E. et al. The Influence of Exposure And Physiology on Microplastic Ingestion by the Freshwater Fish Rutilus Rutilus (Roach) in the River Thames, UK. Environ Pollut. **2018**, 236:188–194. https://doi.org/10.1016/j.envpol.2018.01.044

[38] Botterell, Z. L. R.; Beaumont, N.; Dorrington, T. et al. Bioavailability and Effects of Microplastics on Marine Zooplankton: A Review. Environ Pollut. **2019**, 245:98–110. https://doi.org/10.1016/j.envpol.2018.10.065

[39] O'Donovan, S.; Mestre, N. C.; Abel S, et al. Ecotoxicological Effects Of Chemical Contaminants Adsorbed to Microplastics in the Clam Scrobicularia Plana. Front Mar Sci: **2018**, 5. https://doi.org/10.3389/fmars.2018.00143

[40] Saha, M.; Naik, A.; Desai, A. et al. Microplastics in Seafood as an Emerging Threat to Marine Environment: A Case Study in Goa, West Coast of India. Chemosphere. **2021**, 270:129359. https://doi.org/10.1016/j.chemosphere.2020.129359

[41] Smith, M.; Love, D. C.; Rochman, C. M.; Neff, R.; Microplastics in Seafood and the Implications for Human Health. Curr Environ Heal Reports. **2018**, 5:375–386. https://doi.org/10.1007/s40572-018-0206-z

[42] Yang, S. *et al.*, A Comparative Review of Microplastics in Lake Systems from Different Countries and Regions, *Chemosphere*, **2022**, 286, p. 131806. doi: 10.1016/J.CHEMOSPHERE.2021.131806.

[43] Nel, H. A.; Froneman, P.W. A Quantitative Analysis of Microplastic Pollution Along the South-Eastern Coastline of South Africa. *Pollut. Bull.* **2015**, 274–279. doi: 10.1016/J.MARPOLBUL.2015.09.043.

[44] Matsuguma, Y. *et al.* Microplastics in Sediment Cores from Asia and Africa as Indicators of Temporal Trends in Plastic Pollution. *Arch. Environ. Contam. Toxicol.* **2017**, 73, no. 2, 230–239. doi: 10.1007/S00244-017-0414-9/FIGURES/7.

[45] Masiá, P. *et al.*, Potential Microplastics Impacts on African Fishing Resources," *Sci. Total Environ.* **2022**, 806, p. 150671. doi: 10.1016/J.SCITOTENV.2021.150671.

[46] Akindele, E. O.; Ehlers, S. M.; and Koop, J. H. E. First Empirical Study of Freshwater Microplastics in West Africa Using Gastropods from Nigeria as Bioindicators. *Limnologica.* **2019**, 78, p. 125708. doi: 10.1016/J.LIMNO.2019.125708.

[47] Abdellatif, G.; Mahmoud, A.S.; Peters, R.W.; Mostafa, M.K. Waste Plastics and Microplastics in Africa: Negative Impacts and Opportunities. Annual AIChE Meeting, 2021. Boston, MA, USA. https://www.aiche.org/academy/conferences/aiche-annual-meeting/2021/proceeding/paper/217f-waste-plastics-and-microplastics-africa-negative-impacts-and-opportunities.

131b Process Systems Engineering Tools for Assessing Upcycling Technologies of Plastic Waste

Borja Hernández Blázquez, Oluwadare Badejo, Esun Selvam, Pavel Kots, Dionisios Vlachos, Marianthi Ierapetritou

Global plastic production is estimated at >350 million tons per year and it is projected to increase to over 1200 million tons by 2050.[1,2] Most plastics are discarded after their first use. Only 9% of the United States (US) plastics are recycled and 16% are incinerated.[3] The remaining 75% is disposed into landfills representing an average estimated annual loss of $7.2 billion of market value and 3.4 EJ as embodied energy[4] (equivalent to 3% of the total energy consumption in the US[5]). The most widespread recycling technology is mechanical since it has low processing costs. However, it degrades the polymer properties, and the plastic can only be recycled up to 6 times. Thermochemical recycling techniques are one alternative that can produce virgin monomers (recycling) and added-value products (upcycling).[6] The most widespread thermochemical technologies involve pyrolysis, gasification, and hydrothermal liquefaction. Recently, the intensification with catalysts and new heat transfer technologies have increased the yield of valuable petrol fractions (olefins and aromatics). Furthermore, other technologies like direct hydrocracking and hydrogenolysis have also been developed.[7] These technologies reduce the direct energy consumption operating at mild temperatures but utilize H_2 as a depolymerization agent. Understanding the economic and environmental potential of these and conventional ones is necessary to guide technology selection.

In this presentation, we demonstrate the potential of process systems engineering tools to provide this guidance. We structure the work in three parts:

1. Simulation-based process design of the thermochemical depolymerization technologies with Techno-Eeconomic Analysis (TEA) and Life Cycle Assessment (LCA) for determining the costs and environmental impacts of each of the technologies.
2. A Supply Chain (SC) framework for determining the optimal technology, location, and means of transport. This supply chain integrates the economics and environmental impact analyses of previous depolymerization technologies in parametric and piecewise linear models with the costs and environmental impacts of transportation. The model is formulated as a Mixed Integer Linear Programming problem (MILP) with 1.5 million variables and ~2 million equations.
3. A superstructure optimization determining the optimal use of the pyrolysis products. This tool is formulated as a MILP problem with ~2,000 variables and 1,700 equations.

In the first stage, we compare the recent technologies with conventional technologies. The analysis determines pyrolysis as the most profitable technology and hydrogenolysis supplied with blue hydrogen as the most sustainable. Both are the only two profitable technologies (positive internal rate of return (IRR)), but pyrolysis has a higher IRR if microwave heating is used than hydrogenolysis. At the second stage of the analysis considering the entire SC, pyrolysis is also recommended an economic objective is used. In the design of the SC to minimize the global warming potential, a combination of mechanical recycling, hydrogenolysis, and hydrocracking technologies is recommended. Apart from using the two objectives separately, multi-objective optimization is performed using the epsilon constraint method for determining the Pareto set of solutions. These intermediate points show how pyrolysis is preferred as we are closer to the economic objective, substituting first hydrogenolysis, then mechanical recycling and finally hydrocracking. It should be noted that for these cases, pyrolysis is considered to produce olefins and subsequently lubricants as the final product. However, olefins can be transformed into multiple products. Thus, at the last stage of our analysis, we optimize the superstructure of alternative routes for pyrolysis products. The economic optimization propose a multiproduct refinery with the most valuable products, aldehydes and lubes. Among them, aldehydes are very energy and emission intensive since they need H_2 and they are substituted in the environmental optimization that minimizes the GWP. The production of low-density polyethylene and acetone is preferred instead.

References

1. Rochman, C.M., Browne, M.A., Halpern, B.S., Hentschel, B.T., Hoh, E., Karapanagioti, H.K., Rios-Mendoza, L.M., Takada, H., Teh, S., and Thompson, R.C. (2013). Classify plastic waste as hazardous. Nature *494*, 169-171. 10.1038/494169a.

2. Ellen-Macarthur-Foundation (2016). The New Plastics Economy: Rethinking the future of plastics. Available in: https://ellenmacarthurfoundation.org/the-new-plastics-economy-rethinking-the-future-of-plastics

3. United-States-Environmental-Protection-Agency (2020). Advancing Sustainable Material Management: 2018 Tables and Figures. Assessing Trends in Materials Generation and Management in the United States. Available in: https://www.epa.gov/sites/default/files/2021-01/documents/2018_tables_and_figures_dec_2020_fnl_508.pdf

4. Milbrandt, A., Coney, K., Badgett, A., and Beckham, G.T. (2022). Quantification and evaluation of plastic waste in the United States. Resour Conserv Recy *183*. ARTN 106363. 10.1016/j.resconrec.2022.106363.

5. EIA (2022). Energy Consumption by Sector, Mothly. . Available in: https://www.eia.gov/totalenergy/data/monthly/pdf/sec2_3.pdf Last access: 31st December 2022. .

6. Dogu, O., Pelucchi, M., Van de Vijver, R., Van Steenberge, P.H.M., D'hooge, D.R., Cuoci, A., Mehl, M., Frassoldati, A., Faravelli, T., and Van Geem, K.M. (2021). The chemistry of chemical recycling of solid plastic waste via pyrolysis and gasification: State-of-the-art, challenges, and future directions. Prog Energ Combust *84*. ARTN 100901. 10.1016/j.pecs.2020.100901.

7. Li, H.; Aguirre-Villegas, H. A.; Allen, R. D.; Bai, X.; Benson, C. H.; Beckham, G. T.; Bradshaw, S. L.; Brown, J. L.; Brown, R. C.; Cecon, V. S.; Curley, J. B.; Curtzwiler, G. W.; Dong, S.; Gaddameedi, S.; García, J. E.; Hermans, I.; Kim, M. S.; Ma, J.; Mark, L. O.; Mavrikakis, M.; Olafasakin, O. O.; Osswald, T. A.; Papanikolaou, K. G.; Radhakrishnan, H.; Sanchez Castillo, M. A.; Sánchez-Rivera, K. L.; Tumu, K. N.; Van Lehn, R. C.; Vorst, K. L.; Wright, M. M.; Wu, J.; Zavala, V. M.; Zhou, P.; Huber, G. W. Expanding Plastics Recycling Technologies: Chemical Aspects, Technology Status and Challenges. *Green Chemistry* **2022**, *24* (23), 8899–9002. https://doi.org/10.1039/D2GC02588D.

131c Design and Optimization of Circular Economy Networks: Recycling of Polyethylene Terephthalate As a Case Study

Abdulhakeem Ahmed, Akhil Nair, Carl Laird, Ana I. Torres

Circular network design is an emergent approach amongst various sectors in achieving sustainability. Its applications range from food waste mitigation in the food sector to prolonging material life and reducing reliance on virgin resource utilization in the textiles industry. Studies in the literature highlight the positive environmental and economic impacts of implementing a circular design. In this work, we design circular networks for recycling polyethylene terephthalate (PET). PET is the primary plastic product used in single-use plastics and textiles, thus constituting a large share of created plastic and generated waste. The circular PET supply chain includes recycling by mechanical, chemical, and other physical processes; the vast amount of options makes it difficult to assess the best pathway and the extent of material circularity.

This work proposes a superstructure optimization approach to assess circular PET value chains. The superstructure considers fossil and biomass-based feedstocks for producing virgin PET via the two most common linear pathways. Additionally, we implement circularity with mechanical recycling and several chemical recycling options to recover monomers. The superstructure optimization problem is formulated as a mixed integer linear program (MILP). We obtain the Pareto optimal solutions that minimize the cost and consumption of new material. Preliminary results show that a combination of mechanical recycling and alcoholysis provides the circular network with the lowest operational cost. As expected, material circularity is improved by adding chemical recycling processes at the expense of higher costs.

The presentation will discuss the PET linear and circular superstructures, the proposed MILP model, and the Life cycle analysis of the economically-optimal PET circular value chain.

This work is funded by Carnegie Mellon University, the GEM fellowship, and the Thomas and Adrienne Klopack Graduate Fellowship in Engineering.

131d Cradle-to-Cradle Life Cycle Assessment of Multilayer Plastic Films Used in Sheet Molding Compound Process

Farshid Nazemi, Rachmat Mulyana, Jose M. Castro, Bhavik Bakshi

Multilayer plastic films have a wide range of applications in many industries due to their unique carrier, barrier, and flexibility properties. Multilayer plastic film market is expected to surpass US$210 Bn. This has resulted in a significant generation of this type of plastic film waste, which unlike monolayer plastic films, cannot be treated or recycled using the conventional recycling techniques for monolayer plastic films. This is mainly due to heterogeneity, complexity of structure, and thermodynamic incompatibility of multiple layers of polymers with different chemical composition. Despite the recent efforts to overcome technical barriers for recycling or recovering this type of plastic film, to our knowledge, there is no study that identifies the environmental sustainability and tradeoffs between all possible end of life (EoL) alternatives for treating this type of waste. Therefore, the overarching goal of this study is to evaluate the life cycle environmental impact of multilayer plastic film supply chains, from the extraction of natural resources to waste handling and treatment (cradle-to-cradle). We consider six major scenarios based on the current typical, state-of-the-art and R&D technologies available as EoL treatment options, i.e., landfilling, incineration, pyrolysis, reuse, mechanical recycling (downcycling), and chemical recycling (downcycling). In this work, we are focused on high barrier/carrier plastic films, which are a specific type of multilayer plastic films that are used during the production process of Sheet Molding Compounds (SMC). SMCs are thermoset materials that are used for manufacturing products with high mechanical strength. High barrier/carrier films are used as a carrier that enables a continuous operation for SMC production and as a barrier that prevents the escape of styrene monomers from SMCs, which is crucial for the molding process. While we obtain life cycle inventory (LCI) data from our industry partners, technology vendors, LCI databases, and literature, we run experiments when there exists a gap in available LCI data. For the reuse scenario, we perform experiments for cleaning and thermally stitching the film to enable its reuse in the continuous SMC manufacturing process. In addition, for the mechanical recycling scenario, we perform experiments for pelletizing and molding the film to produce recycled or downcycled products. We use US Environmental Protection Agency (EPA) Tool for Reduction and Assessment of Chemicals and Other Environmental Impacts (TRACI) with Cumulative Exergy Demand (CExD) and economic circularity as complementary metrics to provide insights on environmental impacts, resource use, and circularity efficiency. Life cycle assessment (LCA) results for the landfilling (current typical) scenario indicates that the supply chain of virgin resins used to produce multilayer films contribute the most to the negative environmental impacts. This shows the great opportunity to mitigate these adverse environmental impacts by employing recycling and recovery strategies. Moreover, a preliminary comparative LCA is performed for landfilling, incineration, pyrolysis, and mechanical downcycling scenarios. For the incineration scenario, we assume that the energy generated through the incineration process is used to generate electricity. For the pyrolysis scenario, pyrolysis oil and char are considered as the final products of the pyrolysis process. For the mechanical downcycling scenario, we use the multilayer film waste to produce molded products with real-world applications. A substitution approach is used for the EoL products to evaluate the environmental impacts of each scenario. Comparative LCA results indicate that the mechanical downcycling scenario outperforms other scenarios in terms of the environmental sustainability performance. In this scenario, Global Warming Potential (GWP) and CExD are 50% and 80% less than those of the base case (landfilling) scenario, respectively. Also, economic circularity of the downcycling scenario is 25%, which is the highest among all scenarios. This indicates the advantage of mechanical downcycling over landfilling and energy recovery scenarios (i.e., incineration and pyrolysis) due to the substitution of virgin materials in producing molded products. Nevertheless, we will perform LCA on other possible alternatives for treating multilayer films such as reuse, closed loop mechanical and chemical recycling. By performing a comparative LCA on all EoL scenarios, we can identify the environmental bottlenecks in the supply chain, and identify the environmental tradeoffs between different scenarios, which is a step toward developing sustainable circular economy strategies for multilayer plastic films.

131e Computational Modeling of Dissolution-Based Plastic Recycling: Solubility Prediction, Solvent Screening, and Process Design

Panzheng Zhou, Jiuling Yu, Aurora del Carmen Munguía-López, Kevin Sanchez-Rivera, George Huber, Victor M. Zavala, Reid Van Lehn

Dissolution-based plastic recycling is a promising approach to separate and recover high-quality pure polymer resins from plastic waste by exploiting differences in polymer solubility. Recently, we proposed a dissolution-based recycling process, called Solvent-Targeted Recovery and Precipitation (STRAP), in which resins are sequentially recovered from multicomponent plastic waste by selective dissolution followed by precipitation. The design of STRAP processes for complex waste streams with several components requires careful consideration of solvent selection, operating temperatures, and the sequence of dissolution/precipitation steps. Determining these choices is challenging due to the wide range of solvents and temperatures that could be possibly used for a given set of target polymers. Here, we report a computational scheme to predict polymer solubilities and design STRAP processes. We first employed molecular dynamics simulations and the conductor-like screening model for real solvents to predict temperature-dependent polymer solubilities in various solvent systems. With this approach, we established a computational solubility database for over 10 common polymers in 1007 solvents at multiple temperatures. Solubilities were experimentally measured in selected systems to validate computational predictions. We then developed a tool that automates the selection of solvents for all possible polymer dissolution sequences to identify feasible STRAP processes. We show that this tool can be combined with process simulation models to study the economic and environmental impact of different dissolution/precipitation sequences. Finally, we demonstrate the application of these computational methods via multiple case studies in which pure resins are successfully recovered from polymer mixtures, indicating the capability of using computational tools to guide effective STRAP process design.

[1] Zhou P, Sánchez-Rivera KL, Huber GW, Van Lehn RC. Computational Approach for Rapidly Predicting Temperature-Dependent Polymer Solubilities Using Molecular-Scale Models. *ChemSusChem*. 2021.

[2] Zhou P, Yu J, Sánchez-Rivera KL, Huber GW, Van Lehn RC. Large-scale Computational Polymer Solubility Predictions and Applications to Dissolution-based Plastic Recycling. *Submitted*. 2023.

131f Recycling Waste Plastic to Fuels: Detailed Mechanistic Modelling of LLDPE Pyrolysis

Aswathy K. Raghu, Paulami Majumdar, Linda J. Broadbelt

Plastic pollution has now turned into an acute environmental problem. Currently, about 400 million metric tons of plastic is generated every year out of which less than 10% is recycled[1]. The remaining is dumped in landfills, incinerated (which adds CO_2 into the atmosphere) or ends up as unregulated waste and interferes with the biosphere. Conversion to chemicals (and fuels) by pyrolysis is an effective strategy of recycling waste plastics. Polyethylene is a widely used plastic for packaging applications and forms a significant portion of the waste produced[2]. Linear low-density polyethylene (LLDPE) is a type of polyethylene that is popular for bags, wrap films, etc. owing to its attractive mechanical properties. Studies suggest that thermal pyrolysis of LLDPE yields liquid fractions with diesel-like properties[3].

We present a microkinetic model for the thermal pyrolysis of LLDPE. A microkinetic model is a multiscale mechanistic model which predicts reactor level conversions by taking into account the kinetic information of the mechanistic steps involved. Being a system involving a large number of species and reactions, it is computationally challenging to model every mechanistic step of polymer pyrolysis. Therefore, we employ population balance modelling in which species are divided into classes based on their characteristics and the variation of their molecular weight distributions, with the progress in reaction tracked using the moments. The number averaged and weight averaged molecular weights of the species classes and the reaction mixture can be calculated from these moments. The Broadbelt group has used this method of moments extensively for modelling pyrolysis of other polyolefins, including polypropylene, polystyrene, and high density polyethylene (HDPE). [4–6] In this approach, kinetic parameters are formulated using the Arrhenius relationship, and reaction rates are considered in terms of the moments of species classes, where the classification is done based on features such as unsaturated ends, type of radical (mid versus end), etc. Besides the molecular weight distribution of species classes, specific species of known carbon length can also be tracked to make predictions of product distributions from the model. A semi-batch reactor system is considered. The final model is a differential-algebraic system with differential equations describing the rates of change of moments of species classes and the rates of formation of specific species. Due to the chain length dependency of some of the reactions in pyrolysis, the method of moments has a closure problem which is resolved by providing an algebraic approximation for the third moments. Moreover, the fact that the same reaction could yield different kinds of products necessitates the calculation of likelihoods of their formation using algebraic equations. The Differential Algebraic Equation (DAE) model is written in C and is solved using the solver Double Precision Differential/Algebraic Sensitivity Analysis Code (DDASAC).

LLDPE is made by co-polymerizing ethene and alpha-olefins such as, butene, hexene, octene, etc. As a result, LLDPE contains a large number of short chain branches. The reactions that occur during LLDPE pyrolysis are similar to those during HDPE pyrolysis, but needs several additional equations to solve for, due to the presence of branches. The reaction families considered in the model are chain fission, radical recombination, mid-chain beta scission, radical addition, end-chain beta scission, disproportionation, bimolecular hydrogen abstraction and intramolecular hydrogen shifts, which are the typical free radical reaction types. Mechanistic models based on these reactions have been demonstrated to be able to capture product distribution of HDPE pyrolysis with reasonable accuracy[6] and will serve as a starting point for LLDPE models as well. The species classes are determined based on the characteristics such as chain types (linear/branched), radical types (end-radical, mid-radical, tertiary mid-radical or dead species) and the types of ends (saturated/unsaturated).

Figure 1 shows the number average molecular weight (M_n) as a function of reaction time predicted by a basic LLDPE model compared to an analogous HDPE model. This basic model tracks only the species classes and not any specific species. Isothermal reaction at 673.15 K is considered. The branch characterization of LLDPE in this model is in terms of number of branches/1000 carbons. The average branch length is considered to be 6. Due to the presence of bonds at branch points involving tertiary carbon atoms that are weaker than bonds involving secondary carbon atoms, LLDPE is expected to decay faster than HDPE. The agreement of the model result with the anticipatory behavior is an indication that the preliminary LLDPE model is reasonable. Further observations about LLDPE pyrolysis can be made after modifying the model to capture more details pertaining to specific species. The quantitative predictions of different fractions of products produced by thermal pyrolysis of LLDPE and their validation with experimental results will be presented.

References

1. United Nations Environment Programme (2021). From Pollution to Solution: A global assessment of marine litter and plastic pollution. Nairobi.

2. UNEP. Single-use Plastics: A Roadmap for Sustainability. 2018. (Rev. ed., pp. vi; 6).

3. Al-Salem SM, Chandrasekaran SR, Dutta A, Sharma BK. Study of the fuel properties of extracted oils obtained from low and linear low density polyethylene pyrolysis. Fuel. 2021;304:121396.

4. Kruse TM, Wong HW, Broadbelt LJ. Modeling the evolution of the full polystyrene molecular weight distribution during polystyrene pyrolysis. Ind Eng Chem Res. 2003;42:2722–35.

5. Kruse TM, Wong HW, Broadbelt LJ. Mechanistic modeling of polymer pyrolysis: Polypropylene. Macromolecules. 2003;36:9594–607.

6. Levine SE, Broadbelt LJ. Detailed mechanistic modeling of high-density polyethylene pyrolysis: Low molecular weight product evolution. Polym Degrad Stab. 2009;94:810–22.

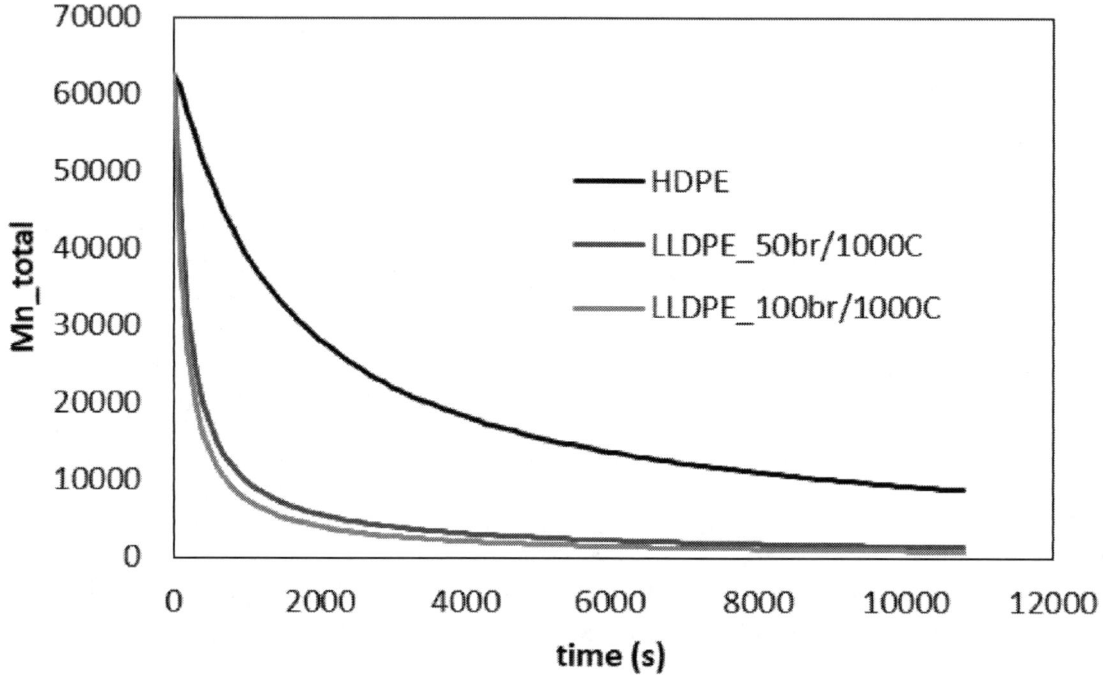

131g Upcycling of Waste PVC-Based Materials to Electrospun Nanofibers for Water Treatment

Atta Ur Razzaq, Milad Esfahani

In recent decades, the increased use of polyvinyl chloride (PVC) in industries and households has led to a surge in PVC waste pollution, which mandates developing solutions for the removal of waste PVC from the environment. We report upcycling, the conversion of waste material to a high-value-added product, of PVC-based products to electrospun fibers (membranes). Two common PVC items, waste PVC pipe and waste PVC pool floats, were chosen and upcycled into electrospun membranes for water treatment. The fabrication process and fiber characteristics, such as morphology, surface charge, and mechanical strength of upcycled fibers, were studied and compared with the same fibers fabricated using research-grade (RG) PVC (commercial PVC powder). In addition, the effect of additives such as calcium carbonate in PVC waste products on the physicochemical properties of upcycled fibers was evaluated. The results showed that upcycling of waste PVC to electrospun fibers is feasible since the upcycled fibers showed similar or superior properties compared to their equivalent fibers from RG-PVC. Finally, the performance of upcycled fibers on the removal of dyes from the water was evaluated. The upcycled fibers from waste PVC pipe and pool float outperformed the RG-PVC fibers in removing methylene blue (MB) from water by showing more than 97 % removal efficiency and adsorption capacity of 195 mg/g compared to RG-PVC with 89 % removal and 179 mg/g adsorption capacity. This study clearly shows the potential of upcycling PVC waste materials from landfill into membranes for enhanced water treatment as a promising solution to the current plastic and water pollution.

131h Recycling of Multilayer Plastic Packaging Materials By Solvent-Targeted Recovery and Precipitation (STRAP)

Kevin Sanchez-Rivera, Aurora del Carmen Munguía-López, Panzheng Zhou, Victor Sanfins Cecon, Jiuling Yu, Zhuo Xu, Shreyas Kolapkar, Kevin Nelson, John Estela, Tim Osswald, Greg W. Curtzwiler, Keith L. Vorst, Ezra Bar Ziv, Reid Van Lehn, Victor Zavala, George Huber

Many plastic packaging materials manufactured today are composites made of distinct polymer layers (*i.e.,* multilayer films) which end up incinerated or in landfills at end of their useful life. We have demonstrated a new strategy called Solvent-Targeted Recovery And Precipitation (STRAP) to deconstruct multilayer rigid and flexible films into their constituent resins using a series of solvent washes that are guided by thermodynamic calculations of solvent-polymer solubility. Three computational methods were used for solvent selection and determining process conditions: Hansen Solubility Parameters (HSPs), Molecular Dynamics (MD) simulations, and COnductor-like Screening MOdel for Realistic Solvents (COSMO-RS). We show that the STRAP process can separate representative polymer resins (HDPE, LDPE, PP, EVOH, PET, PVC, PA6, PA66, and PA 6/66) from post-industrial multilayer films and mixed flexible packages with >95 wt% material efficiency. In this process, a single polymer component was selectively dissolved in a solvent system to be later separated from the insoluble layers and then precipitated by the addition of an antisolvent and/or decreasing the solvent temperature. This resulted in segregated streams of pure dry polymers that can be much more efficiently reused. A techno-economic and life cycle assessment of one example STRAP process with a flexible multilayer film were performed and it was determined that this approach can produce resins at costs comparable to market values and could reduce environmental impacts when compared to film production from virgin resins. In the same example, the FTIR spectra of the separated polymer fractions were comparable to the virgin resins and no significant changes in thermal and molecular properties were observed. Our team was able to produce and characterize a cast film using STRAP recovered PE and currently, we are in the process of scaling up the STRAP technology to a 25 kg/hr development unit which will produce sufficient quantities of material to make and qualify plastic films.

ADVANCED RECYCLING OF PLASTICS IN A CIRCULAR ECONOMY: STATE OF THE ART AND MAJOR CHALLENGES FOR COMPLEMENTARY CHEMICAL RECYCLING TECHNOLOGIES

Philip Biessey, Ruhr University, Bochum, Germany

Introduction

The chemical industry is more and more confronted with social and political expectations to increase process efficiency and to provide new process technologies, allowing for the implementation of the sustainable development goals; for that purpose, circular economy principles in terms of circularity strategies are considered in a process engineering context (Figure 1). These circularity strategies have the common goal of keeping materials as value products in utilization cycles for as long as possible. Specifically in plastics, product re-design is often discussed as an important circularity strategy, aiming at products that can easily be recycled at their end of life ("design for recycling").

Besides product re-design, the avoidance and/ or substitution of crude oil derivatives and other fossil resources are often discussed as circularity strategies to avoid waste generation. Besides that, the circularity strategies include recycling for resource and/ or material recovery at the end of a products life cycle. In the chemical industry, such processes aim at the reintegration of material flows into specific production processes, e.g. the integration of pyrolysis oils in steam crackers as a naphtha substitute [1] or the re-integration of monomers obtained from chemical recycling in polymer production [2]. Regarding plastics in particular, chemical recycling technologies are currently strongly focused: The complementary – rather than competitive [3] – reintegration of material flows from both mechanical and chemical recycling processes aims at substituting crude oil derivatives and allowing for closed material cycles, thus reducing waste volumes in the genuine sense of a circular economy.

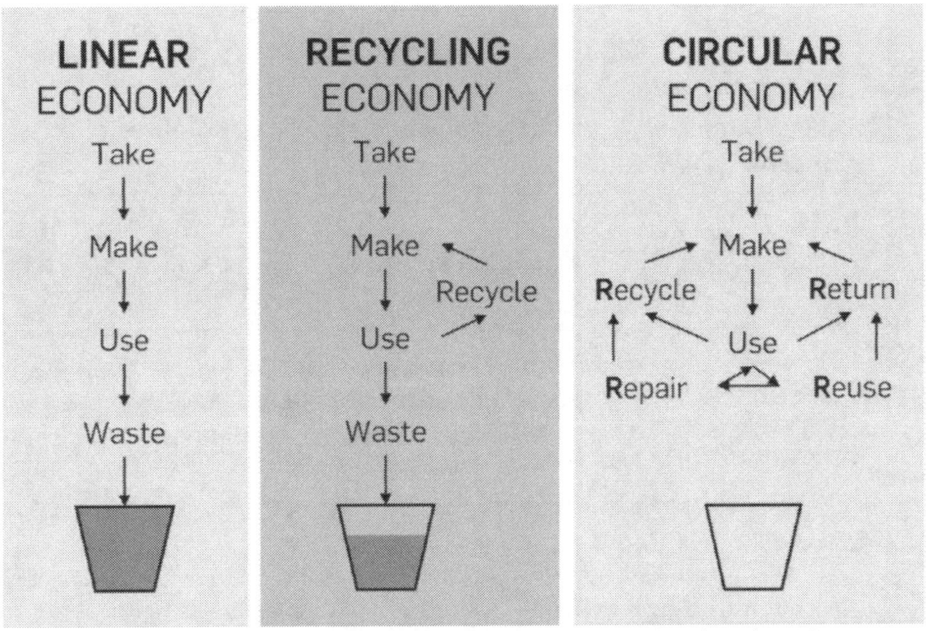

Figure 1 - Linear, Recycling, and Circular Economy Models (based on [4])

Classification of chemical recycling technologies

Chemical recycling processes can be classified according to their basic operating principle into cracking, gasification and solvolysis processes (Table 1). They all have in common, that the chemical structure of the polymer is changed by, e.g., cracking or cleavage mechanisms; this *intended* change in polymer structure differentiates chemical from mechanical recycling.

Solvolysis, which is often referred to as depolymerization[1], is characterized by the cleavage of specific bonds in polymers by using solvents, that are specifically selected according to the targeted bond cleavage. As chemical agents, glycols, amines, methanol and other substances are used, leading to the nomenclature for specific solvolysis processes such as glycolysis, aminolysis, and methanolysis [5]. Solvolysis processes typically aim for monomer recovery; consequently, the most relevant utilization pathways are monomer-based processes such as (re-)polymerization processes. Solvolysis must be differed from solvent-based recycling, which is a purification technique in which the actual structure of the polymer is retained.

In gasification, the feed material reacts with chemical agents such as steam/ oxygen mixtures, or air, to produce a mixture of light hydrocarbons, which are utilized in one of the many different established utilization pathways in the (petro-)chemical industry or in power generation [6].

Table 1 – chemical recycling processes (extended excerpt taken from [7])

	chemical recycling			
	solvolysis	pyrolysis	liquefaction	gasification
working principle	cleavage of specific bonds in polymers by using solvents	thermal and/ or catalytic cracking of polymers in the absence of oxygen	combined pyrolysis mechanism and evaporation of liquefied plastic melt	reaction of plastic waste with gasification agents
chemical agents [a)]	glycols, water/ steam, amines, methanol, enzymes	solid catalysts, e.g., ZSM-5 zeolites		e.g., steam, air, oxygen
(typical) products	monomers	mainly (pyrolysis) oil; gas, solids		light hydrocarbon mixtures, such as syngas
utilization path [b)]	Monomer-based processes, i.e., polymerization	Synfuel, refineries/ steam cracker, feedstock for chemical industry		refineries, synfuel, feedstock for chemical industry, power generation

[1] depolymerization is a somewhat imprecise term, since it is regularly used for all recycling technologies, that change the structure of a polymer; thus, a distinction between the actual operating principles is not possible when using the term "depolymerization"

Cracking processes can be divided into pyrolysis, which is associated with thermal and/ or catalytic cracking of polymers in the absence of oxygen; the other mechanism is the so called liquefaction, which combines pyrolysis and the evaporation of liquefied polymer components in a plastic melt [7]. In both cases, the product consists of three phases: a liquid phase, a gas phase, and a solid phase. The liquid phase can and often does contain two immiscible liquid phases – an oily phase, which is referred to as pyrolysis oil, and an aqueous phase. In most use cases, the oily liquid phase is considered the main product of the process since its composition is often similar to crude oil derivatives such as naphtha or other (petro-)chemical related substances and hydrocarbon mixtures. Due to this similarity, the integration of pyrolysis oil into a steam cracker as a naphtha substitute is often discussed as a promising reintegration route [8]. Nonetheless, it should be noted that different challenges arise in that case in terms of purification processes that are necessary to clean the pyrolysis oil from contaminants (such as heteroatoms) and/ or olefins, which are typically saturated in a downstream hydrotreatment step [6]; consequently, all downstream operations influence the product oil composition obtained from the actual conversion step. Thus, the above-mentioned distinction between pyrolysis and liquefaction is important, since the product characteristics regarding phase fractions and their respective composition vary, depending on the dominating mechanism [7].

Since the purification of pyrolysis oils is challenging and requires the use of both energy and resources, research activities focus on avoiding the formation of, e.g., heteroatoms, or the degradation of certain undesired feed substances in the conversion step by using specific catalysts. These efforts have already shown promising results, e.g., regarding the de-brominating of flame retardants in *polystyrene* (PS) liquefaction [9]. Apart from that, if catalysts are employed in today's pyrolysis and/ or liquefaction processes, typically solid catalysts such as ZSM-5 zeolites are utilized [10]. However, it should be noted that product characteristics of all chemical recycling processes do not only depend on the actual operating principle, but strongly on feedstock characteristics and process conditions (reactor types, temperatures, residence times, catalysts, etc.) as well [10].

Major challenges in chemical recycling

Both the diversity of chemical recycling processes themselves, and – for most of the related process technologies – their low level of technical maturity point out at fundamental challenges for chemical recycling technologies that must be considered to develop a plausible and knowledge-based complementary strategy for chemical recycling technologies.

These challenges are manifold: they range from phenomenological issues, e.g. the investigation of degradation kinetics of complex feed streams in large-scale reactors, to challenges associated with the actual design of resource and energy efficient downstream sequences on industrial scale. There is a lack of reliable methods for the design of separation sequences, since both the transfer of model-based and experimentally obtained results to an industrial scale is challenging due to scaling effects and the complex compositions of, e.g., pyrolysis oils, and their consideration in modelling and evaluation approaches. The determination of product phase compositions is a challenge in its own right, since no analytical standard is available for the characterization of, e.g., pyrolysis oils, making comparisons between different experimental investigations and/ or entire technologies and the transfer of a stream into a process model difficult [11].

Despite their wide range, the challenges in chemical recycling technologies can be classified in three main aspects:

1. *High-level evaluation methods*: The lack of a high-level evaluation method for chemical recycling technologies, allowing for an early identification and selection of technologies that enable promising combinations of feed material, physical and intangible (e.g., legislative) constraints, and desired products.
2. *Process development*: the lack and/ or limitation of methods for (a) the (model-based) integration of the actual conversion step, (b) the following design of downstream sequences, and (c) the evaluation of the entire process.
3. *Reintegration routes*: methods for the identification and evaluation of suitable reintegration routes in terms of closed carbon cycles.

High-level evaluation methods

Knowledge-based methodological approaches for a high-level evaluation of chemical recycling technologies are necessary to identify promising "technology matches" regarding desired products, available feed material, and other technical and non-technical constraints. With such a high-level evaluation method, the number of chemical recycling processes to be considered in detail in terms of process development and process evaluation can be strongly limited. This identification procedure is crucial for focusing the available research resources on promising technologies, rather than applying a watering can-like focus on all possible, or at least on many different technology scenarios regardless of their actual potential for an ecologically and economically feasible realization.

Respective evaluation approaches have been proposed lately that consider technical evaluation criteria on a higher level such as *technology readiness level* (TRL), process sensitivity towards feedstock contamination, range of treatable polymer types and others [12]. This approach allows for the identification and selection of suitable processes, when an actual waste problem in terms of feed characteristics, socio-local waste management issues, desired throughput, etc. is defined; further, it allows for the identification of process limitations, and, thus, for focusing on respective research questions that must be answered to realize a certain technology on relevant scales. Another approach connects the planetary boundaries concept with *life cycle assessment* (LCA)-orientated environmental measures such as *greenhouse gas emissions* (GHG), freshwater use, and others, by deriving limits for the mentioned measures [13]. By that, this approach allows an assessment of whether a particular process or combination of technologies has a potential to be compatible at all with a circular economy and with planetary boundaries, respectively.

Process development

Although research and industrial activities on chemical recycling of plastics have been carried out for decades, there are still fundamental challenges in process development to be addressed and clarified. In particular, suitable approaches for the determination of the real-world process energy demand including actual conversions steps and downstream effort, and appropriate balancing methods are needed or must be further developed to enable a knowledge-based design of entire chemical recycling plants.

As an example, the minimum energy demand of the degradation of plastics such as PS is typically determined using the difference of the educt and products standard enthalpies of formation (e.g., [14]), although it is known that exothermic regions in the degradation of PS exist [15] which might lower the overall energy demand for the degradation step (although this behavior has not been observed in processes with continuous gas phases yet). Further, the energy demand for downstream operations must be considered since it can be in a similar order of magnitude as the actual pyrolysis step, as it was shown by the author and co-workers for the chemical recycling and downstream processing of PS [16]. Additionally, a systematic methodology for the implementation and combination of both experimental data – especially degradation kinetics of the conversion of realistic plastic waste mixtures - and model-based approaches for the design and scaling of reactors and entire processes for chemical recycling technologies are missing, although the availability of a design tool for reactors is crucial for knowledge-based engineering.

Identification and evaluation of product reintegration routes

The application scenarios of chemical recycling process technologies are often limited to "plastics-to-plastics"-orientated approaches such as the derivation of naphtha by pyrolysis oils, or monomer recovery. Depending on yields and product compositions, this perspective is very reasonable, but it can be too narrow when yields and product composition is not considered critically, since, on the one hand, the downstream effort in terms of purification of pyrolysis oils is often underestimated. The purification is typically necessary to get rid of impurities such as heteroatoms and high olefin contents if a re-integration of pyrolysis oils into steam crackers on an industrially relevant level is considered, and is, thus, strongly connected to yields and product composition. On the other hand, the actual idea behind a circular plastics economy, which is in a genuine sense the reduction of waste amounts and the idea of closing carbon cycles, and thus leading to a "waste-to-product" perspective, is threatened to get out of sight when the focus is limited on "plastics-to-plastics" approaches.

Polystyrene pyrolysis and re-polymerization: an example

The pyrolysis of PS is well-known for its high liquid yields and monomer content in the liquid phase; liquid yields of around 90 wt.-% have been reported, with monomer contents in the liquid phase of around 75 wt.-% [2, 17], depending on process conditions and selected catalysts up to 90 wt.-% [18]. Thus, the chemical recycling of PS is considered a promising technology for the reintegration of monomers obtained from feed material, that is – e.g., due to impurities and/ or contaminants – not suitable for mechanical recycling.

Although offering high liquid yields with a high monomer content, a purification of the pyrolysis oil to further increase the monomer content is necessary for the re-polymerization of the obtained monomer to PS. In a recent study, it was shown that the purification of pyrolysis oil obtained from PS pyrolysis in a fluidized bed can carried out using a thermal separation sequence, consisting of several distillation units [16]. The purity requirements for re-polymerization were achieved, and it was shown that the energy demand of the overall process – including pyrolysis, thermal separation, and the actual re-polymerization – was around 1/3 of the conventional, crude oil-based process route, thus strongly indicating a favorable reintegration route for PS waste. Nonetheless, it was found that the energy demand of the thermal separation is roughly in the same order of magnitude as the pyrolysis step, which is mainly because the styrene monomer is a typical medium boiling component in the pyrolysis oil, thus increasing the separation effort. However, it was clearly shown that the energy demand not only of the actual conversion step but the entire process including downstream effort must be considered in process balancing and evaluation.

Based on the above-mentioned study, we carried out an LCA of the pyrolysis and purification step, investigating the global warming potential (among other indicators) as a measure for GHG emissions of the developed separation sequences. In this study, two different purification routes were investigated: the first sequence (scenario "styrene", Figure 2), consisting of two distillation columns and producing styrene monomer and both a light and heavy fraction; this sequence is equal to the one proposed in the study by Merkel et al. [16]. In the other sequence ("mixtures"), the light components are further purified in a third distillation column, allowing the separation of mixtures largely consisting of toluene (light component) and ethylbenzene (heavy component), respectively. Additionally, the PS content in the feed material was varied between 40 – 80 wt.-% in both scenarios, whereas the overall feed composition from Merkel et al. [16] was considered.

The LCA results show (Figure 2) that the overall GWP is negative in every investigated case, except from the "styrene" scenario without further separation and with the lowest PS content in the feed stream. This is due to the relatively high amount of by-product formation and, consequently, the large amount of light fraction ("upper product") that is incinerated and the bottom product treatment ("refinery sludge"), and their respective emissions. In every other case, a negative GWP is obtained, indicating that the developed separation sequences and investigated feed compositions are not only advantageous in terms of the overall energy demand compared to the conventional route, but can also be considered as carbon sinks. Additionally, the GWP is further decreased with increasing purification effort in terms of by-product purification (scenario "mixtures"), since more components are made available for material re-integration, and, consequently, no material streams are incinerated. In all cases, the GWP decreases with increasing PS content in the feed material due to a respective reduction of side-product fractioning. Nonetheless, it should be noted that process scenarios leading to promising results in terms of certain indicators such as energy demand and global warming potential, do not necessarily perform well in terms of other relevant environmental indicators [19]; thus, the LCA results and especially the inter-dependencies between different indicators must be carefully and critically reviewed in the final stage of process evaluation.

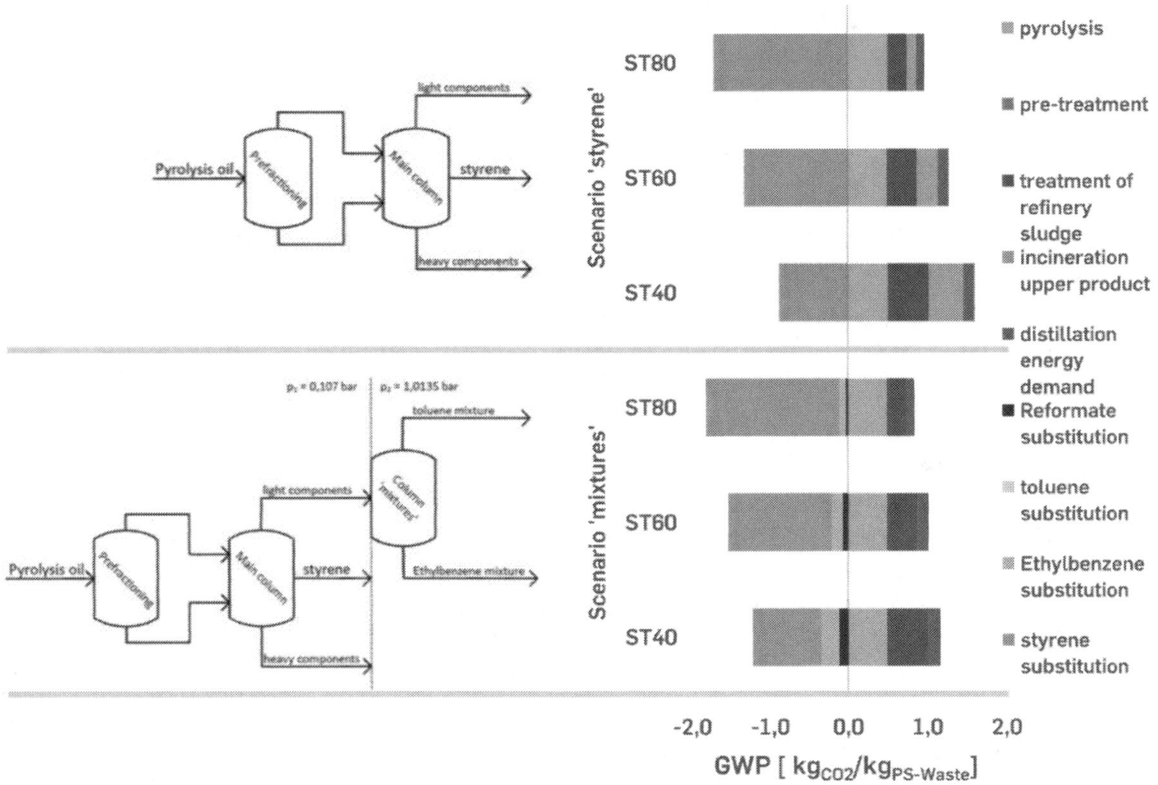

Figure 2 – Global warming potential of different purification routes and feed compositions for polystyrene pyrolysis and purification

Conclusion

Although chemical recycling is intensively discussed and investigated as a complementary technology for a circular plastics economy, there are still important challenges that must be overcome for the implementation of chemical recycling technologies on an industrial scale. It was shown, that three major challenges can be derived: (1), the need for a high-level evaluation method allowing for an early identification of technologies that offer promising combinations of feed material, desired products, and both technical and non-technical constraints; (2), challenges in process development, namely the lack and/ or limitation of methods for a model-based integration of the conversion step, downstream design, and overall process evaluation methods; and (3), the identification and evaluation of energy and resource efficient product reintegration routes, allowing for closed carbon cycles. Existing and emerging solution approaches were discussed using the example of PS pyrolysis and the respective downstream design, product reintegration, and process evaluation by means of LCA, indicating a possible way to overcome the discussed challenges.

References

[1] M. Kusenberg, G. C. Faussone, H. D. Thi, M. Roosen, M. Grilc, A. Eschenbacher, S. de Meester, K. M. van Geem, *The Science of the total environment* **2022**, *838 (Pt 2)*, 156092. DOI: https://doi.org/10.1016/j.scitotenv.2022.156092

[2] W. Kaminsky, M. Predel, A. Sadiki, *Polym. Degrad. Stab.* **2004**, *85 (3)*, 1045 – 1050. DOI: https://doi.org/10.1016/j.polymdegradstab.2003.05.002

[3] I. S. Lase, D. Tonini, D. Caro, P. F. Albizzati, J. Cristóbal, M. Roosen, M. Kusenberg, K. Ragaert, K. M. van Geem, J. Dewulf, S. de Meester, *Resources, Conservation and Recycling* **2023**, *192*, 106916. DOI: https://doi.org/10.1016/j.resconrec.2023.106916

[4] https://thercollective.com/blogs/r-stories/circular-economy-vs-linear-economy (Accessed on October 31, 2023).

[5] I. Vollmer, M. J. F. Jenks, M. C. P. Roelands, R. J. White, T. van Harmelen, P. de Wild, G. P. van der Laan, F. Meirer, J. T. F. Keurentjes, B. M. Weckhuysen, *Angewandte Chemie International Edition* **2020**, *59 (36)*, 15402 – 15423. DOI: https://doi.org/10.1002/anie.201915651

[6] K. Ragaert, L. Delva, K. van Geem, *Waste management (New York, N.Y.)* **2017**, *69*, 24 – 58. DOI: https://doi.org/10.1016/j.wasman.2017.07.044

[7] P. Biessey, J. Vogel, M. Seitz, P. Quicker, *Chemie Ingenieur Technik* **2023**, *95 (8)*, 1199 – 1214. DOI: https://doi.org/10.1002/cite.202300042

[8] M. Kusenberg, M. Roosen, A. Doktor, L. Casado, A. Jamil Abdulrahman, B. Parvizi, A. Eschenbacher, E. Biadi, N. Laudou, D. Jänsch, S. de Meester, K. M. van Geem, *Chemical Engineering Journal* **2023**, *473*, 145259. DOI: https://doi.org/10.1016/j.cej.2023.145259

[9] T. Kolb, A. Schedl, H. Kerschbaumer, N. Niessner, H.-W. Schmidt, B. Wilhelmus, *Chem. Ing. Tech.* **2023**, *95 (8)*, 1314 – 1322. DOI: https://doi.org/10.1002/cite.202200196

[10] H. Li, H. A. Aguirre-Villegas, R. D. Allen, X. Bai, C. H. Benson, G. T. Beckham, S. L. Bradshaw, J. L. Brown, R. C. Brown, V. S. Cecon, J. B. Curley, G. W. Curtzwiler, S. Dong, S. Gaddameedi, J. E. García, I. Hermans, M. S. Kim, J. Ma, L. O. Mark, M. Mavrikakis, O. O. Olafasakin, T. A. Osswald, K. G. Papanikolaou, H. Radhakrishnan, M. A. Sanchez Castillo, K. L. Sánchez-Rivera, K. N. Tumu, R. C. van Lehn, K. L. Vorst, M. M. Wright, J. Wu, V. M. Zavala, P. Zhou, G. W. Huber, *Green Chem.* **2022**, *24 (23)*, 8899 – 9002. DOI: https://doi.org/10.1039/D2GC02588D

[11] M. Kusenberg, A. Eschenbacher, M. R. Djokic, A. Zayoud, K. Ragaert, S. de Meester, K. M. van Geem, *Waste management (New York, N.Y.)* **2022**, *138*, 83 – 115. DOI: https://doi.org/10.1016/j.wasman.2021.11.009

[12] M. Fischlschweiger, *State of the art and future challenges of polymer recycling in the context of circular economy*, (Bio)Process Engineering - A Key To Sustainable Development, Aachen **2022**.

[13] M. Bachmann, C. Zibunas, J. Hartmann, V. Tulus, S. Suh, G. Guillén-Gosálbez, A. Bardow, *Nat Sustain* **2023**, *6 (5)*, 599 – 610. DOI: https://doi.org/10.1038/s41893-022-01054-9

[14] R. Meys, F. Frick, S. Westhues, A. Sternberg, J. Klankermayer, A. Bardow, *Resources, Conservation and Recycling* **2020**, *162*, 105010. DOI: https://doi.org/10.1016/j.resconrec.2020.105010

[15] J. D. Peterson, S. Vyazovkin, C. A. Wight, *Macromol. Chem. Phys.* **2001**, *202 (6)*, 775 – 784. DOI: https://doi.org/10.1002/1521-3935(20010301)202:6<775:AID-MACP775>3.0.CO;2-G

[16] A. Merkel, L. Plessing, G. A. Luinstra, M. Grünewald, P. Biessey, *Chemie Ingenieur Technik* **2023**, *95 (8)*, 1323 – 1331. DOI: https://doi.org/10.1002/cite.202100210

[17] M. Zeller, N. Netsch, F. Richter, H. Leibold, D. Stapf, *Chem. Ing. Tech.* **2021**, *93 (11)*, 1763 – 1770. DOI: https://doi.org/10.1002/cite.202100102

[18] H. Ukei, T. Hirose, S. Horikawa, Y. Takai, M. Taka, N. Azuma, A. Ueno, *Catal. Today* **2000**, *62 (1)*, 67 – 75. DOI: https://doi.org/10.1016/S0920-5861(00)00409-0

[19] R. Meys, A. Kätelhön, M. Bachmann, B. Winter, C. Zibunas, S. Suh, A. Bardow, *Science (New York, N.Y.)* **2021**, *374 (6563)*, 71 – 76. DOI: https://doi.org/10.1126/science.abg9853

297a Fate and Transport of Perfluorooctanesulphonic Acid (PFOS) and Microplastics in Water: Insights for Their Removal By Adsorption

Jasneet Pala, Medha Kasula, Gabriel Barbosa, Milad Esfahani, Heath Turner

Microplastics are becoming a concern for the environment and are gaining significant attention due to their challenging removal process and ability to transport coexisting pollutants. Because of their mobility, small size, and accumulation tendency, microplastics are toxic for terrestrial and aquatic living organisms and humans, causing inflammation and oxidative stress. In addition, the presence of per- and polyfluoroalkyl substances (PFAS) in water causes health effects on humans and animals over time. In this work, we investigated the interaction between these two emerging water contaminates at different conditions (pH, temperature, ionic strength, etc). The interaction of perfluorooctanesulphonic (PFOS) acid and micron size polyethylene (PE) (20 microns) and polypropylene (PP) (17 microns) were investigated via experiment and modeling. The isothermal adsorption of PFOS on the microplastics was measured at different conditions (time, pH, concentration). The results showed that PFOS adsorption on PE and PP was 9% and 13.95% after 24 hours, respectively, and 43.11 ± 0.18% and 35.90 ± 20.94% after one week of contact time. However the extension of contact time from a week to two weeks resulted in lower adsorption (36.17% for PE and 22.62% for PP) which implied desorption of PFOS from these microplastics. In addition, molecular dynamics simulations were employed to model the adsorption mechanism of four different perfluorooctyl PFAS molecules on PE, PP, and PVC microplastic particles by inspecting the potential-of-mean-force of pulling the PFAS molecules from the aqueous bulk phase to the plastic-water interfacial region. The molecular simulation results suggest a pronounced dispersive interaction between the polymer and PFAS molecules for PE and PP; while the dispersive interaction dominates the PVC systems, as compared to PE and PP, a much stronger coulombic PFAS-PVC interaction is found.

297b Highly Selective and Efficient Ammonia Removal and Recovery from Hydrolyzed Urine Using Zinc Oxalate

Brandon Clark, William Tarpeh

Recovering valuable products from wastewater can reduce energy inputs, costs, and environmental damage associated with pollution mitigation and chemical manufacturing but requires highly selective separation materials integrated into effective unit processes. The advantages of selective resource recovery materials are especially apparent within the nitrogen cycle. Roughly 1% of global energy is used by the Haber-Bosch process to convert atmospheric dinitrogen gas (N_2) to ammonia (NH_3), producing carbon dioxide (CO_2) as a byproduct. Anthropogenic sources of reactive nitrogen have unbalanced the nitrogen cycle because 80% of wastewater globally is discharged without treatment, and reactive nitrogen threatens aquatic ecosystems with eutrophication. Moreover, conventional biological wastewater treatment converts 100-fold less reactive nitrogen back to N_2 per year than is generated by Haber-Bosch, necessitating additional techniques that supplement biological treatment. Materials capable of selectively recovering reactive nitrogen from wastewaters without chemical transformation can provide an economic incentive for expanding global environmental remediation and can mitigate dependence on both the Haber-Bosch process and biological wastewater treatment.

Ion exchange can efficiently recover ionized and ionizable reactive nitrogen species like ammonium (NH_4^+)/ammonia (i.e., total ammonia nitrogen or TAN). Ion exchange with fixed-bed column operation exhibits several process advantages for selective nitrogen recovery, including low electricity requirements, flexible scale, and suitability for decentralized applications. However, existing ion exchange resins with fixed anionic moieties are not intrinsically selective for the ammonium cation over other cations. High valence, small hydration radii, and low hydration energies dictate high ion exchange selectivity. However, ammonium is monovalent and has intermediate hydration properties compared to other common wastewater cations like sodium (Na^+), potassium (K^+), magnesium (Mg^{2+}), and calcium (Ca^{2+}).

To overcome the selectivity constraints of ion exchange, we synthesized zinc oxalate from oxalic acid and zinc sulfate. Because divalent zinc (Zn^{2+}) is an ammine-complexing transition metal cation and zinc oxalate is extremely water insoluble ($1.38*10^{-9}$ g/100 mL), precipitated zinc oxalate is an effective aqueous ligand exchanger. While ion exchange involves purely outer-sphere electrostatic attraction, ligand exchange involves a ligand supplying both electrons to a coordinate covalent bond. Selectivity is achieved because zinc binds too strongly to oxalate for competing cations to ion exchange with zinc, but TAN in the neutral ammonia form bypasses ion exchange competition and binds to the zinc itself with its electron lone pair. Additionally, if trace amounts of zinc-ammonia complexes are removed by ligand exchange competition between ammonia and oxalate, then oxalate will dissolve because sodium oxalate, potassium oxalate, and ammonium oxalate are water soluble. Consequently, nearly all that is left in the precipitate is zinc oxalate.

Considering the optimal pH for ammonia adsorption (9-10), hydrolyzed urine, which contains 80% of nutrients found in municipal wastewater but comprises only 1% of wastewater volume, was identified as a promising wastewater candidate. Batch ammonia adsorption tests were conducted in real hydrolyzed urine with zinc oxalate and a commercial tertiary amine resin (AmberLite IRA67, pKa 9) to buffer pH and preserve TAN speciation as ammonia over ammonium without introducing additional cations to the solution. Initial and final ion concentrations were analyzed with ion chromatography (IC). As long as the dose of zinc oxalate was high enough to establish an ammonia to zinc ratio of 2 and below in the system, over 95% of ammonia was removed, and zinc elution was below 0.5%. Furthermore, negligible amounts (i.e., below the detection limit of IC) of the competing cations present in hydrolyzed urine, sodium and potassium, were removed. Above an ammonia to zinc ratio of 2, ammonia began to outcompete oxalate for zinc ligand sites, and the zinc-ammonia complex began to elute. To show effective TAN recovery as well as removal, ammonia-loaded zinc oxalate was placed in 250 mM acetic acid. Over 99% of ammonia was recovered as ammonium acetate, and zinc elution due to oxalate protonation was below 0.2%. Therefore, zinc oxalate is highly selective for ammonia removal in a real wastewater stream, mild acids are effective at efficient TAN recovery, and zinc oxalate is robust against low pH levels. Ultimately, ligand exchange adsorbents can aid in advancing a future vision of a circular nitrogen economy that extracts ammonia pollutants from wastewaters and offsets anthropogenic imbalances to the global nitrogen cycle.

297c Electrified Resource Recovery from Synthetic Anaerobic Digester Centrate: Impact of Anode Type on Operational, Design and Stream Variables

Sana Heydarian, Syed Asad Abbas, Damilola Daramola

Eutrophication of natural water streams is an undesirable effect of nutrient-rich streams disposal into these water bodies and excessive use of fertilizers on agricultural lands[1]. As a prevention act, the post-anaerobic digester liquid stream (centrate) in wastewater treatment plants (WWTP), due to its high loads of nitrogen (N) and phosphorous (P), is not permitted to be discharged into the natural water bodies. Hence, this centrate stream will be looped back to the head of the WWTP facility, and it will increase the nutrient content of the aeration tank influent[2,3]. As stated in several literature studies, aeration tanks are one of the most energy-consuming processes in WWTPs and an increased nutrient load will result in increased energy consumption[4,5]. Furthermore, the natural resources of P are anticipated to be depleted in the near future[2,3].

One approach that can simultaneously address all the aforesaid challenges is recovering nitrogen and phosphorous as a compound called Magnesium Ammonium Phosphate or struvite[1,6]. Struvite is a commercial fertilizer product and has the added benefit of providing a sustainable P resource for the agricultural sector[3]. The process of struvite recovery from wastewater has been performed using several methods including chemical, biological, and electrochemical processes, to name a few[7]. This research focuses on the electrochemical pathway of recovering this valuable fertilizer product because of its sustainability, as renewable energy sources could be used for operating the electrochemical technology, as opposed to non-renewable resources for other methods[8].

In this work, the performance of a planar three-electrode electrochemical cell in a batch configuration with two different anode types has been compared. Materials that have been utilized as the anode are graphite and sacrificial magnesium anode, while synthetic wastewater mimicking the real-world centrate has been used as the electrolyte. It is unknown how these different anodes affect or interact with the other parameters and impact the behavior of the system. Therefore, the aim of this study is to identify and compare the most influential stream, design and operational factors in the functionality of each process. Additionally, nutrient removal capability in the two systems will be compared and recovered solid products will be discussed. Figure 1 represents the main variables in the P percentage recovery for both anode types. Based on this statistical analysis, the most significant factors in the P recovery and the effect directions are distinct in each system. Our results suggest by choosing the optimum set of variables, high P recoveries with good solid product quality can be obtained from both types of systems.

References

(1) Galbraith, S. C.; Schneider, P. A.; Flood, A. E. Model-Driven Experimental Evaluation of Struvite Nucleation, Growth and Aggregation Kinetics. *Water Research* **2014**, *56*, 122–132. https://doi.org/10.1016/j.watres.2014.03.002.

(2) Guštin, S.; Marinšek-Logar, R. Effect of PH, Temperature and Air Flow Rate on the Continuous Ammonia Stripping of the Anaerobic Digestion Effluent. *Process Safety and Environmental Protection* **2011**, *89* (1), 61–66. https://doi.org/10.1016/j.psep.2010.11.001.

(3) Holloway, R. W.; Childress, A. E.; Dennett, K. E.; Cath, T. Y. Forward Osmosis for Concentration of Anaerobic Digester Centrate. *Water Research* **2007**, *41* (17), 4005–4014. https://doi.org/10.1016/j.watres.2007.05.054.

(4) Soares, R. B.; Memelli, M. S.; Roque, R. P.; Gonçalves, R. F. Comparative Analysis of the Energy Consumption of Different Wastewater Treatment Plants. *International Journal of Architecture, Arts and Applications* **2017**, *3* (6), 79. https://doi.org/10.11648/j.ijaaa.20170306.11.

(5) Ghimire, U.; Sarpong, G.; Gude, V. G. Transitioning Wastewater Treatment Plants toward Circular Economy and Energy Sustainability. *ACS Omega* **2021**, *6* (18), 11794–11803. https://doi.org/10.1021/acsomega.0c05827.

(6) Kékedy-Nagy, L.; English, L.; Anari, Z.; Abolhassani, M.; Pollet, B. G.; Popp, J.; Greenlee, L. F. Electrochemical Nutrient Removal from Natural Wastewater Sources and Its Impact on Water Quality. *Water Research* **2022**, *210*, 118001. https://doi.org/10.1016/j.watres.2021.118001.

(7) Balaguer-Barbosa, M. Recovery of Nutrients from Anaerobically Digested Enhanced Biological Phosphorus Removal (EBPR) Sludge through Struvite Precipitation. *USF Tampa Graduate Theses and Dissertations* **2018**.

(8) Wang, Y.; Kuntke, P.; Saakes, M.; van der Weijden, R. D.; Buisman, C. J. N.; Lei, Y. Electrochemically Mediated Precipitation of Phosphate Minerals for Phosphorus Removal and Recovery: Progress and Perspective. *Water Research* **2022**, *209*, 117891. https://doi.org/10.1016/j.watres.2021.117891.

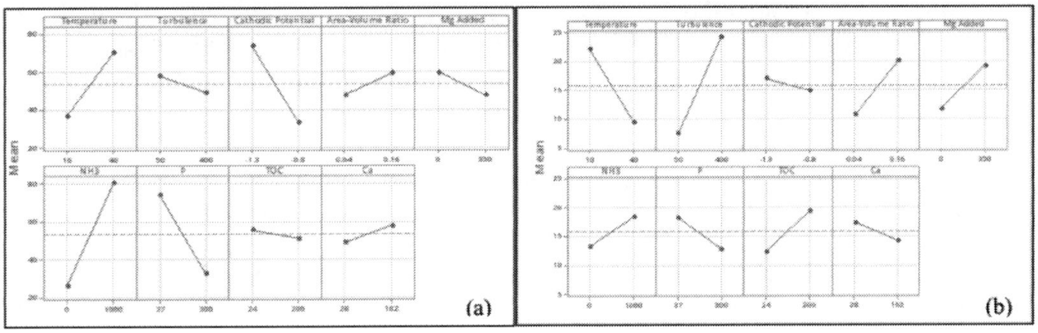

Figure 1. Main effects plot in the P percentage recovery in the system with (a) magnesium anode and (b) graphite anode.

297d Retrofitting of Municipal Wastewater Treatment Plants for Maximizing Resilience Under Cost Limitations: Graph-Theoretic Approach

Jean Pimentel, Akos Orosz, Emmanuel Aboagye, Heriberto Cabezas, Ferenc Friedler, Kirti Yenkie

According to the United Nations (UN), the world population will surpass 9.5 billion people by 2050 (UN, 2022). One of the most serious consequences of the fast population growth is the increase in water pollution. Every year, more than 300 km^3 of municipal wastewater are treated globally (Mateo-Sagasta et al., 2015), and it is expected that this figure will increase alongside with the population. Hence, the capacity of numerous wastewater treatment plants (WWTPs) has been exceeded, especially during rainy seasons when the volume of water to be treated may be significantly enlarged, and high volumes of water remain untreated. Because of this, analysis and retrofitting of WWTP has emerged as a key area in wastewater treatment engineering, and have attracted attention of numerous stakeholders (Bozkurt et al., 2016).

Process Network Synthesis (PNS) refers to the task of defining the structure of a process from a set of plausible building blocks, i.e., defining the basic units comprising the transformation process and their connectivity. The algorithmic approach to PNS has been deployed before for the synthesis of wastewater treatment plants, as it endows the designer with the capacity to explore numerous alternatives (Galan and Grossmann, 1998; Grossmann et al., 2014). Similarly, the task of retrofitting WWTPs can be regarded as a PNS problem where the best improvements to the current structure needs to be determined; such a problem can also be addressed algorithmically by resorting to optimization methods. Previous contributions have addressed this problem of finding the best retrofitting based on cost optimization (Bozkurt et al., 2016); however, determination of the best retrofit should not be based exclusively on economic criteria and additional sustainability metrics should be involved. An especially significant property for this kind of system is its resilience, which accounts for its capacity to recover from failures. However, the comparison of resilience for assorted designs requires a method to account for the effects of both expected and unexpected failures simultaneously. The consideration of unexpected failures is a daunting task, as it requires the enumeration of possible consequences for all possible failures in the process units (Orosz et al., 2022); such an enumeration can be performed by considering the properties of the process structure itself.

The P-graph framework is a graph-theoretic approach for process synthesis that is based on an unambiguous graphical representation and a set of combinatorial axioms and algorithms (Friedler et al., 1992). These elements are employed to exploit the properties of a synthesis problem's structure to enhance its solution, which endows numerous traits for the designers. On one hand, the graphical representation permits a user-friendly depiction of the changes in the process and its manipulation during the solution procedure. On the other hand, the combinatorial algorithms permit the algorithmic construction of a rigorous superstructure and the enumeration of the feasible processes that comprise it. Moreover, they handle the design decisions structurally, thus, numerous infeasibilities can be removed from the outset and no binary decision variables need to be included in the problem (Friedler et al., 1996); hence, the risk of convergence to local optima related to the structure is reduced. Because the optimization is enhanced and the search space is drastically reduced, the framework can efficiently deliver not only the best solution but also the set of n-best alternative designs. Such a set of solutions provides the designers with insightful information concerning the process and may be analyzed to determine the best alternative in light of criteria that are not considered in the model.

In this work, the P-graph framework (Friedler et al., 2022) is employed as a synthesis tool for retrofitting WWTPs that considers sustainability, resilience, and economic indicators. For this, the structure of the existing process is extended with plausible operation alternatives, as shown in Figure 1. Then the optimization is performed to determine the new connections is most suitable for resilience enhancement. For this, combinatorial algorithms are used for the rigorous examination of all possible retrofitting options and the exhaustive enumeration of failure cases in the extended process structure. The examination renders the n-best alternatives for retrofitting that maximize the resilience of WWTP under a limited budget, from which the designers may select the most suitable alternative, thereby yielding insightful information for the decision-makers.

References

Bozkurt, H., van Loosdrecht, M.C.M., Gernaey, K.V., Sin, G., 2016. Optimal WWTP process selection for treatment of domestic wastewater – A realistic full-scale retrofitting study. Chem. Eng. J. 286, 447–458. https://doi.org/10.1016/j.cej.2015.10.088

Friedler, F., Orosz, Á., Pimentel Losada, J., 2022. P-graphs for Process Systems Engineering. Springer International Publishing, Cham, Switzerland. https://doi.org/10.1007/978-3-030-92216-0

Friedler, F., Tarján, K., Huang, Y.W., Fan, L.T., 1992. Graph-theoretic approach to process synthesis: Axioms and theorems. Chem. Eng. Sci. 47, 1973–1988.

Friedler, F., Varga, J.B., Fehér, E., Fan, L.T., 1996. Combinatorially Accerelated Branch-and-Bound Method for Solving the MIP Model of Process Network Synthesis, in: Floudas C.A., Pardalos P.M. (Eds.), Nonconvex Optimization and Its Applications. Kluwer Academic Publishers, Norwell, MA, pp. 609–626.

Galan, B., Grossmann, I.E., 1998. Optimal Design of Distributed Wastewater Treatment Networks. Ind. Eng. Chem. Res. 37, 4036–4048. https://doi.org/10.1021/ie980133h

Grossmann, I.E., Martín, M., Yang, L., 2014. Review of optimization models for integrated process water networks and their application to biofuel processes. Curr. Opin. Chem. Eng., Energy and environmental engineering / Reaction engineering 5, 101–109. https://doi.org/10.1016/j.coche.2014.07.003

Mateo-Sagasta, J., Raschid-Sally, L., Thebo, A., 2015. Global Wastewater and Sludge Production, Treatment and Use, in: Drechsel, P., Qadir, M., Wichelns, D. (Eds.), Wastewater. Springer Netherlands, Dordrecht, pp. 15–38. https://doi.org/10.1007/978-94-017-9545-6_2

Orosz, Á., Pimentel, J., Argoti, A., Friedler, F., 2022. General formulation of resilience for designing process networks. Comput. Chem. Eng. 165, 107932. https://doi.org/10.1016/j.compchemeng.2022.107932

United Nations, 2022. World population prospects 2022: summary of results. United Nations, New York.

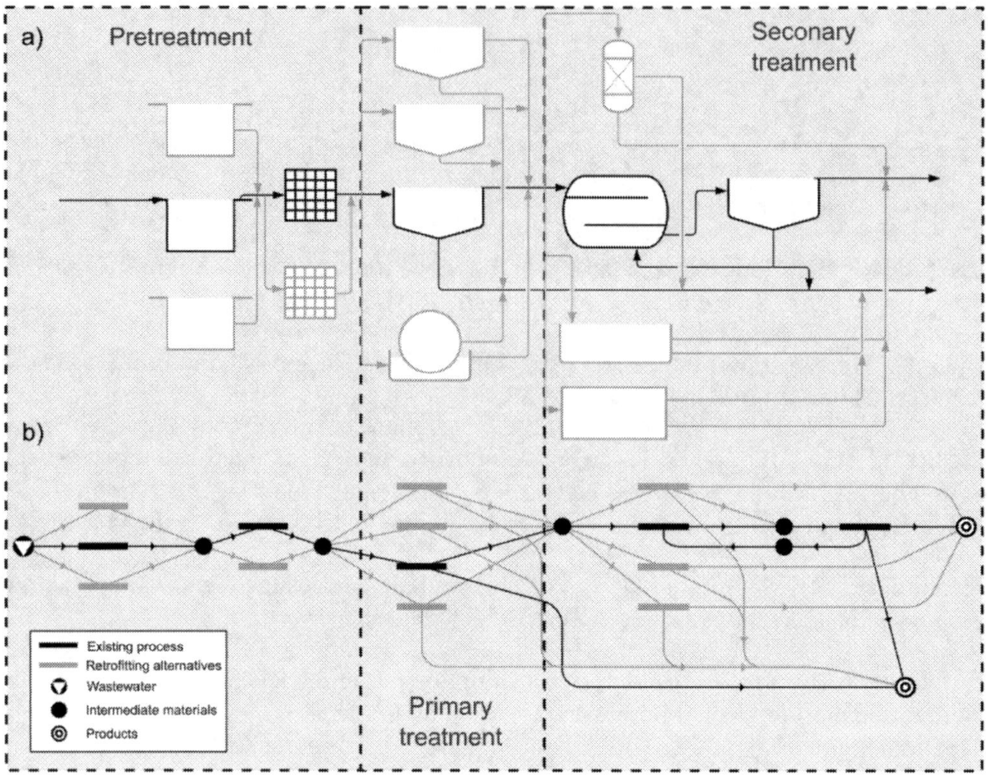

Figure 1. Representation of extended structure for retrofitting illustrating the original WWTP in black, and the extension with plausible retrofitting options in blue for conventional representation (a) and P-graph representation (b).

297e Temperature Swing Solvent Extraction Hypersaline Desalination with Stepwise Release to Advance the Productivity-Selectivity Tradeoff

Kinnari Shah, Ian Billinge, Eliza Dach, Ngai Yin Yip

Hypersaline brines are of growing environmental importance but are presently underserved by conventional treatment methods. Temperature swing solvent extraction (TSSE) is an emerging technology that is able to desalinate hypersaline brines, thereby reducing their volume and facilitating disposal while simultaneously producing fit-for-purpose water to alleviate supply stress. TSSE utilizes a switchable solvent with thermally responsive polarity to extract water from the hypersaline brine while rejecting salts. Low-temperature heat triggers the solvent to switch from a hydrophilic to a more hydrophobic state, causing the product water to disengage from the extract and regenerating the solvent to be recycled into the process. Our previous work revealed a tradeoff between water productivity and water/salt selectivity that constrains the desalination performance of TSSE. In this study, we demonstrate the improved performance of the novel temperature swing solvent extraction stepwise release (TSSE-SR), departing from the nominal productivity-selectivity tradeoff. Thermally switchable solvents of diisopropylamine, triethylamine, and tert-octylamine were evaluated for TSSE-SR desalination of 1.0 M NaCl brines. The introduction of an intermediate temperature step in TSSE-SR allows salt to preferentially disengage over water from the organic solvent extract. TSSE-SR drastically improves salt rejection while minimizing sacrifices in water recovery yield. The potential contribution of the intermediate step to amine loss is assessed and confirmed to be insignificant. Hunter-Nash analysis is conducted on diisopropylamine-water-NaCl ternary phase diagrams to provide a theoretical basis for the enhanced TSSE-SR performance. Finally, TSSE-SR is benchmarked against standard TSSE and the enhancements in productivity and selectivity are quantified. This innovative approach can expand the spectrum of viable solvents for hypersaline desalination to include those that have high water productivity but low selectivity in single-stage TSSE operations.

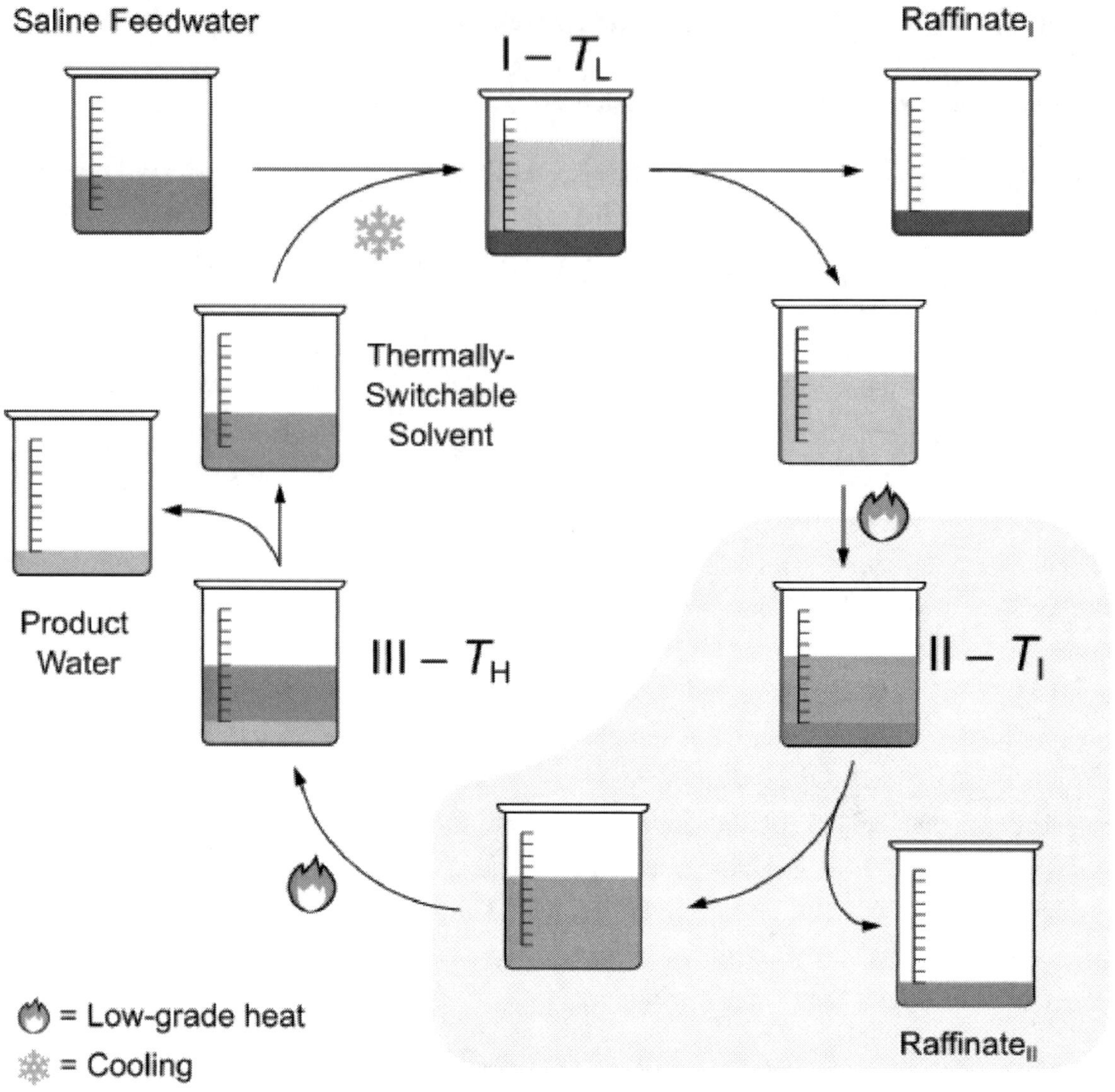

Saline Feedwater
I – T_L
Raffinate$_I$
Thermally-Switchable Solvent
Product Water
III – T_H
II – T_I
Raffinate$_{II}$
= Low-grade heat
= Cooling

188b Adding a Spatial Dimension to Nature Based Solutions in Carbon Neutrality Planning through the Optimization of Ecological Co-Benefits

Michael Charles, Bhavik Bakshi

The United Nations' Sustainable Development Goals (SDGs) emphasize the need for engineering and technology development to approach sustainability holistically, including human health and well-being. This requires expanding existing approaches for carbon mitigation to consider co-benefits and question unintentional consequences in our climate actions. In pursuit of this goal, we present a computational design approach for techno-ecological systems that includes ecosystem co-services in the pursuit of carbon neutrality, and even carbon positivity. Specifically, we consider the health impacts of air quality regulation provided by land-related responses. Our approach builds upon our past research on optimization-based approaches for long-term climate action plans and spatially-explicit industrial site design, proposing a new spatially-explicit operationalization framework for including ecological responses in action-based climate plans. This framework considers health risk reductions and identifies where and when land-based solutions should be implemented, while leaving the 'how' to local ecological experts and land management teams.

The spatial guidance for prioritizing land-based projects provided by this framework enables the identification of areas where pollutant exposure can be reduced and health-related incidences, like hospitalizations and mortality, can be minimized. When these projects should be implemented is dependent on budget constraints and ecological dynamics. We provide a practical case study and results that show the comparison between our previous spatially-explicit approach, which maximized the physical uptake of pollutants and disregarded climate regulation, and our co-benefit approach, which minimizes health risks associated with pollutant exposure and values the benefit of carbon sequestration. The health risk assessments focus on the impacts of criteria air pollutants ozone (O3) and particulate matter less than 2.5 μm in diameter (PM2.5).

Our results display the significance of including population data into techno-ecological and sustainable design methodologies. The inclusion of spatially-explicit health impact assessments in the design approach results in a 77% increase in social economic benefits with only a 6% increase in private costs to industry compared to our previous approach focused only on the physical uptake of ecosystems. The new insight from this approach results in a higher economic value for ecosystem services, promoting investments in restoration, and an increased focus on understanding the health impacts on neighboring populations around industrial sites.

Overall, our work aims to build upon previous research to provide a framework that addresses both land-based climate action and local public health simultaneously. We hope that this approach will lead to better decision-making and investment in land-based projects that not only address climate change but provide local benefits to ecosystems and communities. Although this approach does not address the challenges of measuring, monitoring, and reporting additional carbon sequestration, it does acknowledge improved return of investment on high transaction costs that are often associated with land-based responses. Further research can explore the use of our framework in other locations, including those with different socioeconomic and environmental characteristics. Additionally, the framework can be extended to include other co-benefits, such as biodiversity conservation and water quality improvements, and to assess the potential trade-offs between these co-benefits.

188c Applying Process Design Principles to Glass Recycling for Coastal Restoration

Julie Albert, Franziska Trautmann

The US recycles less glass than other countries due mainly to government policies, lower consumer engagement, and cost. Successful glass recycling facilities are often co-located with bottle manufacturers that recycle old glass into new. However, many communities, particularly those in rural areas and along the coast, lack this infrastructure. Furthermore, recycled glass sand has many potential benefits over natural sand for many applications; for example, it can be made coarser, does not compact as much, and is more resistant to erosion, making it an excellent – and underutilized – material for coastal protection and restoration.

This talk will describe research initiated by two chemical engineering professors and a chemical engineering alumna of their department now being conducted with funding from the National Science Foundation Convergence Accelerator Program, Track E: the Networked Blue Economy. The overarching vision of the *ReCoast Team,* founded in 2021 in response to the NSF's request for proposals, is to create coastal community recycling programs to keep glass out of landfills and instead use it for glass sand products that support coastal restoration, preservation, and resiliency. Through extensive regional economic, social, cultural, and environmental research, the team is ensuring ecological safety and mitigation of land loss.

The focus of the talk will be on correlations between the physical and chemical properties of recycled glass sand with how the glass recycling process is carried out, from glass collection to final product. Key findings include: (1) for small-to-mid-size operations, limiting collection to food and beverage bottles (soda-lime glass) to avoid contaminants like lead (e.g., leaded crystal) and minimize wear-and-tear on machinery from reinforced glasses (e.g., borosilicate) is important; (2) for food and beverage bottles, the parts-per-million presence of various metals that give the bottles color or clarity do not leach based on TCLP analysis by a certified lab; (3) size and angularity of the crushed glass products both increase with wear on the machinery, so operations intending to produce coarse sand for restoration may need to change wearable parts more frequently than manufacturer recommendations or compared to operations selling to different markets. A brief discussion of research demonstrating ecological safety in marsh environments and technoeconomic feasibility will also be presented to provide context for these results.

*Funding: NSF 2137730, 2230769

188d Study of Impacts of Air Velocity, Ambient Temperature and Terrain on Chemical Diffusion in Unplanned Release

Sk. Moumita Manjur, Sidney Lin

Air pollution is now a burning question for the wellbeing and even existence of human in today's world. Chemical industries contribute about 52% to it among all potential sources of air pollution. Any unplanned/accidental release of chemical species can lead to serious loss, injury or even life risk to the community along with contributing to the general air pollution. To ensure a greener environment, unplanned release of air pollutants from industrial process higher than permissible limit must be monitored for public health and safety. A mathematical simulation can provide better understanding over a certain chemicals' diffusion trends in different scenarios and provide guidelines to a proper planning for reduction of pollution along with right decision making on emergency response and rescue action in case of any dire incident of chemical release.

A finite element analysis model of chemical diffusion from chemical storage leakage has been developed using COMSOL Multiphysics®. In this model, the transport of released chemical through air is driven by both diffusion and convection. Concentration profiles and histories of released chemical were calculated using the Fick's 2nd law of diffusion and convection by velocity field of air. Ambient air velocity field is calculated over time using incompressible form of the Navier-Stokes and continuity equations. Physical and thermodynamic properties of the chemical species at different temperature, pressure and composition reported in literatures were used in this mode. Concentration profiles and histories of pollutant species were calculated in different scenarios including: i) without air flow, ii) with air flow at different speed and direction, iii) with obstacles in terrain. The calculated results are analyzed for impacts of terrain, air velocity, and air temperature on the propagation of chemical species with time for an unplanned release situation.

188e Characterization and Modeling of the Thermal Degradation of Fluoropolymers into the Gaseous Evolution of per- and Poly-Fluoroalkyl Substances (PFAS)

Enoch Nagelli, Simuck Yuk, Victor A. Jaffett, Justin Hunter, Bryan A. Lagasse, Taylor N. Miller, Jacob A. Nauman, Patrick B. Dawson, Augustine H. Kim, Joseph D. Taptich, Kwabena K. Davis, Robert Peters, Pamela L. Sheehan

Fluoropolymers have the potential to thermally break down into long chain per- and polyfluoroalkyl substances (PFAS) and/or polychlorofluoro- substances to the environment. [1] Recently, the U.S. Environmental Protection Agency (EPA) released a technical update on the use of incineration to manage PFAS waste streams by incineration at high temperatures, upwards of 1000 to 1400 °C. [2] The EPA's Contaminant Candidate List now includes several PFAS subtypes with thousands of other distinct chemicals for identifying additional PFAS that could potentially require future regulation under the Safe Drinking Water Act. [3] In order to understand the environmental impact, it is important to characterize and study the gas phase thermal breakdown fragmentation and desorption products that evolve during the combustion of fluoropolymers. Research on the efficacy of incineration is limited, but recent models predict that the combustion of C_1 and C_2 PFAS results in the formation of the products of incomplete combustion (PICs) such as CF_2 and CF_3 radicals [4]. In our research study, we analyzed the thermal breakdown and degradation products of polytetrafluorethylene (PTFE), polyvinylidene fluoride (PVDF), and polyvinylidene hexafluoropropylene (PVDF-HFP) using Thermogravimetric Analysis from 20 to 1000 °C, coupled to Mass Spectrometry (TGA-MS). The TGA-MS data provides a signature for possible gaseous degradation products that evolved at specific temperatures. The separation of individual byproducts and isolation of gaseous byproducts remains a challenge experimentally, however, atomistic modeling can lead to predictive chemical reaction pathways and thus provide insight into the initial fragmentation of the fluoropolymers during thermal breakdown.

Building on the gas phase characterization from TGA-MS results, process modeling with ChemCad and Aspen Plus was utilized to study the mass and energy balance for the combustion of the fluoropolymers of interest (such as polytetrafluoroethylene (PTFE) and polyvinylidene fluoride (PVDF) with hexafluoropropylene (HFP)). The process is modeled in a Gibbs Reactor in atmospheric combustion conditions with varying oxygen (air) flow rates and temperatures, ranging in temperatures from 250 to 1000 °C. For example, the process feed stream was specified to study the breakdown of PVDF with its constituent monomers which include HFP to form perfluorobutyl ethylene (PFBE), chlorotrifluoroethylene (CTFE), and perfluoropropyl vinyl ethers (PPVEs) reacting with varying air flow rates of stoichiometric oxygen (O_2) and nitrogen (N_2). In addition, Density Functional Theory (DFT) calculations were also used to gain additional mechanistic understandings on the thermal degradation of relevant fluoropolymer structures. Overall, this study is focused on the instantaneous characterization of the gaseous reaction products from the combustion of fluoropolymers using mass spectrometry. The mechanistic modeling provides context to the kinetic and energetic pathways for the potential thermal breakdown occurring in the TGA-MS studies. Experimental and modeling results together inform the process design modeling for large scale energy and mass balance calculations.

KEYWORDS: Per- and polyfluoroalkyl substances (PFAS), Gas Phase Characterization, Fluoropolymers, Thermal Gravimetric Analysis Coupled Mass Spectrometry, Thermal Degradation Analysis, Combustion, Process Design, Aspen Plus, CHEMCAD.

CONTACT: **Dr. Enoch A. Nagelli, Dept of Chemistry and Life Science**, United States Military Academy, West Point, New York 10996. **Email:** enoch.nagelli@westpoint.edu **TEL:** 845-938-3904.

REFERENCES

1. David A. Ellis, Scott A. Mabury, Jonathan W. Martin and Derek C. G. Muir "Thermolysis of fluoropolymers as a potential source of halogenated organic acids in the environment" *Nature* 412, 321–324 (2001). https://doi.org/10.1038/35085548.

2. U.S. Environmental Protection Agency. (2020, February). *Per- and Polyfluoroalkyl Substances (PFAS): Incineration to Manage PFAS Waste Streams.* https://www.epa.gov/chemical-research/technical-brief-and-polyfluoroalkyl-substances-pfas-incineration-manage-pfas and https://www.epa.gov/chemical-research/status-epa-research-and-development-pfas#contamination.

3. U.S. Environmental Protection Agency. (2022, October). EPA Issues Final List of Contaminants for Potential Regulatory Consideration in Drinking Water, Significantly Increases PFAS Chemicals for Review https://www.epa.gov/ccl/contaminant-candidate-list-5-ccl-5

4. Jonathan D. Krug, Paul M. Lemieux, Chun-Wai Lee, Jeffrey V. Ryan, Peter H. Kariher, Erin P. Shields, Lindsay C. Wickersham, Martin K. Denison, Kevin A. Davis, David A. Swensen, R. Preston Burnette, Jost O.L. Wendt, and William P. Linak (2022) "Combustion of C1 and C2 PFAS: Kinetic modeling and experiments" *Journal of the Air & Waste Management Association*, 72:3, 256-270, DOI: 10.1080/10962247.2021.2021317

188f Machine Learning Modeling of Electrochemical Recovery of Phosphorus from Municipal Wastewater in a Batch Reactor

Lawrence Ajayi, Syed Asad Abbas, Jason Trembly, Damilola Daramola

Phosphorus is an irreplaceable macronutrient necessary for plant growth and other industrial processes, whereas when in excess in water bodies can lead to eutrophication. Recovery of phosphorus from wastewater became important as demand for fertilizer and supply of non-renewable rock phosphate increase and decrease respectively[1]. Electrochemical recovery of phosphorus from wastewater has been established to be efficient, modular and robust with little or no chemical dosage[2,3]. This approach can also be used to achieve a process that enables easy automation with little or no human intervention. However, the efficiency of electrochemical driving force depends partly on the water chemistry[1] which varies based on the type of wastewater and treatment method. To achieve an efficient automated process, process parameters need to concurrently change based on the composition of the wastewater and the intended percentage recovery[4]. Therefore, it is important to develop a model that takes both process and stream parameters as input variables to determine the percentage recovery of phosphorus as an output variable.

This research study aims to develop an experimental data-driven model to predict the recovery of phosphorus using various machine learning algorithms. One major advantage of machine learning is that it doesn't require understanding of the science behind the process[5] nor does it make any presumption about the data[6]. Hao *et al.* developed a model for removal of phosphorus in an electrochemical fixed bed reactor packed with magnetite[4], however, process variables such as temperature and turbulence that could affect the reaction kinetics and mass transfer respectively were not considered in the model development. Recently, the use of sacrificial magnesium anode as opposed to an inert anode with magnesium salt[2,7] has been found to facilitate phosphorus recovery, especially when struvite (a slow release fertilizer) is the desired precipitate[8]. Therefore, different anode types (sacrificial versus non-sacrificial) will also be used as an input variable for the model development.

Preliminary data has shown recovery of phosphorus to be directly and inversely related to the initial concentration of ammonia and phosphorus respectively in the wastewater being treated as indicated in the attached figure. This initial data, which varies the concentration of major ions in the wastewater and process parameters including cathodic potential, size of the electrode, temperature, and turbulence, will be used for model development. Machine learning algorithms such as multiple linear regression, random forest and support vector machine will be used to train, test, and validate different models. These various models and feature importance tests will be assessed for precision and ranking of parameters with the most significant impact on the recovery efficiency. These results will be presented at the meeting.

References:

(1) Kékedy-Nagy, L.; English, L.; Anari, Z.; Abolhassani, M.; Pollet, B. G.; Popp, J.; Greenlee, L. F. Electrochemical Nutrient Removal from Natural Wastewater Sources and Its Impact on Water Quality. *Water Res.* **2022**, *210*, 118001. https://doi.org/10.1016/j.watres.2021.118001.

(2) Belarbi, Z.; Daramola, D. A.; Trembly, J. P. Bench-Scale Demonstration and Thermodynamic Simulations of Electrochemical Nutrient Reduction in Wastewater via Recovery as Struvite. *J. Electrochem. Soc.* **2020**, *167* (15), 155524. https://doi.org/10.1149/1945-7111/abc58f.

(3) Lei, Y.; Zhan, Z.; Saakes, M.; van der Weijden, R. D.; Buisman, C. J. N. Electrochemical Recovery of Phosphorus from Acidic Cheese Wastewater: Feasibility, Quality of Products, and Comparison with Chemical Precipitation. *ACS EST Water* **2021**, *1* (4), 1002–1013. https://doi.org/10.1021/acsestwater.0c00263.

(4) Hao, J.; Zhang, J.; Li, X.; Qiao, M.; Zhao, X. Machine Learning Facilitates the Application of Electrochemically Induced Precipitation for the Removal of Phosphorous. *ACS EST Water* **2023**, *3* (2), 616–625. https://doi.org/10.1021/acsestwater.2c00631.

(5) Cao, B.; Adutwum, L. A.; Oliynyk, A. O.; Luber, E. J.; Olsen, B. C.; Mar, A.; Buriak, J. M. How To Optimize Materials and Devices via Design of Experiments and Machine Learning: Demonstration Using Organic Photovoltaics. *ACS Nano* **2018**, *12* (8), 7434–7444. https://doi.org/10.1021/acsnano.8b04726.

(6) Nageshwari, K.; Senthamizhan, V.; Balasubramanian, P. Sustaining Struvite Production from Wastewater through Machine Learning Based Modelling and Process Validation. *Sustain. Energy Technol. Assess.* **2022**, *53*, 102608. https://doi.org/10.1016/j.seta.2022.102608.

(7) Belarbi, Z.; Trembly, J. P. Electrochemical Processing to Capture Phosphorus from Simulated Concentrated Animal Feeding Operations Waste. *J. Electrochem. Soc.* **2018**, *165* (13), E685. https://doi.org/10.1149/2.0891813jes.

(8) Tan, X.; Yu, R.; Yang, G.; Wei, F.; Long, L.; Shen, F.; Wu, J.; Zhang, Y. Phosphate Recovery and Simultaneous Nitrogen Removal from Urine by Electrochemically Induced Struvite Precipitation. *Environ. Sci. Pollut. Res.* **2021**, *28* (5), 5625–5636. https://doi.org/10.1007/s11356-020-10924-8.

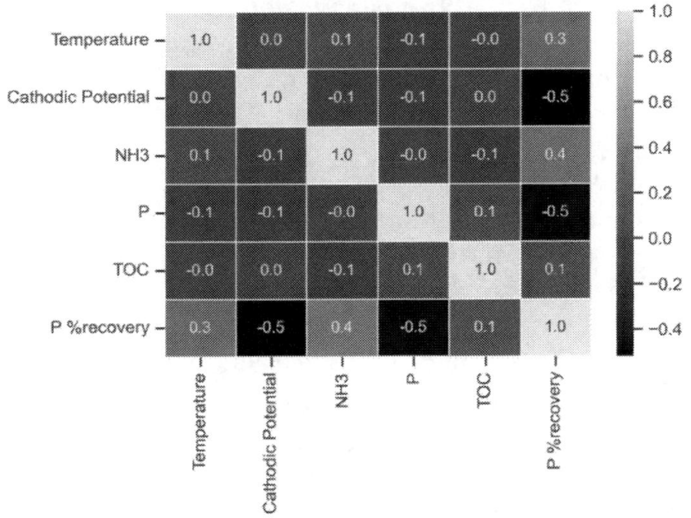

Figure 1: Correlation matrix of P %recovery and experimental variables based on electrochemical P recovery from synthetic municipal wastewater.

169a Removing Naturally Occurring Radioactive Materials (NORMs) from Produced Waters of Oil and Natural Gas Industries: From Camd to Experimental Proof-of-Concept

Ubadire Onyemaobi, Narendra Boppana, Rajib Mukherjee, Urmila Diwekar

CAMD is the reverse use of the group contribution method to generate molecules with desirable properties. The group contribution method (GCM) is a technique to estimate and predict thermodynamics and other properties from molecular structures. This approach dramatically reduces the amount of data needed. Instead of requiring information on thousands or millions of components' properties, data for only a few dozen or a few hundred groups are needed. A GCM uses the principle that some simple chemical components' structures are always the same in many different molecules. The smallest common constituents are the atoms and the bonds. For example, all-organic components are built of carbon, hydrogen, oxygen, nitrogen, halogens, and sometimes sulfur. Together with single, double, and triple bonds, only ten atom types and three bond types are necessary to build thousands of components. The next slightly more complex building blocks of components are functional groups, which are themselves built from a few atoms and bonds. The problem is a combinatorial optimization problem, which is a difficult optimization problem. We present a new optimization framework to solve real-world problems of CAMD for various separations encountered in chemical, pharmaceutical, and environmental industries. This talk will present the methodology and an important real-world case study.

The geologic formations that contain oil and gas deposits also contain naturally-occurring radionuclides, which are referred to as Naturally Occurring Radioactive Materials (NORM). Oil and gas fracking brings NORM to the surface in a concentrated form, which could pose a radiation safety hazard. Further, NORM levels have the potential to climb over time as the fluid being extracted begins to contain proportionally greater amounts of brine. Therefore, the majority of produced water contains significant levels of NORM. In 2011, GTI carried out a Techno-economic Assessment of Water Management Solutions project (2011) supported by a consortium of 23 companies. This consortium identified many priority industrial challenges for the pre- and post-crossover stages of a shale gas development area's water-based life cycle and identified NORM removal as one of the highest priority research areas. They also ascertained that there is currently no commercial product on the market to remove NORM from concentrated produced and flowback waters selectively. In this work, we have used CAMD to design novel clay-based adsorbents for removing 99.9% removal of NORM from produced water. With NSF funding, we then synthesized the adsorbents in the laboratory to show the effectiveness of CAMD. The new adsorbents perform an order of magnitude better than existing adsorbents for NORM removal. We carried out these experiments using produced water from Permian Basin. The concentration of NORM (Radium) in this water was more than 4000 pCi/g. According to regulations, the acceptable quantity of NORM for groundwater is 30 pCi/g. With our novel adsorbents, we could remove NORM from this water to the limit undetectable.

Keywords: Adsorbents, NORM, CAMD, Produced water.

References:

1. Benavides, P. and Diwekar, U., 2015. Optimal design of adsorbents for NORM removal from produced water in natural gas fracking. Part 1: Group contribution method for adsorption. Chemical Engineering Science, 137, pp.964-976.
2. Benavides, P. and Diwekar, U., 2015. Optimal design of adsorbents for NORM removal from

produced water in natural gas fracking. Part 1: Group contribution method for adsorption. Chemical Engineering Science, 137, pp.964-976.

3. Doshi, R., Mukherjee, R. and Diwekar, U., 2018. Application of Adsorbate Solid Solution Theory To Design Novel Adsorbents for Arsenic Removal Using CAMD. ACS Sustainable Chemistry & Engineering, 6(2), pp.2603-2611.

169b Efficient Degradation of Hydrogen Sulfide in Groundwater Using a Chemical-Free Photocatalyst

Mary Worbington, Ella Sheets, Kevin McPeak

About 15% of Americans do not have access to public water systems and must use private wells for drinking water.[1] These private wells are unregulated, and the user is responsible for maintaining and treating their water supply. Current treatment options for private water sources, such as UV disinfection or reverse osmosis, can be expensive, complex, or ineffective against certain contaminants. Hydrogen sulfide is one such contaminant that occurs naturally in groundwater and private wells due to the activity of anaerobic sulfur bacteria. The rotten-egg smell of even low concentrations of H_2S is a source of disgust for users of private wells. Photocatalytic advanced oxidation processes (P-AOPs) offer a chemical-free, lower-cost, and complete solution to treating the various contaminants found in private water sources.

In a previous work, we demonstrated that the P-AOP we developed effectively degrades model textile dyes, pharmaceuticals, pathogens, and chloro-hydrocarbons.[2] In this work, we report our system's effectiveness at treating H_2S-contaminated water. The photocatalyst consists of commercial titania nanoparticles, P25, that have been immobilized on quartz fibers, resulting in a high surface area photocatalyst on UV-transparent supports. The high photocatalyst surface area is advantageous for the mass-transfer-limited degradation reactions, and the UV-transparent supports maximize photon management in our reactor.

We test the efficacy of our reactor system in degrading H_2S under illumination from a UVC lamp. Our system achieves an electrical energy per order (E_{EO}) of < 1 for H_2S. We also investigate the effect of hydroxyl radical and hole scavengers, TOC, and pH on photocatalytic performance by performing tests using a synthetic water source prepared with interfering compounds and water from a Baton Rouge wastewater treatment plant. Finally, we propose improvements needed to scale our reactor system for practical application in the treatment of private wells.

(1) DeSimone, L. A. Quality of water from domestic wells in principal aquifers of the United States, 1991–2004. *US Geological Survey* **2009**, 139.

(2) Daniel E. Willis, E. C. S., Madhusudan Kamat, Sarah K. Glass, MaCayla J. Caso, Mary Worbington, Tochukwu Ofoegbuna, Samuel D. Snow, Kevin M. McPeak. Efficient Chemical-Free Degradation of Waterborne Micropollutants with an Immobilized Dual-Porous TiO_2 Photocatalyst. *ACS ES&T Engineering* **Under Review**.

169c Peracetic Acid Activated By Reduced Graphene Oxide Catalytic Membrane for Micro-Pollutants Removal from Agricultural Wastewater

Erda Deng, Haiqing Lin

Emerging recalcitrant micropollutants (MPs) with high toxicity are frequently detected in various water sources, including 2,4-dichlorophenoxyacetic acid (2,4-D) and its derivatives which are one of the most used herbicides in the world. Recently, peracetic acid (PAA) has been applied as a potent yet cost-effective oxidant. Developing PAA activation methods based on the direct electron-transfer (DET) mechanism are essential to the rational design of robust PAA oxidation systems targeting water decontamination. 2-D carbon materials with high electrical conductivity like graphene oxide (GO) and reduced GO (rGO) could mediate the DET activation of PAA. Herein, we demonstrated, for the first time, an rGO membrane activated PAA catalytic oxidation filtration system realizing MPs removal. We first investigated the oxidation efficacy of the rGO, PAA, and MPs mixture system and evaluated the MPs removal ratio in batch experiments. The rGO/PAA systems demonstrated over 80% removal of 2,4-D compared to only 20% removal using only PAA. The PAA and MPs mixture filtrating through an rGO membrane demonstrated over 50% removal of 2,4-D. Additionally, the rGO membrane systems exhibited non-chemical add-in into the feed, reducing the need for follow-up purification steps. Moreover, a 50-nm-thick rGO modified Nylon membrane used to purify wastewater containing 2,4-D in a crossflow system exhibited more than 30% catalytic removal of 2,4-D. This high removal efficiency for agricultural organic contaminants suggests its potential application in degrading MPs.

169d Optimizing the Membrane Composition Surrounding *E. coli* Amtb Ammonium Transporter Proteins for Highly Ammonium-Selective Protein-Loaded Biomimetic Membranes

Brandon Clark, William Tarpeh

Recovering valuable products from wastewater can reduce energy inputs, costs, and environmental damage associated with pollution mitigation and chemical manufacturing but requires highly selective separation materials integrated into effective unit processes. The advantages of selective resource recovery materials are especially apparent within the nitrogen cycle. Roughly 1% of global energy is used by the Haber-Bosch process to convert atmospheric dinitrogen gas (N_2) to ammonia (NH_3), producing CO_2 as a byproduct. Anthropogenic sources of reactive nitrogen have unbalanced the nitrogen cycle because 80% of wastewater globally is discharged without treatment, and reactive nitrogen threatens aquatic ecosystems with eutrophication. Moreover, conventional biological wastewater treatment converts 100-fold less reactive nitrogen back to N_2 per year than is generated by Haber-Bosch, necessitating additional techniques that supplement biological treatment. Materials capable of selectively recovering reactive nitrogen from wastewaters without chemical transformation can provide an economic incentive for expanding global environmental remediation and can mitigate dependence on both the Haber-Bosch process and biological wastewater treatment.

In biology, membrane transporter proteins are responsible for moving atoms and molecules through cell membranes with an extremely high degree of selectivity. Therefore, these proteins are attractive options for incorporation into synthetic membranes that also aim to be highly selective towards one or a few targets. The most well-studied example is aquaporins, selective and highly permeable water transport proteins, and biomimetic aquaporin-embedded membranes have already been commercialized for water desalination. However, a major challenge is ensuring that the bio-membrane layer shows good stability, holds the maximum concentration of proteins for high permeability, and preserves correct protein folding. The design of such a layer may be unique for each protein according to each protein's binding characteristics.

E. coli AmtB ammonium transporter protein (EcAmtB) is a promising candidate for an ammonium-selective biomimetic membrane because its crystal structure and transport mechanism have been extensively characterized, and it has proven to be selective for ammonium outside of a living cell ($K_{d,Na+} / K_{d,NH4+}$ of roughly 1700). However, it has also recently been reported that EcAmtB does not transport ions well in the absence of POPG, a glycerol-terminated phospholipid found in *E. coli* native membranes. Reports have investigated multiple binding domains for both POPG and POPE, an ethanolamine-terminated phospholipid also native to *E. coli*. Both seem to greatly stabilize the protein and cause correct folding into an effective transporter. A difficulty of this reliance is that phospholipids are often considered to be less stable in synthetic membranes compared to synthetic copolymers. Hence, it is imperative that the membrane composition dependence of EcAmtB is extensively investigated before a final membrane is constructed.

To this end, solid supported membrane-based electrophysiology (SSME) is used to analyze ammonium permeability and selectivity in multiple proteoliposomes containing EcAmtB and various membrane compositions, including both phospholipids and synthetic copolymers. Preliminary results with proteoliposomes containing binary phospholipid compositions POPE/POPG, POPC/POPG, POPE/PA, and POPC/PA suggest that POPE may be an essential phospholipid for EcAmtB. The replacement of POPG with phosphatidic acid (PA) resulted in a marginal worsening in K_m, but the replacement of POPE with the choline-terminated POPC resulted in negligible ammonium transport at initial ammonium concentrations from 1 mM to 50 mM.

Ultimately, transporter protein-loaded biomimetic membranes can potentially aid in advancing a future vision of a circular nitrogen economy that extracts ammonium pollutants from wastewaters and offsets anthropogenic imbalances to the global nitrogen cycle.

169e Harnessing Green Nanosilver Supported on Reduced Graphene Oxide for Photocatalytic Degradation of Methylparaben

Sadaf Khan

Methylparaben (MeP) is a hazardous water pollutant found in the effluent from cosmetic industries. It has been reported to cause disruption in various organisms' endocrine or hormonal systems, and its extensive use has led to its increased discharge into water bodies, posing a significant threat to the environment and human health. Therefore, there is a pressing need to develop effective techniques to remove MeP from contaminated water sources.

Plasmonic nanoparticles have emerged as a promising pathway for the removal of pollutants from water treatment processes due to their ability to perform direct visible-light-driven photocatalysis. In this study, a hybrid catalyst of silver nanoparticles on reduced graphene oxide (rGO/AgNPs) was fabricated using a biobased in-situ reduction process and employed as a visible light photocatalyst to treat MeP. The use of green reducing agents during the synthesis process reduces the overall cost and environmental impact of the process. The synthesized catalyst demonstrated significantly enhanced adsorption and photocatalytic degradation of MeP than rGO or AgNPs, individually. The conditions that influence the kinetics of the photocatalytic degradation by rGO/AgNPs were comprehensively studied using response surface methodology (RSM), including solution pH, catalyst dose, persulfate concentration, MeP initial concentration, and time. The catalyst showed good stability and reusability and could induce up to 97.6% MeP degradation efficiency.

Furthermore, the mechanism of the photocatalytic reaction was investigated by scavenger tests and density functional theory (DFT) study. The results indicated that hydroxyl radicals (•OH) played a crucial role in the photocatalytic degradation of MeP. Additionally, the DFT calculations revealed that the adsorption and electron transfer mechanism of MeP on the rGO/AgNPs surface was more favorable than that on the individual components. This study provides a novel approach for the development of low-cost photocatalysts for cosmetic wastewater treatment utilizing visible light irradiation. The biobased synthesis of rGO/AgNPs and the insights into the photocatalytic mechanism could pave the way for the future development of eco-friendly and effective catalysts for the removal of MeP from contaminated water sources.

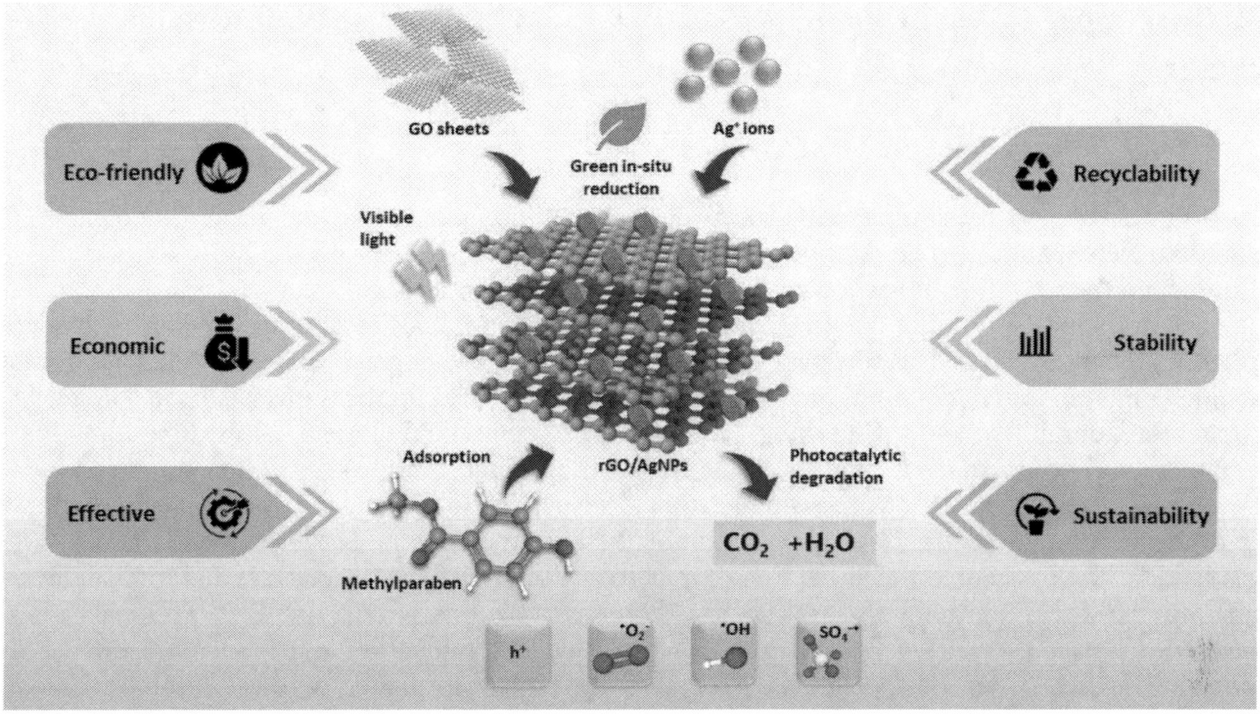

GO sheets
Ag+ ions
Green in-situ reduction
Eco-friendly
Recyclability
Visible light
Economic
Stability
Adsorption
rGO/AgNPs
Photocatalytic degradation
Effective
Sustainability
Methylparaben
CO2 +H2O
h+
•O2−
•OH
SO4•−

169f The Degradation of Phenol in the Proposed Mist Flow Photocatalytic Reactor

Shoma Kato, Yasuki Kansha

Industrial wastewaters contain phenolic compounds that are toxic in low concentrations and can persist in the environment. The removal of these compounds is necessary for the prevention of pollution and the recycling of wastewater. However, traditional biological treatment for the treatment of those compounds is difficult due to the toxicity of those compounds to microorganisms. Advanced oxidation processes that involve the generation of radicals have been proposed for the degradation of phenolic compounds. The use of photocatalysts for water treatment applications has been studied for their stability, durability, and low toxicity. However, photocatalytic treatment can be limited by light irradiation conditions and mass transfer. To address these limitations, the use of mist generated by ultrasound has been proposed. Mist is small droplets suspended in air, and it provides benefits such as increased surface area and shorter light penetration distance in photocatalytic treatment. Prior investigations have explored the utilization of mist in a reactor utilizing ultraviolet and ultrasound irradiation. However, the generated was saturated on top of the liquid bulk, and the additional generated mist was not utilized. In this study, a reactor that can transport the generated mist using air flow was proposed. After transporting the generated mist, it was irradiated with UV light for the promotion of photocatalytic reactions. Using the reactor, the extent of the degradation of the model pollutant phenol was investigated in the mist droplets. The results indicate that the irradiation time of the mist droplets was shorter than the previous reactors, but the degradation of the pollutants was improved. The results provide important insight into the improvement of the irradiation conditions of the photocatalyst for promoting photocatalytic reactions.

169g A Novel, Low Energy Approach to Nutrient Removal

Carlos Borras

Algal blooms remain a key environmental issue in the United States due to high levels of phosphorus from industrial runoff. In most cases, water treatment facilities are responsible for dumping large amount of phosphorus into lakes and waterways in their effluent streams. Phosphorus can also attach to soil particles, so even when disposed of as a solid, phosphorus will eventually enter a water source. Algal blooms are not only unsightly and have offensive odor, but can also develop neuro and hepatotoxins (a danger to humans and aquatic life) if left unchecked. Our patented technology is designed to remove phosphorus from water sources, reducing the overall phosphorus concentration to environmentally sustainable levels. This process utilizes aluminum-copper galvanic cells that generate a spontaneous reaction requiring no external power source. These cells are "powered" by the differences in the Fermi potentials of the selected electrodes. Phosphorus removal is achieved through one of two possible mechanisms. Either high concentrations of aluminum generated on the surface of a sacrificial aluminum electrode causes the precipitation of aluminum phosphates, or hydrolysis of aluminum ions to aluminum oxide/hydroxide generate a polymeric compound with a high phosphorus adsorption capacity. Phosphorus removal efficiency can be improved through the addition of hydrogen peroxide to improve the ion release efficiency. This process is easily scalable and has shown phosphorus removal levels of over 99% from water sources containing 1 – 6 ppm of phosphorus.

This process was evaluated and selected for water treatment in an industrial wastewater emergency in April of 2021. A former fertilizer facility on the west coast of Florida developed a significant leak in one of its lined water storage compartments. For 9 months, under the supervision of the Florida DEP, this process produced discharge water with environmentally acceptable amount of phosphorus and prevented any release of untreated water. By the end of June 2023, this process should be operating in one of the surface water canals around Lake Okeechobee that carry approximately 100 metric tons of phosphorus into Lake Okeechobee, with a maximum flow of up to 6,000 GPM.

169h Modeling Water Absorption of Hydrogel Coated Seeds

Tori Phillips, Jean-Francois Louf

Plant species that grow naturally in semi-arid environments have developed various survival strategies to capture water. One of them is to produce mucilage: a hydrogel seed coat that absorbs soil moisture and helps maintain hydration during drought events. Inspired by these mucilage-producing seeds, hydrogel coatings have been developed to improve drought resistance in other seed species. Some field tests showed encouraging results, such as increased growth by 16% in wheatgrass, 50% in cucumber, 77% in corn, and 100% in pea shrub. However, other tests showed conflicting results as the seed coatings reduced germination rates, especially with thin seed coats (hydrogel mass < 50% of seed mass). Unfortunately, the reason behind the dichotomy of these germination rates is unknown. To investigate the effect of a hydrogel seed coat on water transport, we designed artificial seeds made of clay and coated them with an alginate hydrogel. We then measured the associated imbibition rate and examined how the permeability variation in the flow direction in this spherical geometry affects the dynamics of water moving into the clay core. We found that the hydrogel coating did not delay the imbibition of the clay core and could be captured by the Lucas Washburn theory. We then successfully confronted our model with experiments conducted on coated soy seeds. Further, we tested the germination rate of our coated soy seeds and observed that (1) the seeds coated in the hydrogel had fewer yields, but (2) the ones that germinated had a more extensive root system, and (3) the hydrogel-coated seeds that did not germinate were beginning to decay, indicating an overly wet surrounding environment. Together, our results further help to elucidate the reasons behind the conflicting yields reported when coating seeds with hydrogel.

156a Evaluation of Hollow Fiber Membrane Modules for CO2 Direct Air Capture and Utilization in Low-Purity CO2 Processes

Vitor Renan Vitor Gama, Oishi Sanyal, Fernando Lima

The detrimental impact of greenhouse gas emissions on the climate has motivated the surge of carbon dioxide capture and utilization (CCU) strategies. In particular, Direct Air Capture (DAC) of CO_2 is being widely considered as an effective way to capture CO_2 directly from the atmosphere, addressing historical emissions. High throughput gas separation membranes with high CO_2/N_2 selectivities have an advantage of enabling continuous process operations and therefore must be considered as a standalone or in hybrid form for DAC[1].

This presentation will address a staged hollow fiber membrane process for concentrating atmospheric CO_2 (420 ppm) to a product stream with CO_2 concentration of <10%. Some applications of such low-purity CO_2 streams will be highlighted, including utilization in a photobioreactor. This study is conducted in Aspen Plus, where the membrane separation process scenarios are simulated, and considerations for energy demand and cost of each case are analyzed using the Process Operability framework[2,3].

The operability framework maps the feasible regions of operation for the scenarios proposed, taking into consideration the available input conditions while seeking to maintain the process within desired energy efficiency and net-negative carbon emission goals. As the use of gas separation membranes for DAC has not been widely explored yet, this work is expected to provide better understanding of the process bounds, regarding the techno-economic feasibility and the overall process being "carbon-negative".

[1] - Garry S.P. Taifan, Christos T. Maravelias. Generalized optimization-based synthesis of membrane systems for multicomponent gas mixture separation. Chem. Eng. Sc. 2022. https://doi.org/10.1016/j.ces.2022.117482.

[2] - Gazzaneo, V., Lima, F. V. Multilayer Operability Framework for Process Design, Intensification, and Modularization of Nonlinear Energy Systems. Ind. Eng. Chem. Res. 2019. https://doi.org/10.1021/acs.iecr.8b05482.

[3] - Gama, V., Dinh, S., Alves, V., Dantas, B., Bishop, B. A. and Lima, F. V. Modeling and Process Operability Analysis of a Direct Air Capture System. In Proceedings of the 13th IFAC Symposium on Dynamics and Control of Process Systems, including Biosystems (DYCOPS). 2022.

156b Advanced Structured Material for Direct Air Capture

Mustapha Soukri, Paban Sitaula, Daniel O'Nolan, Lindsey Chatterton

Advanced Structured Material for Direct Air Capture

Mustapha Soukri, Paban Sitaula, Lindsey Chatterton, and Daniel O'Nolan

Research Triangle Institute, 3040 E Cornwallis Rd, Durham, North Carolina, 27709, USA

ABSTRACT:

As fossil fuels remain a major source of energy throughout the world, developing efficient negative emission technologies, such as direct air capture (DAC), which remove carbon dioxide (CO_2) from the air, becomes critical for mitigating climate change.

DAC has been intensively studied in recent years, and a variety of adsorbent materials have been developed to directly capture CO_2 from the air. Among them, amine-based materials are attractive adsorbents for DAC and are widely studied due to their promising performance under humid conditions, lower amine volatilization rates compared to amine solvents, and the strong affinity of amine groups for CO_2. While the amine-containing solid adsorbents are tolerant to moisture in the air, it is generally known that physisorbents are not effective adsorbents for CO_2 capture in the presence of moisture due to competitive adsorption between CO_2 and H_2O, which yields high sorbent regeneration costs.

We have developed an amine-based sorbent for DAC *via* a reductive amination, cross-linking reaction of polyamine and polyaldehyde compounds. The sorbent adsorbed CO_2 from air up to 5.0 wt.% at 25°C, 75% relative humidity, over 100 cycles. The CO_2 adsorption capacities from simulated air were determined using an advanced gravimetric sorption analyzer and a packed-bed reactor (PBR) respectively. We have then integrated this sorbent to a novel contactor that is optimized for wind-driven operation. Creating an amine sorbent film for DAC on a structured substrate such as a Stainless-Steel plate surface has been developed.

We will present the development results of the sorbent performance under simulated DAC conditions. We will also highlight the effects of common contaminates in air as well as the sorbent kinetics. Furthermore, we will present the outcome of the coating effort of this material on the surface of Stainless-Steel plate including the choice of the binder resins, drying techniques, solvents for coating formulation, viscosity of the coating formulation, adhesion promoters, and surface pretreatments which all play an important role in creating a uniform and strongly bound film to the substrate.

156c Optimal Design and Operation of a Hydrogen Generation and Storage System for Utilization in the Peaker Plant

Md Emdadul Haque, Ijiwole Solomon Ijiyinka, Pavitra Senthamilselvan Sengalani, Zhili Feng, Debangsu Bhattacharyya

There are considerable efforts worldwide for reducing the use of fossil fuel for energy production. While renewable energy sources are being increasingly used, fossil fuel still contribute about 80% of the energy used worldwide. As a result, the level of CO_2 is still increasing fast in the atmosphere currently exceeding about 410 ppm. Various research efforts are ongoing to reduce CO_2 build up in the atmosphere. One approach is utilization of hydrogen as a fuel in power plants, and this can reduce CO_2 emission by 30-40% [1]. Hydrogen can be used as an energy storage medium producing it during the time when electricity is in excess while utilizing it later when needed. There are different options for hydrogen storage, among them storing it as a compressed gas is currently the cheapest and most practical option especially for distributed storage. [2]. However, capacity and maximum operating pressure of storage must be optimal for cost-efficient storage of hydrogen [3,4]. There are three leading approaches for utilization of stored H_2 for power generation- co-injection with natural gas in an existing natural gas combined cycle (NGCC) power plant, firing of H_2 by itself in a turbine, and use it in a fuel cell. Co-injection of H_2 with natural gas facilitates use of existing gas turbines in NGCC plants and peaker plants. Peaker plants are generally operated when the electricity is in high demand and therefore plays a key role in maintaining the resiliency and reliability of the electric grid. Currently no studies on optimal design and operation of hydrogen generation and storage facilities integrated with peaker plants could be found in the literature.

This research presents an integrated model of hydrogen production, storage and utilization for the peaker plants. The system includes electrolyzers for producing hydrogen, hydrogen compression and cooling system, hydrogen storage, hydrogen expansion and heating system and an aeroderivative gas turbine for power generation. Two types of electrolyzers, namely alkaline and proton exchange membrane (PEM) electrolyzers, are evaluated. To store hydrogen as a compressed gas, an optimum storage vessel is designed. A model of the vessel is developed by considering geometry of the vessel that takes into account the design thickness of the wall by considering the stress and corrosion allowance. A cost model of the hydrogen storage vessel is developed and compared with the in-house data. Optimal design and operation of the hydrogen storage system and the peaker plant are done by maximizing the net present value (NPV) of the integrated system. Due to the dynamics of the storage and generation systems, a dynamic optimization problem ensues that is solved using Python/PYOMO. The NPV optimization problem is solved for 14 regions by considering their respective locational marginal price (LMP) of electricity [5] with varying carbon taxes. A clustering algorithm [6] is used to reduce the number of representative days to be considered for optimization for a year-long optimization. Results show that the optimal size and design pressures of the hydrogen storage vessels and the design and operating conditions of the peaker plant differ not only from region to region based on the LMP profile, but also can differ for the same region due to the difference in carbon tax.

Bibliography

1. Nugroho, A.P.; Kenshi, I.; Atsushi, K. Impact of hydrogen fuel for CO_2 emission reduction in power generation sector in Japan. *Energy Procedia*, **2017**, 105, 3075 – 3082.
2. Bernad, C.T.; Somtochukwu, G.N. Hydrogen Production, Distribution, Storage and Power

Conversion in a Hydrogen Economy - A Technology Review. *Chemical Engineering Journal Advances*, **2021**, 8, 100172.
3. Office of Energy Efficiency & Renewable Energy. Hydrogen Storage. https://www.energy.gov/eere/fuelcells/hydrogen-storage (accessed on March 20, 2023).
4. Department of Energy. Onboard Type IV Compressed Hydrogen Storage System- Cost and Performance Status. DOE Hydrogen and Fuel Cells Program Record. https://www.hydrogen.energy.gov/pdfs/19008_onboard_storage_cost_performance_status.pdf (accessed on March 20, 2023)
5. Scott, I. J., Carvalho, P. M. S., Botterud, A. & Silva, C. A. Clustering representative days for power systems generation expansion planning: Capturing the effects of variable renewables and energy storage. *Energy*, **2019**, 253, 113603.
6. Sun, Y.; Wachche, S.; Mills, A.; Ma, O.; Meshek, M.; Buchanan, S.; Hicks, A.; Roberts, B. 2018 Renewable Energy Grid Integration Data Book. *NREL*, **2020**, Golden, CO.

156d Structural Optimization and Modeling of Carbon Capture through Adsorption

Mayra Gonzalez Ramirez, Tiras Lin, Thomas Roy, Thomas Moore, Du Nguyen, Pratanu Roy, Larry Biegler, Grigorios Panagakos

The CO_2 emissions associated with human economic activity are known to cause undesired changes to the earth's climate. An essential step in mitigating these emissions is their separation from flue gas mixtures from point sources, e.g., power and industry plants, or from air. Thus, Carbon Capture and Sequestration (CCS) is one of the most promising technologies developed to alleviate trends in global warming and detrimental effects of CO_2 on human ecosystems. Furthermore, industrial stakeholders and policy makers are particularly interested in developing these technologies on a larger scale to achieve sustainability goals and agendas.

In this work, we focus on developing novel computational tools and reactor designs to model and scale-up the adsorption process, mainly known for its effectiveness in treating dilute mixtures. More precisely, we target a process called Direct Air Capture (DAC) which uses adsorption for removing CO_2 directly from air. In greater detail, we will develop novel, optimal reactor designs for the Pressure Swing Adsorption process (PSA) and Temperature Swing Adsorption (TSA) in CO_2 capture through a powerful, versatile, and flexible mathematical framework called Topology Optimization. Moreover, we wish to leverage the detailed modelling and simulations to gain insights regarding the scale-up of the device.

In the past, some useful insights into optimizing and scaling-up microfluidic reactors for first-order kinetics coupled with mass and momentum transport have been developed [1]. However, these studies have not been applied to more complicated kinetics e.g. coupled reaction-diffusion for adsorption process. Another challenge in deriving optimal structures for adsorption is the inherent dynamic nature of the process and the not well understood mass transfer between solid material and fluid. To model the mass transfer, we get inspiration from work done for PSA and TSA process optimization using trust region methods [2]. In these models, the mass transfer has been treated as a linear driving force coupled to the other governing and constitutive equations.

To achieve these goals, we set the coupled partial differential equations (PDEs) problem in structural optimization format and expand the computational framework developed for optimizing the first order reactor, to include kinetics from the literature pertaining to state-of-the-art materials for DAC e.g. [3]. The pressure drop along the PSA or TSA device is the natural objective function to minimize. An iterative approach is applied, where we solve the coupled PDEs by the finite elements method and in each iteration the algorithm generates a computationally updated design. After the system converges, the potential scalability of the system is evaluated. The results from the topology optimization of catalytic reactors will be a steppingstone for large-scale carbon capture through adsorption.

References

[1] F. Okkels and H. Bruus, “Scaling behavior of optimally structured catalytic microfluidic reactors,” Physiscal Review E, 2007.

[2] J. Uebbing, L. Biegler, L. Rihko-Struckmann and S. Sager, “Optimization of pressure swing adsorption via a trust-region filter algorithm and equilibrium theory,” Computers and Chemical

Engineering, 2021.

[3] R. Hughes, G. Kotamreddy, A. Ostace, D. Bhattacharyya, R. L. Siegelman, S. T. Parker, S. A. Didas, J. R. Long, B. Omell and M. Matuszewski, "Isotherm, Kinetic, Process Modeling, and Techno-Economic Analysis of a Diamine-Appended Metal–Organic Framework for CO2 Capture Using Fixed Bed Contactors," Energy Fuels, vol. 35 , no. (7), p. 6040–6055. , 2021.

*This work was performed under the auspices of the U.S. Department of Energy by Lawrence Livermore National Laboratory under Contract DE-AC52-07NA27344.

156e Scale-up of CO_2 Capture from Powdered Sorbent to Sorbent-Washcoated Monolith for Direct Air Capture

Ashok Ummireddi, Soumitra Payra, James Akinjide, Joo-Youp Lee

A large portion of CO_2 emissions in ~5.2 gigatons (Gt) is released in relatively small quantities from distributed sources emitted each year in the U.S. Therefore, for such emissions, point source CO_2 capture is not feasible and direct air capture (DAC) is an indispensable part of a diversified portfolio of technologies to mitigate U.S. greenhouse gas emissions. Successful DAC technologies require the separation of high purity CO_2 (e.g., >95%) with high selectivity toward CO_2, low energy requirements, minimum chemical and thermal degradations, reliability, long lifetime, etc.

Adsorption-based CO_2 separation for DAC requires low pressure drop and cost-effective supports that are coated with or contain sorbent particles. In this study, a DAC system was scaled up from sorbent in a powdered form to sorbent-washcoated monolith structure. A performance comparison between the two different forms was made between 20 and 35 °C. Adsorption kinetic model plays a key role in interconnecting the two different systems. A CO_2 adsorption kinetic expression for our CO_2 sorbent particles was determined in a continuously flowing reactor. The mass transfer of CO_2 in the porous sorbent bed can be modeled by solving the convection-diffusion equation within the porous sorbent. Intraparticle diffusion of CO_2 inside the sorbent is modeled with adsorption kinetics. CO_2 adsorption isotherm data is obtained for 100-400 ppm CO_2 in ambient air at temperatures up to 35 °C using chemisorption. The adsorption kinetic model is described by coupling air flow with external and intraparticle diffusion. The phase of sorbent particles washcoated onto monolith surfaces was characterized for the morphology and pore data to link to the washcoated phase.

Our CFD model is combined with external mass transfer, CO_2 adsorption kinetics, and pore diffusion. This combined model will predict the performance of volumetric CO_2 productivity in terms of monolithic geometry (i.e., cell size, cross-sectional area, and length), temperature, inlet wind speeds and directions. The model predictions based on this scale-up principle will provide the key design information on overall CO_2 capture efficiency, volumetric productivity, and throughput in terms of dimensions of monoliths including pitch sizes.

156f Molecular and Process Modelling of Selected Novel Solvents for CO_2 Capture

Fareeha Shadab, Ismail I.I. Alkhatib, Lourdes F. Vega

Today the debate regarding climate change is serious as CO_2 levels are at record-breaking highs, with 40% of emissions coming from fossil fuels. Without significant action, CO_2 concentrations are projected to reach 570 ppmv by 2100, causing a rise in global surface temperatures and sea levels, highlighting the need for sustainable strategies to reduce emissions and achieve net-zero state. CO_2 capture *via* absorption using aqueous amine solvents remains the most widely used industrial process for this purpose. However, a significant technical barrier is the high energy demand for solvent regeneration such as the benchmark 30 *wt.*% aqueous monoethanolamine (MEA), increasing the cost of capture, nowadays capped at new US DOE limits of $30 per ton. To overcome this, it is necessary to search for novel amines with desirable properties that can improve the economics of the capture process. However, evaluating the performance of new amines is challenging due to limited experimental data necessary for detailed techno-economic evaluation, and the absence of a practical framework to determine the most suitable solvent.

Given the recent advancement in molecular modeling tools, molecular-based equations of state provide a suitable platform for the design of modelling screening frameworks. To better enable the implementation of alternative amines for CO_2 capture, a comprehensive understanding of their properties is essential. The main objective of this work is to develop a robust solvent screening tool using the soft-SAFT molecular-based equation of state (EoS) to obtain relevant thermodynamic properties for assessing novel solvents for CO_2 capture. A collection of 40 alternative novel amines were included in this work, belonging to six families: (1) aminoalcohols, (2) alkylethanolamines, (3) dialkylaminoalcohols, (4) cyclic amines, (5) diamines, and (6) multiamines. The soft-SAFT molecular models of these amines were developed to calculate the thermophysical properties of pure amines such as density, vapor pressure, heat capacity and the heat of vaporization. Soft-SAFT also allows the calculation of other thermodynamic properties such as viscosity and surface tension by coupling the equation with appropriate theories, such as free volume theory for viscosity and density gradient theory for surface tension. The capability of soft-SAFT to describe the phase behaviour is a determining factor for the optimization of absorption/desorption process for CO_2 capture. Alongside properties for the aqueous mixtures of amines inclusive of density, vapor-liquid equilibrium and viscosity, we calculated the CO_2 absorption capacities at relevant process conditions.

This work highlights the predictive power of soft-SAFT EoS in cases where no experimental data was available. The modeling was extended to calculate two key performance indicators, namely, cyclic capacity ($\Delta\alpha$) which is the amount of CO_2 effectively separated from the capture plant and the regeneration energy (Qregen) which is linked to the reboiler duty of the stripper column and is the energy required to desorb the CO_2 from the rich amine stream, as screening criteria to determine the most optimal solvent. In our search for the most efficient replacement amine to the golden standard MEA, both 2-Diethylaminoethanol (DEEA) and Diethylenetriamine (DETA) showed promising potential in terms of a high cyclic capacity ($\Delta\alpha$) and low regeneration energy (Q_{regen}), shown from **Figure-1**. Results presented here bolster the significance of using Soft-SAFT molecular equation of state as a practical tool in the search for novel amines for CO_2 capture.

For a holistic analysis of carbon capture, the impact of aqueous viscosity on the mass transfer coefficient and the reaction kinetics on the reaction rate of CO_2 solubility are two important parameters that can be further studied at the target process conditions of 313K for 30 wt% amine. Moreover, the integration of these thermodynamic models with detailed process modeling such as ASPEN HYSYS to examine the techno-economic feasibility can further aid the search for novel solvents by providing a strong base to calculate and predict relevant thermodynamic properties.

Figure-1 Caption: Scatter plot of the Regeneration energy (Q_{regen}) with cyclic capacity ($\Delta\alpha$) of 36 novel amines.

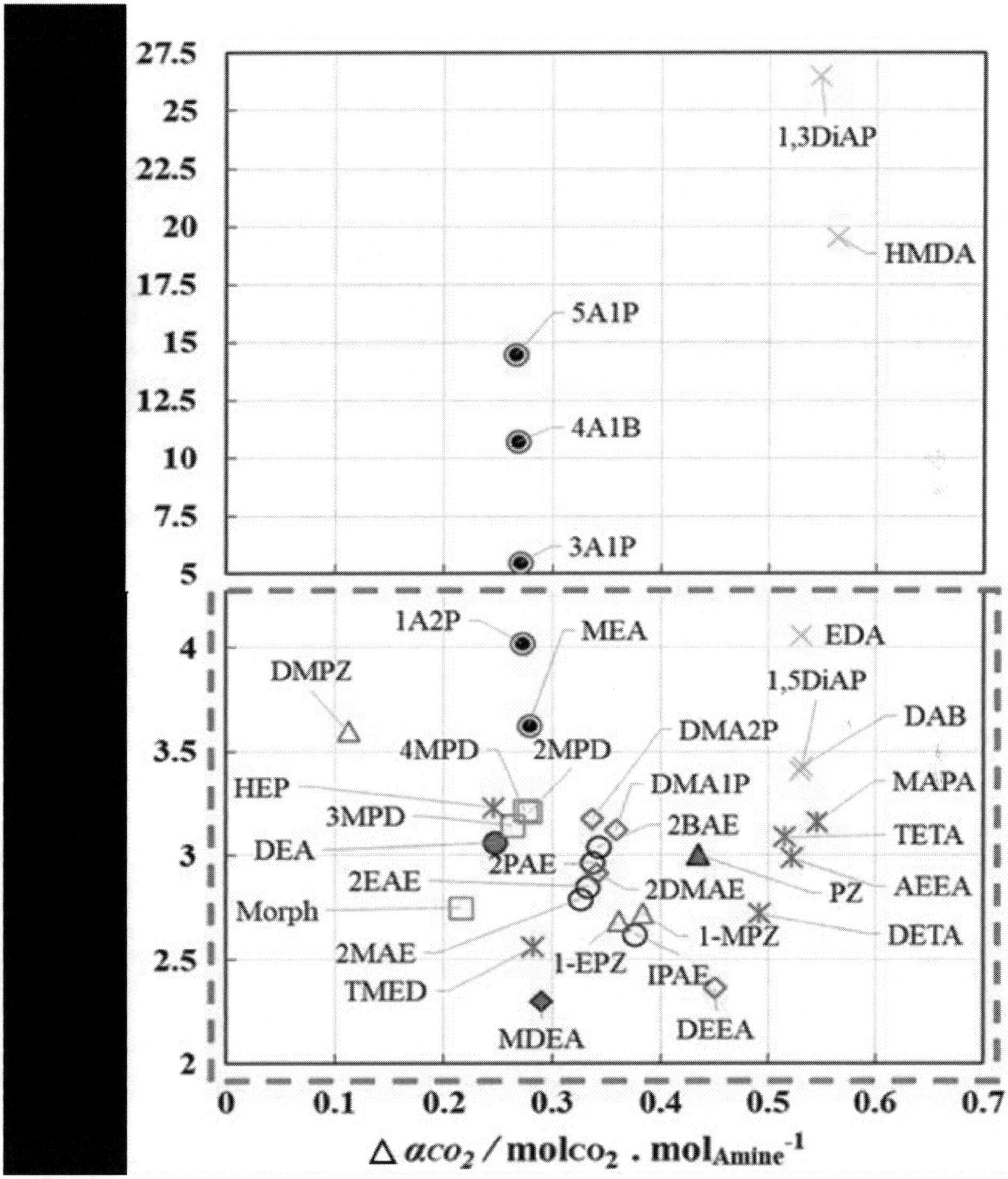

227a Application of Supercritical Carbon Dioxide to Polymer Recovery from Multilayer Plastic Packaging: Experimental Thermodynamic Phase Behaviors

Yoorae Noh, Yucheng Peng, Virginia Davis, Edward W. Davis, Bryan Beckingham

Multilayer plastic packaging possesses barrier properties to extend the storage stability of packaged content and is essential to our daily lives. However, millions of tonnes of polymeric material waste are generated and disposed of annually during manufacturing and/or after use, provoking the importance of packaging polymer reclamation and reuse. In order to address the growing packaging waste burden across the globe, developing novel strategies to reclaim valuable polymers without degrading material properties is critical. Supercritical carbon dioxide ($scCO_2$) utilized technology is one of the novel recovery processes for both post-industrial and post-consumer multilayer food packaging wastes. Unfortunately, there is still a lack of understanding of the systematic phase behavior of $scCO_2$ solvent interchangeable with commercial packaging polymer materials. Specifically, a solvent system with a mixture of $scCO_2$ and co-solvents should be developed to separate poly(ethylene co-vinyl alcohol)/polyamide from polyolefin-based multilayer plastic packaging. In this presentation, I will share the experimental thermodynamic phase behaviors of the solvent mixtures ($scCO_2$ + co-solvent) with food packaging polymer components. Our experimental design and results can be researched to optimize conditions for the polymer separation and recovery. Co-solvents are selected using Hansen solubility parameter calculations for efficient separation of poly(ethylene co-vinyl alcohol)/polyamides from polyolefins in $scCO_2$. By implementing a high-pressure variable-volume reactor system coupled with a $scCO_2$ injection system, the phase behavior of the solvent mixtures and demixing point of the targeted virgin polymers under the solvent mixtures are examined with various temperature and pressure conditions. The results will provide a fundamental understanding of the conditions (including temperature, pressure, and co-solvent composition) that lead to the separation and reclamation of the target polymer under $scCO_2$. Our proposed mechanism will provide a novel pathway to recover high-value polymer materials.

227b Process Advances and Kinetics Models for Hydrolytic Depolymerization of Post-Consumer Polyethylene Terephthalate (PET)

Patricia Pereira, Ashley Caiola, Angelo Testa, Kevin Reasons, Jianli Hu, Phillip E. Savage, Christian Pester

Chemical recycling of waste plastics provides a conceptual path to recover and reuse small molecules via depolymerization or decomposition of the plastics. Polyethylene terephthalate (PET) is used in large volume to make single-use packages for the consumer market (e.g., bottled water). PET is synthesized via condensation reactions, so it can be depolymerized via hydrolysis. This presentation explores advanced process concepts for the hydrothermal chemical recycling of PET waste into one of its monomers, terephthalic acid (TPA), by using microwave reactors and fast hydrolysis.

Microwave methods can reduce reaction times compared with conventional thermal heating processes. We compared TPA yields from microwave-assisted PET hydrolysis with conventional thermal heating with and without potential catalysts (zeolite HY, acetic acid, zinc iodide, and TPA). TPA yields with microwave hydrolysis (70%) with added TPA exceeded that from conventional thermal heating (41%) at 200 °C for 1 h. The TPA yield from hydrolysis with zeolite HY was negligible at 200 °C for 1 h with conventional thermal heating, but with microwave heating, it was 25%.

Fast hydrolysis involves rapid non-isothermal heating (e.g., 10 °C/sec) for a short time (e.g., 60 sec), which promotes plastic depolymerization while limiting byproduct formation. We optimized fast hydrolysis by measuring TPA and oligomer yields at different heating rates (3 to 12 °C/sec), reaction times (15 sec to 3 min), and PET/water mass ratios (1/4-1/10). TPA yields above 80% were obtained in less than one minute, with 100% PET conversion.

Finally, we developed a reaction network and kinetic model to describe the catalytic and non-catalytic hydrolysis of PET. The model can describe isothermal and fast hydrolysis and provides predictions for PET disappearance and formation of the various reaction products.

227c Trojan Horse Implants for Triggered Depolymerization of Polyethylene Terephthalate (PET)

Dhananjay Dileep, Jack Bonde, Aleksei Ananin, Michael J. Forrester, Lauren Burton, Matthew Laws, Shiva Karimidehkordi, George Kraus, Eric Cochran

Polyethylene terephthalate occupies the lion's share in the realm of plastics, with an annual production capacity exceeding 32MMTs. This quotidian presence can be attributed to its perfect mishmash of desirable thermomechanical, chemical, barrier, and optical properties due to its robust chemical structure comprising rigid hydrophobic terephthalic units spaced by aliphatic glycols. This structure provides thermomechanical ease in processing with excellent resistance to most solvolytic attacks, fortified by the terephthalic units. Chemical resilience manifests as recalcitrance; however, it has been observed that this initial barrier to depolymerization can be overcome by relatively inconspicuous tweaks to the polyester backbone, significantly alleviating the energetic demand for chemical recycling. The premise of the study revolves around mirroring molecular reactivity at macromolecular scales by introducing chemically reactive moieties nicknamed trojan horses as a comonomer in the chain. These moieties are then activated through different mechanisms to initiate depolymerization. Mass transfer barriers on the bulk and the molecular scales greatly diminish with molecular weight as the trojan horse snips the chains stochastically along the backbone resulting in oligomers. These oligomers have a high chain end concentration and are more susceptible to solvolytic scission at milder conditions resulting in a depolymerization cascade that culminates into monomers. The depolymerized monomers and trojan horse cocktail can be repolymerized to produce polymer sans deterioration in performance properties. Several trojan horse candidates have been considered, each initiating depolymerization uniquely. Some of the mechanisms explored in the study focus on accentuated solvolysis, neighboring group participation, and oxidative hydrolysis. The trojan horses can be recycled or valorized into commercial products. Alleviated energetic demands for chemical recycling are expected to result in a paradigm shift in the industry towards molecular recycling, making atom economy viable at realizable scales.

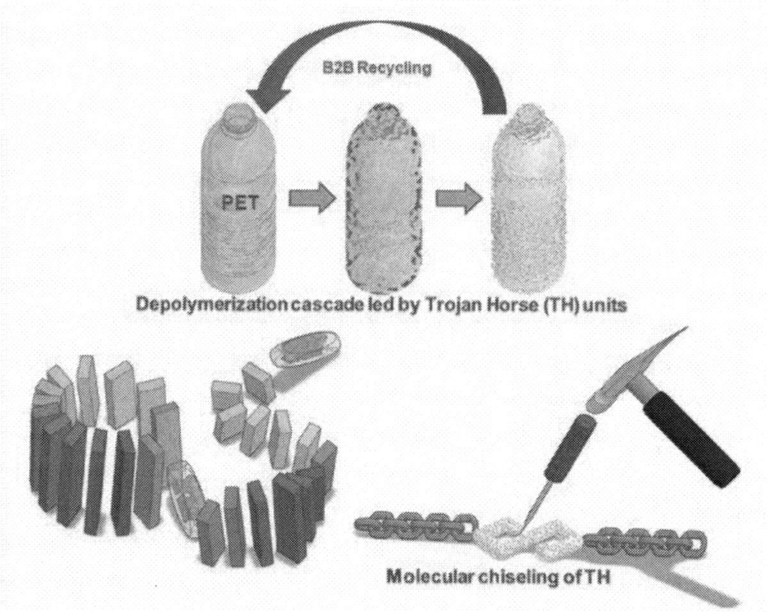

227d Technical Evaluation of Microbial Degradation of Polyethylene Terephthalate (PET) at a Laboratory Level.

Yesenia Castaño Castro, Camilo A. Suarez-Mendez, Carlos A. Jiménez-Junca, Sonia Ruiz

Currently, the global production of plastic is around 359 million tons annually, and it is estimated that around 200 million tons of plastic are distributed in water sources, landfills, and other natural environments [1]. Among the wide range of plastic materials available today, polyethylene terephthalate (PET) is one of those with increasing production in the last 20 years, and it is expected to quadruple in the next 50 years [2]. However, the systems to contain, control, and recycle it have not followed the same trend, mainly because it is estimated that 1 million bottles are produced every minute worldwide, of which only 41% are recycled [3]. There are different alternatives for PET degradation, such as chemical, thermal, and physical alternatives, but these are not alternatives that cover a high demand such as landfill material. Moreover, they may include harmful substances or high energy consumption, and finally, they may affect the physicochemical properties of the polymer [1]. Therefore, research on PET degradation through biological means has been developed worldwide. In Colombia, the Bioprocesses and Reactive Flows research group (BIOFRUN) at the National University of Colombia is conducting technical evaluation of microbial degradation of Polyethylene Terephthalate (PET) at a laboratory level, working with a filamentous fungus and a bacterium, based on the hypothesis that achieving higher biomass production and performing the respective hydrolytic enzyme induction will result in greater enzyme production, with long-chain fatty acids being recommended inducers [4]. In the research conducted by the BIOFRUN group, it has been found that at the reactor level, agitation, a pH around to 9, oxygen saturation (around 40%), and a temperature between 20-24°C favor the production of biomass in small pellets, which reduces transfer problems in the reactor. Additionally, it has been found that PET, as the sole source of carbon, induces the hydrolytic enzyme, with degradation products such as terephthalic acid, bis-(2-hydroxyethyl) terephthalate (BHET), and monohydroxyethyl terephthalate (MHET) quantified in parts per million (ppm), which is why experiments are currently being conducted on PET pretreatments that favor the degradation of PET into its basic monomers and short-chain oligomers.

[1] V. Tournier et al., "An engineered PET depolymerase to break down and recycle plastic bottles," Nature, vol. 580, pp. 219–219, 2020.

[2] I. Tiseo, "Global plastic production 1950-2020," 2022. [Online]. Available: https://www.statista.com/statistics/282732/global-production-of-plastics-since-1950/. [Accessed: 16-Jan-2022].

[3] S. Laville and M. Tayor, "A million bottles a minute: world's plastic binge as dangerous," The Guardian, London, 2017.

[4] D. Castro-Ochoa, C. Peña-Montes, and A. Farres, "Producción y caracaterísticas de cutinasas: Una alternativa interesante para biocatálisis a nivel industrial," Rev. Espec. en Ciencias Químico-Biológicas, vol. 1, pp. 16–25, 2010.

227e Efficient Polystyrene Depolymerization Enabled By Inorganic Nano-Sheet Catalyst

Hassam Mazhar, Mamdouh A. Al-Harthi

The applications of plastics are diverse and widespread, spanning from industrial to domestic utilities. However, global annual plastic production surpassed 760 billion pounds per year in 2019, leading to a rapid buildup of waste and putting immense pressure on waste management systems. The negative implications of plastics on the environment and their inadequate recyclability have resulted in anti-plastic sentiments. It is alarming that approximately 85% of produced plastic is dumped in landfills or incinerators, posing a significant threat to ecosystems due to the release of microplastics and hazardous materials.

The United States is actively promoting plastic recycling due to various reasons, including the implementation of China's National Sword policy, brand owner commitments, and anti-plastic sentiments.

Polystyrene, a widely used polymer, had global production of about 15.6 million metric tons in 2019. PS caters to a vast range of industrial and domestic applications. PS stands out as the only commodity polymer that exhibits the potential for depolymerization into its styrene monomer. Catalytic pyrolysis and solvent-assisted pyrolysis have also been employed to overcome some of the challenges associated with PS depolymerization. Other methods, like microwave-assisted pyrolysis, vacuum conditions, and high pressure, have also been investigated as strategies to achieve PS degradation. However, these methods carry their own set of challenges, such as lower % styrene yields, the formation of a large number of byproducts, the requirement for high temperatures to achieve pyrolysis, and longer reaction times.

In this work, we present catalytic pyrolysis of PS using an inorganic nanomaterial as a catalyst. The depolymerization reactions were carried out under a vacuum with s few weight percent of the recyclable catalyst. The abstract shows preliminary results, as shown figures. Detailed experimental results referring to catalyst characterization and the product analysis through NMR and GC-MS will be presented at the conference.

The salient features of the catalyst based on the preliminary results were as follows

1. High monomer product recovered
2. High percentage of styrene monomer in the product
3. Low coke formation during recycles
4. High reaction rate
5. Easy catalyst separation
6. Easy catalyst regeneration
7. Low-cost catalyst

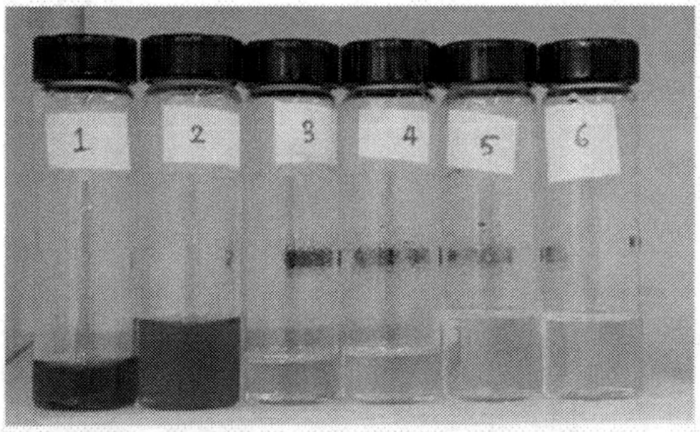

Figure 1 : Depolymerized PS product

Figure 2 : coke deposition after reaction

Sample No	Description of Figure-1
1	Thermal pyrolysis of PS (Without any catalyst) *low purity, low recovery, low % styrene*
2	Depolymerization of PS in the presence of material A (A= developed in our lab) *low purity, high recovery, low % styrene*
3	Depolymerization of PS from **literature** *High purity, low recovery, high % styrene*
4	Depolymerization of PS in the presence of **material A** + literature material *High purity, low recovery, high % styrene*
5	Depolymerization of PS in the presence of **material B** (B= developed in our lab) *High purity, high recovery, high % styrene*

227f Systematic Study of the Rejection and Fouling Behavior of Different Microplastics in the Presence of Natural Organic Matters.

Himangshu Mondal, Milad Esfahani

Micro & nano plastics (MNPs) (such as polystyrene, polyethylene, and polypropylene) are an emerging class of pollutants of this era. MNPs is generated from the abrasion of bigger particles or produced commercially and enter the human body through food and water. Microplastics are cytotoxic and cause unnatural cell death, mutations, weak immune response, cancer, etc. Therefore, it is crucial to remove microplastics from waste/drinking water. MNPs can be removed from water through different strategies such as adsorption, coagulation, and flocculation. However, they lack efficient removal. The membrane-based filtration is a promising technology for the removal of NMPs in the water treatment. However, the rejection and fouling behavior of microplastic is not comprehensively studied especially in the presence of other water components. In this study, we systematically investigate the removal efficiency and fouling behavior of polystyrene, polyethylene, and polypropylene microplastic particles using the ultrafiltration membranes. In addition, the rejection and fouling behavior of the mentioned NMPs will be evaluated in the presence of humic acid (as the model foulant of natural organic matter). The commercial Polyethersulfone membranes and UF membranes fabricated using the phase inversion method were used. We employed the crossflow filtration to determine the flux, fouling and rejection behavior of mentioned membranes. The UV-VIS spectroscopy method and fluorescence spectroscopy method was employed for determining the percentage removal of microplastics. The result shows that the polypropylene (20 ppm) solution did not result in fouling after five hours of filtration with 25 LMH/Bar permeance.

227g Calculated Physicochemical Properties of Glycerol-Derived Solvents to Drive Plastic Waste Recycling

Ademola Soyemi, Tibor Szilvasi

Plastics serve modern life in several ways, but plastic use comes with plastic waste that ends up in landfills or in the natural environment. In response, efforts are being intensified to increase plastic waste recycling on the regulatory and technological levels. A promising method is solvent-based recycling which uses a solvent to selectively dissolve plastic material which can then be recovered and reprocessed into upcycled products. Most known applications of solvent-based recycling disclosed in industrial patents and in academic literature make use common solvents which may pose potential health and environmental risks that can limit their applicability to industrial scale use. Therefore, there is a need to develop 'green' solvents for solvent-based plastic waste recycling. For this purpose, we propose the functionalization of glycerol, a natural product from biomass as well as a byproduct of biodiesel production, that can produce new and potentially less harmful solvents capable of dissolving plastic waste. In this work, we have developed a database of ~10,000 glycerol-based solvents and calculated their physical properties such as the dipole moment, density, melting point, LogS, LogP, Hansen Solubility Parameters (HSPs) using a combination of Density Functional Theory, machine learning and Hansen theory. We show that the physical properties can be tuned based on the attached functional groups, and based on the HSPs of the solvents and selected plastics in addition to the melting point of the solvents, we give recommendations for solvents that can carry out selective plastic dissolution for complex plastic waste feedstock

Reference: *Ind. Eng. Chem. Res.*, 2023. accepted DOI: 10.1021/acs.iecr.2c04567

227i Controlled depolymerization as a novel strategy for the chemical recycling of poly(bisphenol A carbonate) waste materials

Elangeni Ana Gilbert, Javier Fernando Guastavino, Diana Estenoz, Santiago Eduardo Vaillard

Introduction

The amount of waste derived from non-renewable sources is increasing annually becoming a serious environmental problem [1]. In particular, poly(bisphenol A carbonate) (PC-BPA) has a growing annual worldwide production that exceeds 6 million tons [2]. The controlled depolymerization of PC-BPA with polyhydroxylic-alcohols (polyols) allows not only the recovery of BPA, but also the generation of new compounds with carbonate functionality. The structure of the compounds obtained is determined by the depolymerizing agent used (Figure 1).

Objectives

The mail goal of this work is to study the controlled depolymerization of PC-BPA using a poly-alcohol derived from renewable sources as depolymerizing agent for the production of BPA and cyclic carbonates.

Materials and Methods

Pieces of PC-BPA waste were reacted at 90°C for 2 h in a 15 mL round-bottomed flask equipped with a magnetic stir bar, using an organic solvent with a renewable poly-alcohol and an organic base as catalyst. The volatile compounds were recovered by distillation under reduced pressure. Both polymeric precursors (BPA and cyclic carbonate) were obtained as white powders and characterized by 1H-NMR.

Results and Discussion

The depolymerization of PC-BPA waste was studied using a poly-alcohol derived from renewable sources as the depolymerizing agent. The reaction was carried out using different organic solvents and varying the reagents molar ratio in order to maximize the yield of the cyclocarbonate compound. The reaction products were isolated and structurally characterized by spectroscopic techniques (Figure 2). The process allowed the recovery of BPA, and the polymeric precursor with cyclocarbonate functionality (70%) could be used to obtain new polymers, such as others PC and polyhurethanes. In this way, the recovery of BPA would prevent its disposal in landfills and, therefore, its arrival to aqueous sources.

Conclusions

The controlled depolymerization of PC-BPA waste was carried out successfully using a polyhydroxylic-alcohol derived from renewable sources as the depolymerizing agent. Moreover, BPA and cyclic carbonates were obtained with excellent yield avoiding the use of phosgene and its derivatives. The reaction involves only moderate temperature and a single purification stage. In addition, the depolymerization can be carried out by varying the depolymerizing agents in order to obtain different tailor-made polymeric precursors with cyclocarbonate functionality for the synthesis of new materials.

References

1. Coates, G.W., & Getzler, Y.D. (2020). Nat. Rev. Mat., 5(7), 501-516.

2. Do, T., Baral, E.R., & Kim, J.G. (2018). Polymer, 143, 106-114.

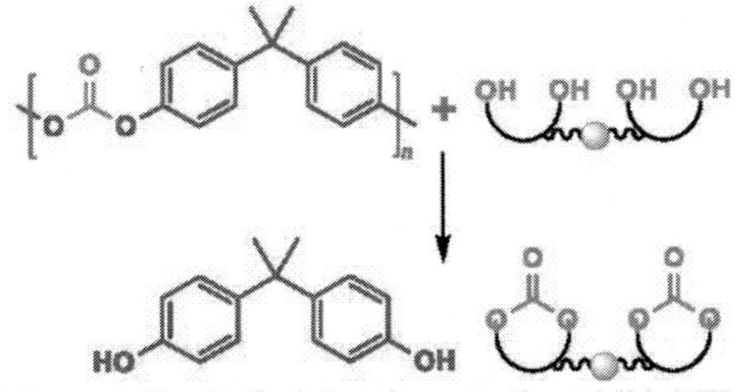

Figure 1. Controlled depolymerization of PC-BPA.

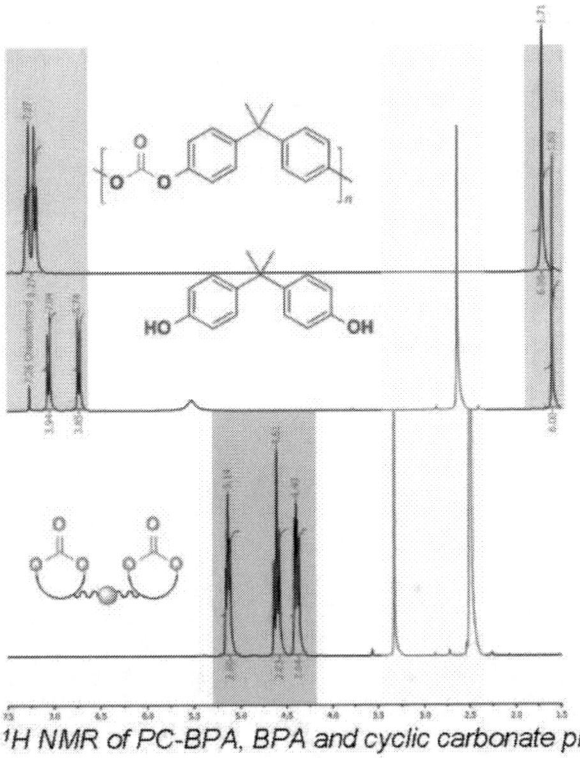

Figure 2. ^{1}H NMR of PC-BPA, BPA and cyclic carbonate precursor.

176a Techno-Economic Assessment of Food Waste to Biofuel By Integrated Hydrothermal Liquefaction with Microbial Electrolysis Cells

Shuyun Li, Lesley J. Snowden-Swan, Juan A. Lopez-Ruiz, Jinyue J. Jiang, Zhiyong J. Ren

According to the United States Department of Agriculture (USDA), over 30% of the U.S. food supply is wasted or lost. A significant amount of this food waste ends up in landfills, contributing to greenhouse gas emissions and other environmental issues. Therefore, it is appealing to convert food waste into green energy in an environmentally friendly and efficient manner. Hydrothermal liquefaction (HTL) combined with microbial electrolysis cells (MEC) can effectively convert food waste into biocrude oil while treating the HTL aqueous phase to produce hydrogen for biocrude upgrading. In this study, a high-fidelity HTL/MEC process model was developed based on experimental data. A comparative techno-economic assessment (TEA) was then conducted to compare the performance of the HTL process with the traditional anaerobic digestion (AD) process for food waste at different plant scenarios. System-level analyses show that MEC can provide the necessary hydrogen for biocrude upgrading with 90% chemical oxygen demand (COD) removal efficiency. Additionally, the treatment cost of the HTL aqueous phase by MEC is comparable to the direct sewer discharge fee for high COD and NH3. The minimum fuel selling price (MFSP) for a 500 dry ton per day HTL/MEC process is $4.3/gasoline gallon equivalent (GGE). When compared to AD pathways in terms of levelized cost of disposal, the results show that AD pathways have a cost advantage at the community scale due to lower capital expenses. However, the integrated HTL/MEC process is cost-competitive with AD above 50 TPD scale.

176c Evaluation of Fuel Properties of Hydrothermal Carbonization-Derived Shrimp Shell Hydrochar

Swarna Saha, Md Tahmid Islam, Josh Calhoun, Toufiq Reza

Shrimp, known as the most consumed seafood, generates a significant quantity of waste which is mostly accounted from its own weight. In order to transform shrimp shell waste into value added product in a sustainable and environment friendly way, hydrothermal carbonization (HTC) process was employed to produce energy-dense hydrochar which can be used as an alternative to fossil fuel. This study focuses on evaluating the fuel, combustion properties, and physical characteristics of shrimp shell hydrochar produced under a wide range of HTC temperature (110°C-290°C) conditions. Although mass yield declines with increasing HTC temperature, higher heating value (HHV) (26.9–27.9 KJ/kg) and energy densification ratio (1.09-1.13) demonstrated a rising tendency. Thermogravimetric analysis demonstrates that increasing the temperature yields more thermally stable hydrochar than raw shrimp shell and CHNS analysis results confirm the presence of 40-43% carbon in the hydrochars. Combustion characteristics evaluated from thermogravimetric analysis reports high comprehensive combustibility index (Si) (3.39× 10-7) and lower maximum weight loss temperature (Tm) (370°C) which ensures vigorous combustion reactivity of the hydrochar prepared at 180°C. Whereas lower burnout index (Db) (0.43× 10-3 %/min3) for the hydrochar at 290°C makes it suitable for ideal ignition and burning. Analysis of results demonstrate hydrothermal carbonization to be a sustainable method of converting shrimp shell waste into potential fuel substitute.

176d Assessment of Combustion Residual Leachate: Treatment Standards, Wastewater Volumes and Critical Mineral Recovery

Chad Able, Alison Fritz, Eric Grol

Combustion residuals and the resulting leachate from storage sites represent a large and variable volume of wastewater in the United States (U.S.). Constituents in this wastewater are regulated at the federal and local levels – federal emissions standards include total suspended solids (TSS), oil/grease, arsenic, and mercury for new and existing sources of leachate, whereas local emissions standards can include pH and a selection of 11 additional constituents, including copper, iron, selenium, mercury, and fluoride. Heterogenous regional standards and variable leachate volumes make quantifying treatment needs for leachate difficult. To address this issue, this work aggregates regional standards in the U.S. and assesses the treatment technologies necessary to meet these standards. Next, using data from the Environmental Protection Agency (EPA) and the Electric Power Research Institute (EPRI), the combustion residual leachate volume from landfills and impoundments is calculated on a national level. The annual wastewater discharge volume of a single landfill with combustion residual leachate can exceed 200 million gallons. In the U.S., landfill leachate is found to represent an annual volume of 7.1–11.3 billion gallons. Closing surface impoundments is a second source of leachate discharge that is determined to yield between 220 and 280 billion gallons of leachate in the U.S. over a three-year closure period.

While treatment of these large volumes of leachate will be costly, recovery of critical minerals offers an opportunity to offset treatment costs and improve critical mineral supply reliability. This work presents critical mineral recovery potential both as the available mass of a given mineral and the expected market value for 17 critical minerals available from EPRI's Combustion Product Information (CPInfo) database. Of the available minerals, magnesium has by far the highest concentration, reaching up to 2,000 ppm in landfill leachate and 1,300 ppm in impoundment leachate. Magnesium recovery from water is an industrially established process that can be readily implemented. Applying average magnesium concentrations and leachate volumes, if magnesium was recovered from all landfills it would represent a $39.2 million market– from all impoundments, this would represent a $650 million market. This supply would account for 9.1% and 151%, respectively, of 2021 U.S. magnesium consumption.

SULFONATED COCONUT SHELL BIOCHAR AS AN ACID CATALYST FOR THE HYDROLYSIS OF WASTE CELLULOSIC BIOMASS

Ogechukwu J. Igboke,[1] Temitope Orimolade,[2] Gavin H. Prevatt,[2] Oludare J. Odejobi,[1,] and Sitaraman Krishnan[2,*] [1]Obafemi Awolowo University, Ile-Ife, Nigeria, [2]Clarkson University, Potsdam, NY, United States*

ABSTRACT

The valorization of coconut shell waste by conversion to acid catalyst particles can circumvent the need for its disposal by CO_2-generating combustion. In this paper, the physicochemical characteristics of sulfonic acid functionalized biochar, derived by the carbonization of coconut shell, are discussed and correlated to its catalytic activity. Coconut shell biochar was obtained by carbonization at a temperature in the range of 400–700 °C and sulfonated by treatment with concentrated sulfuric acid. The BET surface area of the catalyst decreased more than a hundredfold to ≅1–4 m^2/g after sulfonation. The total acid density, determined by the Boehm titration method, increased with the carbonization temperature. The sulfonic acid was also separately quantified. Since a relative excess of concentrated sulfuric acid (5 to 50 mL/g-biochar) was used in the sulfonation step, the sulfonic acid density of the biochar would be expected to be independent of the sulfuric acid amount. However, interestingly, it decreased with the amount of sulfuric acid. The biochar solid acid catalyst was used to hydrolyze corncob, and the reducing sugar concentration in the hydrolysate was measured. The sugar yield increased with the sulfonic acid density of the catalyst particles.

INTRODUCTION

The conversion of biochar to biocatalyst has become of significant interest due to applications in catalysis and as adsorbents and functional materials. Biochar is preferred over other carbon materials, including carbon black and coal coke, in synthesizing catalysts due to its distinct developed porous structure and prominent surface area with the distribution of numerous trace elements and functional groups [1, 2]. It can be obtained from natural renewable feedstocks, for example coal, wood, apricot stones, hazelnut shells, almond shells, pistachio shells, walnut hard shells, coconut shells, as well as various industrial and agricultural waste products [1-6].

Amongst numerous sources of raw material for biochar preparation, coconut shell biochar has gained much interest as activated carbon for functional materials, adsorbents, and catalysts. Abundant carbon and minimal ash content contribute to the efficacy of using coconut shell biochar for biocatalyst development [7]. It has other desirable properties, such as an inherent coarse-grained structure, hardness, and ready availability [6]. Coconut farming, a significant source of agricultural waste in Nigeria, generates approximately 11 million tons of coconut shells annually [1, 2].

Carbonization or slow pyrolysis is the standard method employed during the thermochemical conversion of coconut shells to biochar. The biochar carbonization temperature is a critical parameter due to its direct effect on the pH, fixed carbon and ash content, and surface functionalities [1]. Literature reports that the optimal carbonization temperature for agricultural and forestry residues is 500–900 °C [1, 2]. Furthermore, activators such as alkali metal hydroxides and carbonates, zinc chloride, and phosphoric acid produce porous high-surface-area biochar particles [2, 3, 5]. Although chemical activation may impart better functional properties, it could be associated with the generation of excessive chemical waste and operational safety concerns. For these reasons, it is essential to develop techniques to optimize biochar properties through physical means, such as finding the optimal carbonization temperature, which varies depending on the biomass source.

Sulfonated coconut shell biochar is promising as a biocatalyst for valorizing agricultural cellulosic waste by breaking β-1,4-glycosidic bonds yielding sugars [2, 4, 8]. Many of the properties of biochar, including the specific surface area of the particles, the porosity, and surface acid density, are sensitive to the process conditions used for their production, and these properties influence their catalytic performance. In this study, coconut shell was carbonized between 400 and 700 °C to determine the optimum temperature for biochar synthesis without chemical activation. The biochar particles were reacted with concentrated sulfuric acid to functionalize them with acidic groups with catalytic activity. The sulfonated biochars were characterized in terms of their chemical composition, surface area, crystallinity, and surface acid density, and some of these properties were correlated to their hydrolytic activity on corncob.

METHODS

Preparation of Raw Coconut Shell and Corncob

Raw coconut shells and corncob were collected from the Obollo-Afor market in the Southeastern Part of Nigeria. The coconut shells and corncobs were washed with tap water and sundried. Then, the dried samples were ground via mechanical comminution, and the resulting powders were sifted with 10-mesh and 100-mesh sieves to obtain approximately ≤2 mm and ≤0.149 mm of coarse coconut shell and corncob powders, respectively.

Carbonization of Coconut Shell

Moisture and other volatiles in the coconut shells were removed by placing 500 g of the cleaned and pre-dried sample inside a muffle furnace maintained at 150 °C for 12 h. 50 g of the coconut shell samples were heated at different final carbonization temperatures ranging from 400 to 700 °C in a non-oxidative environment, as discussed in Igboke et al. [8], for approximately 1 h. Then, they were allowed to cool completely in the furnace before unloading. The resulting biochars were labeled BC*x*, where *x* (= 400, 450, 500, 550, 600, 650, or 700) indicates the carbonization temperature in degrees Celsius.

Sulfonation of Biochar

The biochar was ground again to reduce the size to approximately ≤0.149 mm powder. Five grams of the biochar powder was mixed with concentrated sulfuric acid (98 %, 50 mL) in a 250 mL conical flask. The mixture was stirred for 20 min at 150 rpm to homogenize before transferring to a ceramic crucible. The crucible was heated in a muffle furnace at 100 °C for 8 h. After heating, the mixtures were allowed to cool and washed with warm distilled water (temperature ≤ 80 °C) until no sulfate was detected in the extract. The sulfonated biochar was dried at 105 °C for 8 h and stored in an air-tight container for analysis. The sulfonated biochars were labeled SBC*x*.

Evaluation of the Sulfonated Biochar Performance by Hydrolysis of Corncob

The prepared corncob powder was hydrolyzed by mixing 1.0 g of biomass with 10 mL water and 1.0 g of sulfonated biochar catalyst. This mixture was stirred for 20 min at 150 rpm to homogenize and autoclaved at 121 °C for 1 h. Then, the mixture was allowed to cool and filtered. The filtrate was analyzed for the concentration of reducing sugars (in glucose equivalents) using UV-vis spectroscopy [8]. The procedure was repeated for all the prepared sulfonated biochar to determine the reducing sugar concentration in each hydrolysate.

Effect of Sulfuric Acid to Biochar Ratio on the Acid Density of the Sulfonated Biochar

Four grams of the coconut shell biochar powder that produced the highest glucose yield in the preceding experiment was mixed with concentrated sulfuric acid in a 250 mL conical flask in concentrations ranging from 5 to 50 mL of concentrated H_2SO_4 per gram of biochar to determine the optimum reactant amounts for the sulfonation process. Each mixture was transferred into a ceramic crucible and heated in a muffle furnace at 100 °C for 8 h. The mixture was allowed to cool and washed with warm distilled water until no sulfate was detected in the extract. The sulfonated biochar was dried at 105 °C for 8 h and stored in an air-tight container.

Characterization of Biochar and Sulfonated Biochar

The sulfonic acid and total acid densities of the sulfonated biochar catalysts were determined by an acid-base titration method and the Boehm titration, respectively [8]. Thermogravimetric analysis (TGA), FTIR spectroscopy, X-ray diffraction (XRD) analysis, scanning electron microscopy (SEM), energy-dispersive X-ray spectroscopy (EDX), and BET adsorption analysis were performed as described in Igboke et al. [8].

RESULTS AND DISCUSSION

Coconut Shell Biochar

Figure 1 shows that the biochar yield, defined as the fraction of the biomass remaining as a solid biochar, decreased with an increase in the carbonization temperature. The reduction in the yield of the biochar with increased temperature is consistent with previous reports [9-11]. The residual mass and biochar yield are higher at lower carbonization temperatures due to the accumulation of degradation products on the carbonaceous structure [12]. These degradation products volatilize at higher temperatures.

The carbonization temperature is also expected to affect the uptake and diffusivity of the sulfonating agent within the biochar particles. Biochars produced at higher temperatures have been reported to exhibit lower permeability of the sulfonating agent [12]. Thus, the catalytic activity of the sulfonated biochar is expected to show a non-monotonic dependence on the carbonization temperature [2, 13], and the carbonization temperature corresponding to the peak catalytic activity depends on the source and composition of the starting biomass [14]. In the present study, the optimum carbonization temperature for the sulfonated biochar production from coconut shells was found by measuring the reducing sugar yield in corncob hydrolysis using the solid acid catalyst.

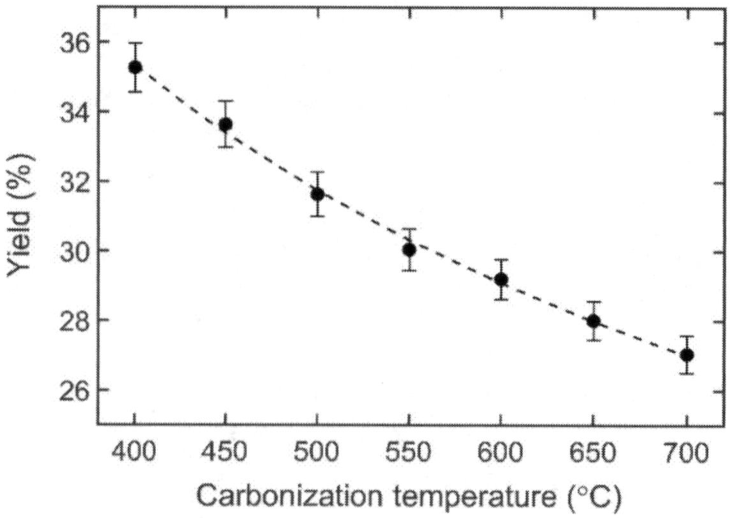

Figure 1: Effect of carbonization temperature on biochar yield

Effect of Carbonization Temperature on Reducing Sugar Production

Figure 2 shows the effect of carbonization temperature on the reducing sugar yield of the sulfonated biochar catalysts in the hydrolysis of corncob. The yield decreased with an increase in the carbonization temperature in the range of 400 to 500 °C, and increased after that up to 650 °C. A correlation between the reducing sugar yield and the sulfonic acid density of the catalyst particles was observed (Figure 2). The reducing sugar yield was proportional to the sulfonic acid density on the particle surface.

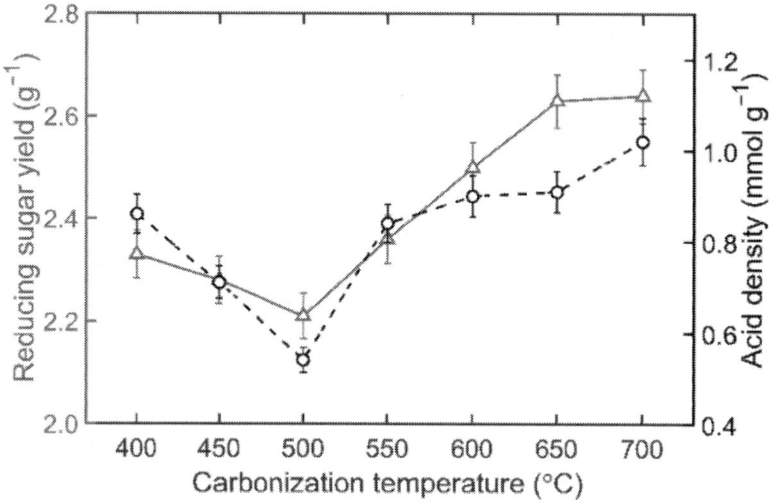

Figure 2: Effect of carbonization temperature on the reducing sugar yield and surface acid density of the sulfonated biochars prepared using different carbonization temperatures.

Characteristics of Biochar and Sulfonated Biochar Catalysts

All the biochar particles, particularly those prepared at lower temperatures, showed infrared spectroscopy bands in the 1600–1580 cm^{-1} range, corresponding to aromatic C=C stretching vibrations [8]. The relative intensity of this band decreased with an increase in the carbonization temperature, consistent with the fact that defect-free graphene is not expected to show a resonant infrared absorption frequency [15]. The sulfonated biochars also showed prominent bands near 1024 cm^{-1}, attributed to asymmetric S=O stretching vibrations, and 1151 cm^{-1}, corresponding to symmetric O=S=O stretching vibrations.

EDX spectroscopy showed an increase in the sulfur to carbon molar ratio from about 0.88±0.25% in the sulfonated biochar, SBC400, to about 1.96±0.34% in SBC700. In terms of wt%, the average S concentration was 1.73±0.52% in SBC400 and 3.98±0.77% in SBC700. da Luz Corrêa et al. [16] reported a sulfur concentration of 3.14 wt% in sulfonated biochars prepared using murumuru kernel shell, comparable to that in SBC700. (The wt% concentrations can be converted to mmol S per gram of biochar using the multiplicative factor of 10/32.07.) The resulting molar values were 0.53±0.16 mmol/g in SBC400 and 1.24±0.24 mmol/g in SBC700. These values, determined by EDX, are in reasonable agreement with the $-SO_3H$ densities of 0.86 mmol/g and 1.02 mmol/g found by titration of the acidic groups with a base (see Figure 2). The biochar's oxygen content increased following sulfonation (from 6.38±1.68 wt% in BC700 to 18.1±2.4 wt% in SBC700), indicating that sulfonation with strong sulfuric acid promotes the production of oxygenated groups as well. The S/O molar ratio increased from 3.5±1.4% in SBC400 to 10.2±3.0% in SBC700

The acid density of the sulfonated biochars varied between 0.54 mmol/g for SBC500 to 1.02 mmol/g for SBC700 (Figure 2). These results are in agreement with the observation that the $-SO_3H$ concentration in carbon-based sulfonated catalysts generally range between 0.1 and 4.9 mmol/g [17]. The highest concentration of sulfonic acid groups was obtained for the coconut shell carbonized at 700 °C. A similar result was found by Kabayo et al. [12], when coffee husks carbonized at different temperatures were sulfonated with sulfuric acid. The lowest acid density associated with SBC500 could result from the accumulation of nonvolatile degradation products in the structure of the carbonaceous material that inhibits complete sulfonation.

Powder X-ray diffraction analysis of the biochars using an X-ray wavelength of 0.154 nm showed the characteristic diffraction peaks at 22.8° and 44.4°, indicating an amorphous, disordered carbon with a relatively low degree of graphitization and the existence of disorderly multi-layered carbon sheets [8, 18, 19] in all the samples analyzed. The specific surface area of the biochar samples, determined using nitrogen adsorption measurements, increased significantly with an increase in the carbonization temperature. It was about 1.3 m^2/g in BC400 and about 430 m^2/g in BC700. However, the specific surface areas were 1.5–3.5 m^2/g after sulfonation.

Figure 3 shows the SEM images of BC600 and SBC600 at different magnifications. The SEM images of BC700, and the best-performing sulfonated biochar catalyst, SBC700, are shown in Figure 4. Uneven heterogeneous particles with porosity in some areas of the particle surface are seen. Pore formation is generally due to the decomposition and volatilization of the biomass constituents [20]. However, these pores can get partially filled due to partial oxidation and condensation reactions mediated by the sulfonating agent [21]. The particles in SBC700 exhibited flakes with a non-porous surface, possibly due to the ash formed during carbonization.

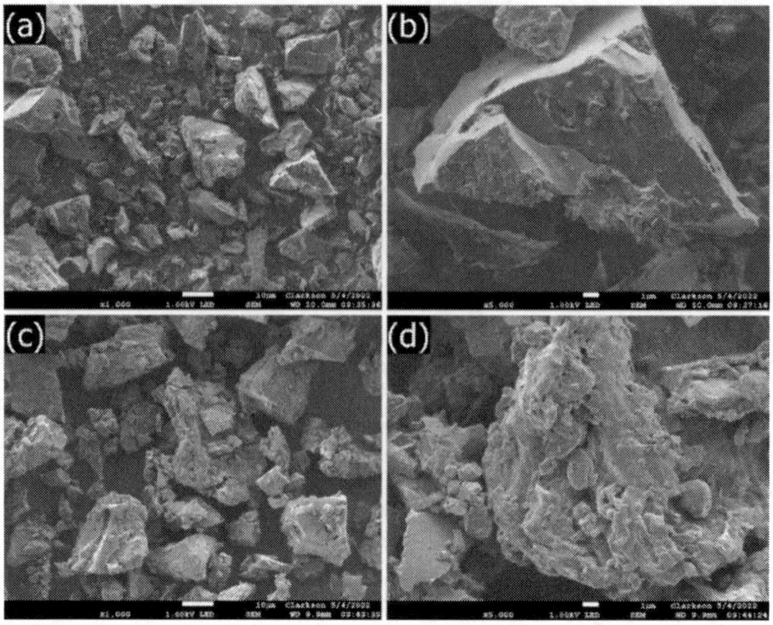

Figure 3: SEM images of BC600 and SBC600. The scale bars represent 10 μm in (a) and (c) and 1 μm in (b) and (d), respectively.

Figure 4: SEM images of BC700 and SBC700. The scale bars represent 10 μm in (a) and (c) and 1 μm in (b) and (d), respectively.

Effects of Biochar to Sulfuric Acid Ratio on Acid Density of Sulfonated Biochar

The volume of concentrated sulfuric acid per unit mass of the BC700 biochar was varied in the sulfonation reaction, and its effect on the surface acid density of the sulfonated biochar was studied. Figure 5 shows the results of the total surface acid density of the sulfonated biochar. The acid density decreased with an increase in the amount of acid used during the reaction. An excessive sulfuric acid likely alters the pore structure at the outer surface of the catalyst particles, making it difficult for further penetration of the acid inside the particle to sulfonate the inner pore surface. Sulfuric acid in amounts ranging from 10 to 50 mL per gram of biochar has been used in the literature [7, 21-23]. However, our results show that as low as 5 mL/g is sufficient to get a relatively high acid density on the particle surface.

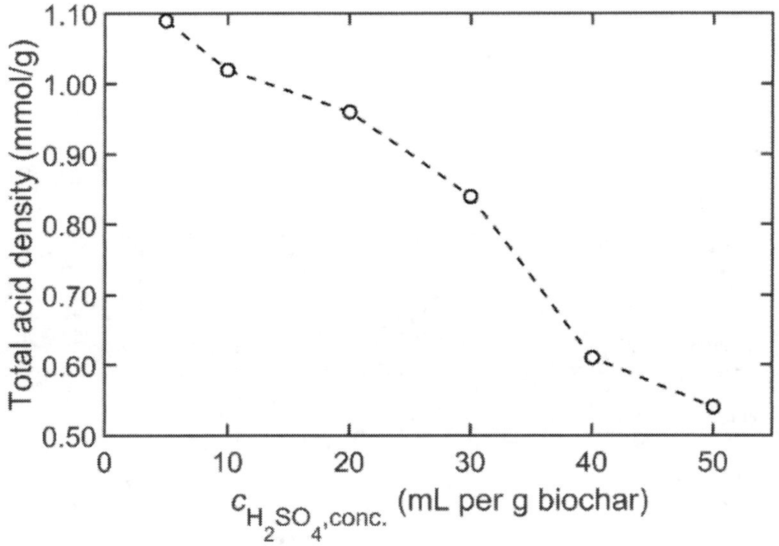

Figure 5: The effect of the amount of sulfuric acid used in the sulfonation step and the acid density of the sulfonated biochar particles.

CONCLUSION

This study found the optimum carbonization temperature and sulfuric acid concentration for preparing solid acid catalysts from coconut shell biomass. Detailed physicochemical characterization and catalytic activity (measured in terms of hydrolysis of corncob to reducing sugars) of the biochar particles are reported. An excess of concentrated sulfuric acid, commonly used in the literature for the sulfonation of biochar, is not necessary. In fact, it is detrimental to the surface acid density of the catalyst particles. There is a significant prospect in producing carbon-based solid acid catalysts from coconut shells, a highly abundant natural waste.

REFERENCES

1. Manya, J.J., *Pyrolysis for Biochar Purposes: A Review to Establish Current Knowledge Gaps and Research Needs.* Environmental Science & Technology, 2012. **46**(15): p. 7939-54.
2. Cheng, F. and X. Li, *Preparation and Application of Biochar-Based Catalysts for Biofuel Production.* Catalysts, 2018. **8**(9).
3. Das, D., D.P. Samal, and M. BC, *Preparation of Activated Carbon from Green Coconut Shell and Its Characterization.* Journal of Chemical Engineering & Process Technology, 2015. **06**(05).
4. Fraga, A.d.C., et al., *Biomass Derived Solid Acids as Effective Hydrolysis Catalysts.* Journal of Molecular Catalysis A: Chemical, 2016. **422**: p. 248-257.
5. Sanni, E.S., et al., *Determination of Optimum Conditions for the Production of Activated Carbon Derived from Separate Varieties of Coconut Shells.* International Journal of Chemical Engineering, 2017. **2017**: p. 1-16.
6. Li, B., et al., *Aqueous Phase Reforming of Biocrude Derived from Lignocellulose Hydrothermal Liquefaction: Conditions Optimization and Mechanism Study.* Renewable Energy, 2021. **175**: p. 98-107.

7. Endut, A., et al., *Optimization of Biodiesel Production by Solid Acid Catalyst Derived from Coconut Shell Via Response Surface Methodology.* International Biodeterioration & Biodegradation, 2017. **124**: p. 250-257.
8. Igboke, O.J., et al., *Composition and Morphological Characteristics of Sulfonated Coconut Shell Biochar and Its Use for Corncob Hydrolysis.* Waste Biomass Valorization, 2023. **14**: p. 3097-3113.
9. Fan, X., et al., *Sorption Mechanisms of Diethyl Phthalate by Nutshell Biochar Derived at Different Pyrolysis Temperature.* 2022: p. 107328.
10. Azzaz, A.A., et al., *Olive Mill By-products Thermochemical Conversion Via Hydrothermal Carbonization and Slow Pyrolysis: Detailed Comparison between the Generated Hydrochars and Biochars Characteristics.* 2022. **10**(2): p. 231.
11. Pütün, A., et al., *Pyrolysis of Hazelnut Shells in a Fixed-Bed Tubular Reactor: Yields and Structural Analysis of Bio-Oil.* 1999. **52**(1): p. 33-49.
12. Kabayo, S.M., et al., *Preparation and Characterization of Solid Acid Catalysts Derived from Coffee Husks.* International Journal of Chemical Science, 2019. **3**: p. 5-13.
13. Yu, J.T., A.M. Dehkhoda, and N. Ellis, *Development of Biochar-Based Catalyst for Transesterification of Canola Oil.* Energy & Fuels, 2010. **25**(1): p. 337-344.
14. Sakhiya, A.K., A. Anand, and P. Kaushal, *Production, Activation, and Applications of Biochar in Recent Times.* Biochar, 2020. **2**(3): p. 253-285.
15. Pitchiya, A.P., et al., *Graphene-Enhanced Ion Transport in Dual-Conducting Composite Films of Polyacetylene and an Imidazolium Iodide Ionic Liquid.* Langmuir, 2023. **39**(19): p. 6767-6779.
16. da Luz Corrêa, A.P., et al., *Preparation of Sulfonated Carbon-Based Catalysts from Murumuru Kernel Shell and Their Performance in the Esterification Reaction.* RSC Advances, 2020. **10**(34): p. 20245-20256.
17. Kang, S., J. Ye, and J. Chang, *Recent Advances in Carbon-Based Sulfonated Catalyst: Preparation and Application.* International Review of Chemical Engineering, 2013. **5**(2): p. 133-144.
18. Inbaraj, B.S., K. Sridhar, and B.H. Chen, *Removal of Polycyclic Aromatic Hydrocarbons from Water by Magnetic Activated Carbon Nanocomposite from Green Tea Waste.* Journal of Hazardous Materials 2021. **415**: p. 125701.
19. Huang, Y.-F., et al., *Microwave Pyrolysis of Lignocellulosic Biomass: Heating Performance and Reaction Kinetics.* Energy, 2016. **100**: p. 137-144.
20. Azad, D., R. Pateriya, and R. Sharma, *Chemical Activation of Pine Needle and Coconut Shell Biochar: Production, Characterization and Process Optimization.* International Journal of Environmental Science and Technology, 2023: p. 1-16.
21. da Luz Corrêa, A.P., et al., *Preparation of Sulfonated Carbon-Based Catalysts from Murumuru Kernel Shell and Their Performance in the Esterification Reaction.* The Royal Society of Chemistry, 2020. **10**(34): p. 20245-20256.
22. Nura, G., *Synthesis and Optimization of Sugarcane Bagasse Derived Sulfonated Biochar Based Solid Acid Catalyst.* 2021, ASTU.
23. Mallick, A., M. Mukhopadhyay, and S. Ash, *Synthesis, Characterization and Performance Evaluation of a Solid Acid Catalyst Prepared from Coconut Shell for Hydrolyzing Pretreated Acacia Nilotica Heartwood.* Journal of The Institution of Engineers (India): Series E, 2020. **101**(1): p. 69-76.

176f Developing Novel Technologies for Produced Water Management

Benaiah Anabaraonye, Jonas F. Sundberg, Simon Ivar Andersen

Produced water (PW) is a significant challenge in oil and gas operations. Globally, on average, for every volume of oil produced, three volumes of water are produced. The complexity of produced water makes handling a real challenge. Produced water typically comprises various hydrocarbons, production chemicals such as scale and corrosion inhibitors, biocides, demulsifiers, and H_2S scavengers. Discharging untreated produced water to the sea poses several significant environmental risks. For these reasons, new technologies are needed. At the Danish Offshore Technology Centre (DOTC), we have designed a multi-dimensional program with a focus on reducing the environmental impact of hydrocarbon extraction in the North Sea. Reducing environmental impact requires dedicated sensing technologies coupled with control algorithms. Therefore, various sensing and monitoring technologies are being developed.

In this work, we present an overview of novel technologies being developed at DOTC to address specific produced water management challenges in the North Sea. First, we discuss the development of a fast, low-cost, and user-friendly capillary electrophoresis (CE) technology to quantify toxic chemicals in the produced water. Monitoring is crucial in any reduction of the discharge and understanding the concentrations of toxic chemicals in the water phase is critical to quantifying the environmental impact factor. We show that low concentrations (ppm-level) of a commercial biocide can be detected in both de-ionized water and complex brine systems.

Next, we discuss the development of metal-organic framework (MOF) based biomimetic oxygen scavengers. The complete removal of oxygen reduces the occurrence of corrosion in the injection wells and significantly reduces the need for oxygen-scavenging chemicals. Metal-organic frameworks are a class of hybrid materials prepared via the self-assembly of inorganic nodes connected by organic linker molecules. This results in extended network solids with a high internal surface area and porosity. This combined with their high tailorability makes them excellent materials for sorption-based processes. We are developing a new metal-organic framework incorporating cobalt complexes in the structure. By doing so, molecular oxygen can be chemically bound via an electron transfer process between cobalt sites and oxygen. This yields a material with very high O_2 to N_2 selectivity, exceeding what is possible using traditional, physisorption-based sorbents. In addition, the process is reversible, and oxygen can be removed by temperature or vacuum swing. The line-of-sight is to replace chemical scavengers with adsorption towers on the platform.

Finally, we present the development of a novel MOF-based BTEX (benzene, toluene, ethylbenzene, and xylenes) sensor. The sensor is based on a MOF-functionalized fiber probe. First, the MOF selectively bind, and extracts dissolved BTEX from produced water, resulting in a local enrichment of the target analytes. Secondly, the interaction will be probed using Raman spectroscopy, a technique capable of producing unique fingerprint spectra and resolution of complex mixtures. Thus, the sensor will be capable of distinguishing individual BTEX components with high sensitivity. The line-of-sight is developing a fiber probe that allows online and or remote sensing.

The technologies being developed at DOTC provide new solutions that can lead to a significant reduction in the environmental impact of produced water management. These results have direct applications in other wastewater management applications.

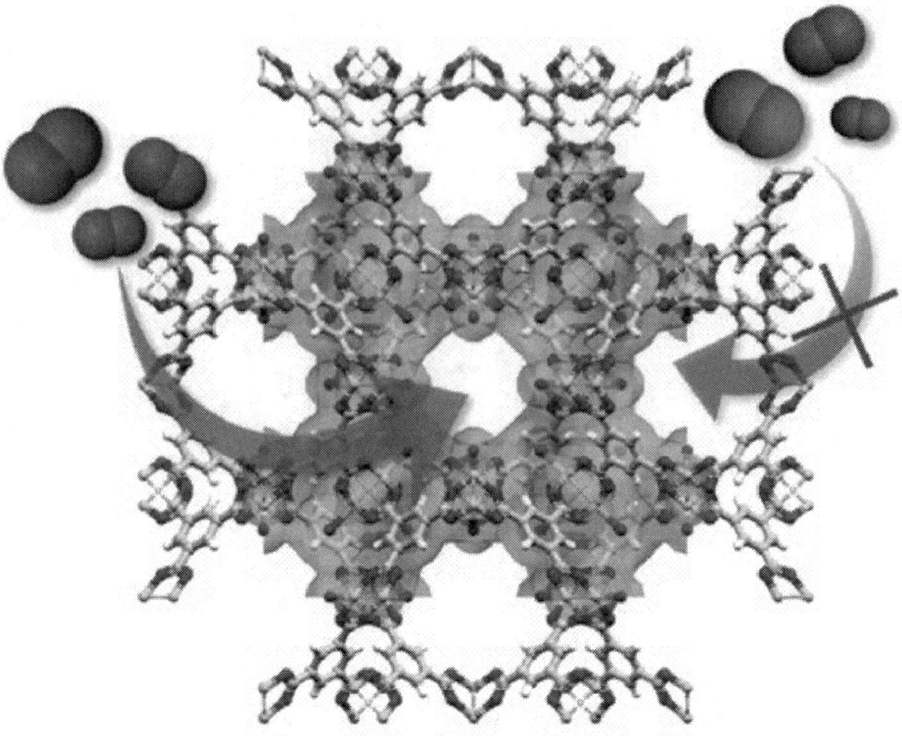

AUTHOR INDEX

AIChE
120 Wall Street, FL 23
New York, NY 10005-4020

ISBN 978-1-7138-9288-5

9 781713 892885